SOLID
STATE
MATERIALS

Springer Science+Business Media, LLC

SOLID
STATE
MATERIALS

EDITORS
S. RADHAKRISHNA
A. DAUD

Springer Science+Business Media, LLC

EDITORS

S. Radhakrishna
Institute of Advanced Studies
University of Malaya
Kuala Lumpur, Malaysia

A. Daud
Physics Department
University of Malaya
Kuala Lumpur, Malaysia

Copyright © 1991 Springer Science+Business Media New York
Originally published by Narosa Publishing House, New Delhi in 1991
Softcover reprint of the hardcover 1st edition 1991

Exclusive distribution in Europe and North America (including Canada and Mexico) by
Springer-Verlag Berlin Heidelberg New York London Paris Tokyo
Hong Kong Barcelona Budapest

For all other countries exclusive distribution by
Narosa Publishing House, New Delhi

All export rights for this book vest exclusively with the publishers. Unauthorized export
is a violation of Copyright Law and is subject to legal action

This book has been produced from the Camera Ready text
provided by the Editors/Contributors

ISBN 978-3-662-09937-7   ISBN 978-3-662-09935-3 (eBook)
DOI 10.1007/978-3-662-09935-3

Preface

Solid State Materials have become the driving force for the pace of todays society. In practically all aspects of Science and Technology materials play an important role. It will be no exaggeration to say that we are in the midst of a materials revolution. Materials scientists have responded to the needs of the industry from time to time. Several solid state devices are based on material modification and tailor-made production. As examples one can quote luminescent dosimeters made from doped LiF crystals, compact rechargeable solid state batteries made from lithium or silver ions embedded in suitable materials, tunnel diodes made from heavily doped silicon, microwave Gunn oscillators made from doped GaAs, semiconductor visible LASER diodes made from heterojunction GaAs-GaP, light emitting diodes made from doped semiconductors, thermoelectric coolers made from bismuth telluride, highly sensitive infrared detectors made from HgCdTe and a host of other devices. These devices have provided the impetus for intensified research into different aspects of Materials Science.

This International workshop on Solid State Materials has been organized to provide hands on experience in the fast changing field of materials to scientists and technologists who are actively involved in developing various areas of Materials Sciences. We would like to express our gratitude to the authors who have taken the trouble to provide manuscripts of their talks in time for us to meet the deadline set by the press.

The organizing committee would like to express its grateful appreciation to The Ministry of Science, Technology and Environment, Government of Malaysia, UNESCO and the Asian Physical Society for sponsoring the workshop and supporting the participation of several distinguished speakers and participants from various countries. We sincerely appreciate the effort of Narosa Publishing House/Springer-Verlag in bringing out the volume in time.

Kuala Lumpur, Malaysia S. RADHAKRISHNA
5 August 1991 A. DAUD

Contents

Solid State Ionic Materials

Limiting Processes in Lithium Solid State Cells

Steen Skaarup

Physics Laboratory III, Technical University of Denmark
DK-2800 Lyngby, Denmark

The voltage during the course of discharge of a lithium solid state electrochemical cell depends on a large number of physical and chemical properties of the components and interfaces. The processes likely to be of most importance and their dependence on the current density are described, with the intention of providing a quantitative basis for the design of cells and for conditions of experiments. The emphasis is on defining and comparing *characteristic times* for the processes involved.

INTRODUCTION

The discovery of crystalline, glassy and polymeric materials with high alkali metal ion conductivities has inspired much research into the fundamental properties of solid electrolytes and solid state insertion electrodes. At the same time, attempts have been made to develop rechargeable batteries based on the unique properties of these materials. Several of these systems are at present close to commercialization [1].

Whether the aim is to develop a battery system for practical use or to do research into materials properties of just one of the components, it is important to be able to estimate which processes are likely to be limiting in the experimental system studied. This is because even a very simple solid state electrochemical cell may be composed of complex components: Electrolyte may be added to the cathode in order to enhance the ionic conductivity, special forms of carbon may also be added in order to provide sufficient electronic conductivity. The nature and geometry of the complex interfaces between these components as well as the grain sizes and shapes influence the response of the cell to an applied current or voltage. Therefore the result may depend on a mixture of thermodynamic, kinetic and macroscopic physical properties [2]. In order to interpret the experiments, it is necessary to have at least a qualitative or semi-quantitative knowledge of which processes are important under the particular experimental conditions. This article seeks to identify some useful general principles in the design and interpretation of experiments.

LIMITING FACTORS DURING CHARGE AND DISCHARGE

Electronic Conductivity

When using metallic lithium as negative electrode material, the voltage drop from the electronic resistance of this component will nearly always be negligible. This is not the case for many of the compounds used for positive electrodes, such as transition metal sulphides and oxides. Whereas the well studied TiS_2 has excellent electronic conductivity, most vanadium oxides, e.g. V_6O_{13}, have electronic conductivities in the range 10^{-2} to 10^{-1} $\Omega^{-1}cm^{-1}$. This could be satisfactory for thin electrodes, but in most cases, there is a strong change when lithium is inserted in the structure [3]. For most of the investigated vanadium oxides, the change is a decrease to values of only $\sim 10^{-4}$ $\Omega^{-1}cm^{-1}$. These values are not bulk material properties, but have been measured on compressed powders (in a manual press at 70 Mpa). The values obtained in this way do not have the same fundamental interest, but may be more relevant to the use of the compound in a battery electrode which will often consist of compressed particles.

If the electronic conductivity is too low, some form of conducting material must be added in order to avoid large IR-drops from the positive electrode component. Usually, conducting carbon is added, but the precise nature of this carbon is crucial: Some special varieties (e.g. Shawinigan Russ and Ketjen Black) have structures that enable them to form a conducting three-dimensional network at low mass percentages (5-15%), whereas ordinary graphite will be inefficient.

The electronic conductivity of the electrolyte is only interesting if it is high enough to cause self-discharge of the cell.

Ionic Motion

If the electronic conductivity is assumed to be sufficiently large, the ionic motion through the system will be the main limiting factor.

Figure 1 shows a schematic solid state battery during discharge. When passing through the cell, the Li^+ ion will experience successive barriers to motion:

– Transport in the anode.
– Transfer across the anode/electrolyte interface.
– Transport in the separator electrolyte.
– Transfer across the electrolyte/cathode interface.
– Transport in the cathode.

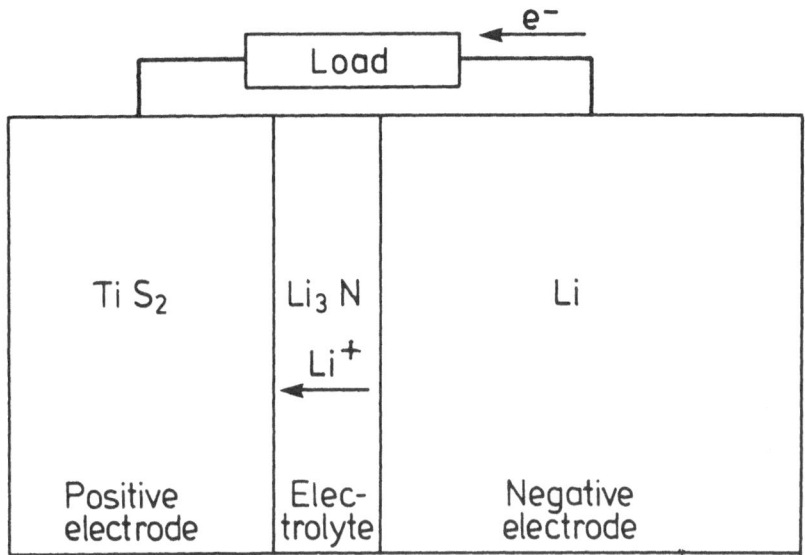

Figure 1. Schematic lithium solid state battery during discharge.

Each process contributes to the total overvoltage measured when current is passed through the cell, and each has a characteristic dependence on the current density and the temperature. In some cases, the current dependence will simply be Ohms law, but the ionic transport in the electrodes can show a more complex behavior. When designing a battery or planning/interpreting a cell experiment, it is crucial to have a clear view of which process or processes are influential. This is only possible if the experiment is carefully controlled in a quantitative way - e.g. the masses of the active materials and the thicknesses of the components must be accurately known. In planning an experiment, the goal will often be to have just one important kinetic contribution, whereas practical battery designs are likely to be limited by several processes simultaneously.

A useful approach is to think in terms of characteristic times which can be defined for some of the processes. They depend on the dimensions of the component (e.g. thickness of the electrolyte) and on the ionic conductivity. The importance of the process will then be determined by the ratio between the characteristic time and the stoichiometric discharge time T_D, which is the time theoretically required to discharge all of the limiting component of the active material. In lithium batteries, the positive electrode material will often be the limiting component. If the characteristic time of a diffusion process is significantly lower than T_D, the corresponding process will not contribute appreciably to the overvoltage. If it is significantly higher, the process will strongly limit the discharge process, leading to a low utilization of active material. If all characteristic times are small compared to T_D, the total overvoltage will be small, and the discharge will follow the thermodynamic equilibrium curve, unless electronic and/or ionic ohmic resistances are large.

Diffusion in the Lithium Electrode

As a solid state cell is discharged, lithium ions near the interface to the electrolyte passes into the electrolyte. This leaves vacancies or larger voids in the lithium electrode. If these voids are not replenished at a sufficient rate, the contact area between the phases will diminish, and the resistance rise [4]. If both phases are hard, the replenishment is performed by selfdiffusion of lithium atoms from the bulk electrode. If the current density is above a critical level, the voltage will drop rapidly after a characteristic time. The relation between current density, I, and characteristic depletion time, t_d, is:

$$I\sqrt{t_d} = \frac{F\sqrt{\pi}}{2} \, C_{Li}\sqrt{D_{self}} \tag{1}$$

F is Faradays constant, C_{Li} the lithium concentration (0.0764 mol/cm^3), and D_{self} the diffusion coefficient. The relation assumes semi-infinite diffusion, e.g. $t_d < t_{Li} = d_{Li}^2/D_{self}$ in the lithium (d_{Li} is the thickness, t_{Li} the electrode time constant).

The selfdiffusion can easily be a limiting factor at room temperature and realistic current densities. Figure 2 shows results for three temperatures. (-20 °C is included because a commercial "room temperature" battery must be able to perform also at lower temperatures, although not at full capacity). Using a 3 hour discharge at 25 °C, the maximum current density is only about 0.5 mA/cm^2.

There are three ways around the problem: One is to apply some form of external pressure to the cell. At room temperature and 1 mA/cm^2 the required pressure is large: About 500 kg/cm^2. The lithium selfdiffusion increases rapidly with temperature [5], and an alternative solution is to raise the temperature (if this is not impossible because of other considerations). At 100 °C, the required pressure can be produced by the inclusion of small springs in the experimental cells. The third solution is to use at least one soft component, that is able to adjust to the geometrical changes. The polymer electrolytes are excellent in this respect, and are able to retain interfacial contact with only a minimum of pressure, perhaps formed by using vacuum packing techniques.

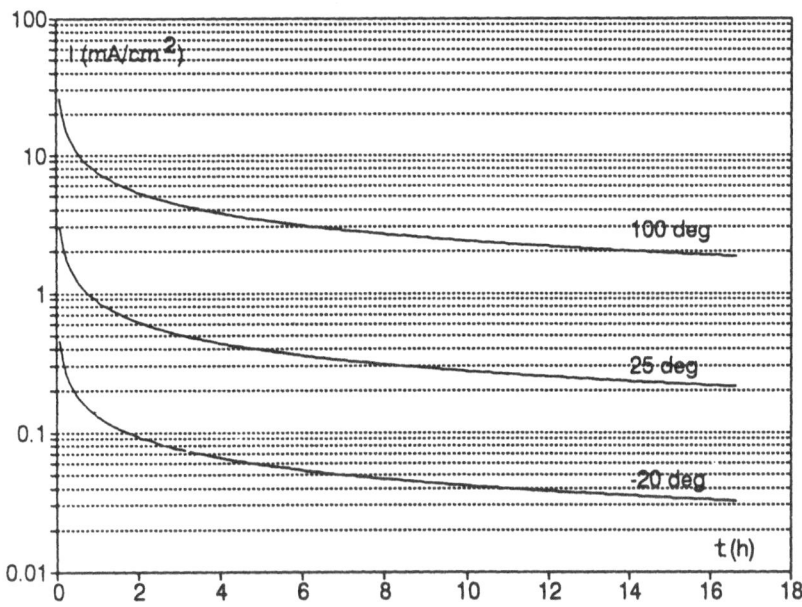

Figure 2. Limiting current densities as a function of the time to reach lithium depletion at the Li/electrolyte interface at the three temperatures indicated.

Electrode/electrolyte Charge Transfer

A satisfactory interface that allows ions to pass between two solids can be difficult to establish, and when used in a rechargeable cell, must satisfy the even more stringent demand of being able to retain contact during the sometimes large volume changes accompanying the cell reaction.

The best way to estimate the magnitude of the charge transfer resistance is by a combination of DC and AC experiments. In cells without reference electrodes, it is preferable to use symmetrical cells so that only one kind of interface is present. Figure 3 shows the AC impedance spectrum of a cell with lithium on both sides. The small semicircle at low frequencies is attributed to the charge transfer resistance, which in this case is estimated to ~25 Ω*cm^2. In this way, the formation of <u>blocking layers</u> can be followed. They are formed if the two phases are not thermodynamically stable. When using polymer electrolytes, such blocking layers can form on the lithium metal. These layers may be broken down by passing current through the interface, causing "delayed action" of the voltage - therefore DC cycling should be used as a supplement to the AC measurements.

Ionic Conduction in the Electrolyte

If the separator electrolyte is a solution of a lithium salt in a non-crystalline solvent such as propylene carbonate or a polymer, both positive and

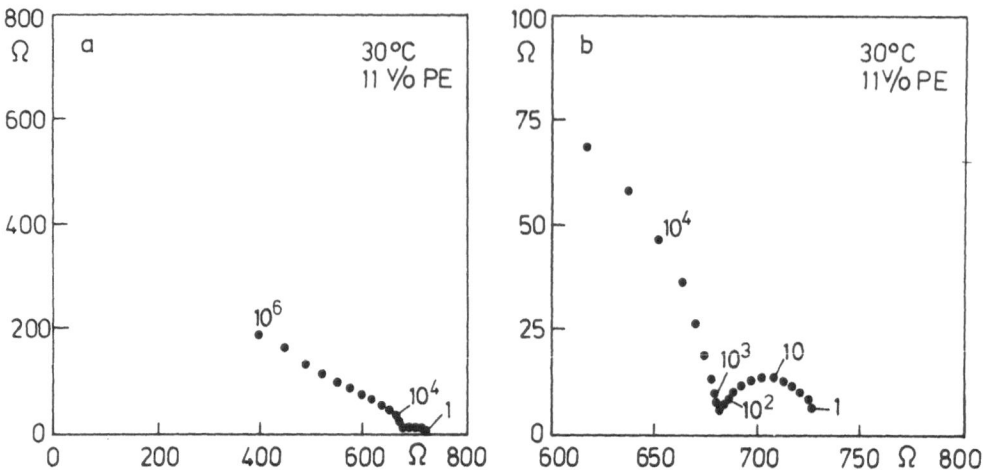

Figure 3. AC-impedance of a symmetrical Li/lithium sulphide glass + poly-ethylene/Li cell. Ordinate: Imaginary part. Abscissa: Real part.
b) Magnified part of a). Frequencies indicated on figure.

negative ions will move when a current is passed. This will cause a concentration gradient to develop: The section of electrolyte closest to the positive electrode will be depleted of cations, whereas the concentration adjacent to the negative electrode will be elevated [6]. This concentration gradient will contribute an overvoltage that can be calculated by the Nernst equation. The diffusion of cations will counteract the formation of the concentration gradient so that a steady state value is reached. If the current density is above a critical value, I_*, the concentration at the positive electrode will eventually reach zero, and the discharge stops because of <u>salt depletion</u> in the electrolyte. If the dissolved salt is mono/monovalent as in most conducting polymers, the limiting current is:

$$I_* = \frac{4FC_eD_+}{d_e} = \frac{4RT}{F} \frac{\sigma t_+}{d_e} \qquad (2)$$

C_e is the salt concentration, D_+ the cation diffusion coefficient, d_e the separator electrolyte thickness (or the radius of a cathode pore containing electrolyte), σ the total ionic conductivity and t_+ the cation transport number.

If the current density is smaller than I_*, the cation concentration will reach a stationary profile, but will never reach zero, whereas for values larger than I_*, the electrolyte depletion will stop the discharge after a <u>critical time</u>, t_*. The achievable current density is proportional to t_+—this points to an advantage for true solid electrolytes, where only cations move. For liquid and polymeric electrolytes, the Li^+ transport number is usually in the range 0.2-0.5. The depletion phenomenon places a limit on the current density for a given ratio between conductivity and thickness. As shown in equation 2, the important parameter is the absolute value of the fraction of the current carried by the cations. This means that t_+ can be accepted to be small if the total conductivity is large enough. The time to reach the salt depletion limit can be expressed as:

$$t_* = \frac{\pi}{16}\left(\frac{I_*}{I}\right)^2 \frac{d_e^2}{D_s} = \frac{\pi R T C_e \sigma t_+}{2(1-t_+)I^2} \tag{3}$$

D_s is the salt diffusion coefficient ($D_s = 2(1-t_+)D_+$). $d_e^2/D_s = t_e$ is the <u>time constant</u> of diffusion of the electrolyte. When using equation 3, it must be remembered that it is only valid for current densities above the critical value I_*.

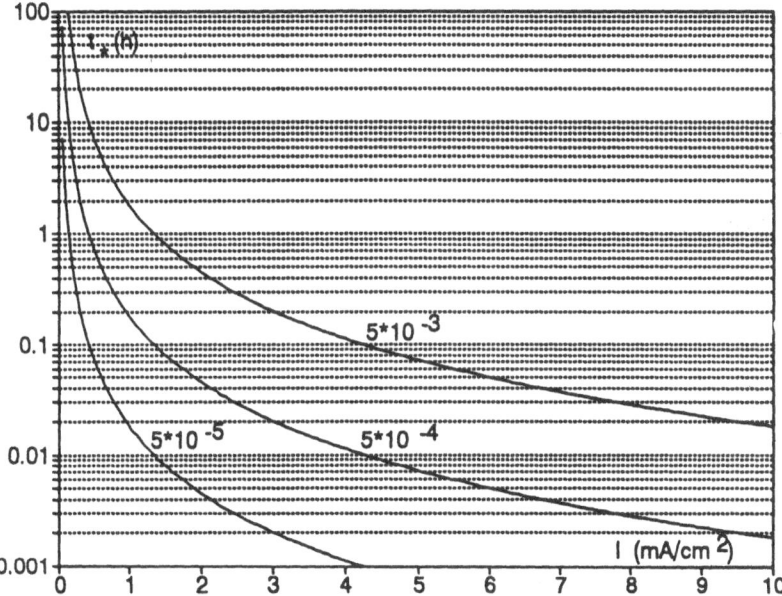

Figure 4. Critical times for reaching salt depletion in the electrolyte for three different total conductivities in $\Omega^{-1}cm^{-1}$ ($t_+ \equiv 0.25$, $C_e \equiv 1$ mol/l).

Figure 4 shows the results for three values of the conductivity in the range which is realistic for battery applications. The equation can be used to estimate the ability of the electrolyte to be discharged both for regular uses over several hours, and for short term, high power applications. An electrolyte corresponding to the upper curve in figure 4 will be able to carry a current of 10 mA/cm^2 for 0.02 hours (72 seconds).

When using liquid organic or polymer electrolytes for lithium batteries, electrolyte depletion is a realistic possibility as a limiting factor, especially at lower temperatures. It may not always be possible to make the electrolyte so thin that the depletion can be avoided at higher power drains. True solid ionic conductors with ($t_+ = 1.0$) avoid this phenomenon—also because the ionic concentration is usually very large ($C_e = 111.5$ mol/l in Li_3N).

Ionic Transport in the Positive Electrode

If the electronic conductivity of the positive electrode is large enough, the discharge of the cathode will be limited primarily by the ionic motion. The discussion will focus on the use of insertion materials. The important parameters for battery cathode applications are the maximum degree of lithium intercalation, C_{max}, and the diffusion coefficient of Li^+ ions, D_{Li}. As lithium moves into the cathode from the electrolyte, the concentration increases initially at the interface only, and is then transported by diffusion into the bulk of the electrode. This creates a concentration gradient, whose steepness is determined by the balance between the current density and how fast the ions are transported away from the interface by diffusion. Under certain simplifying assumptions, (e.g. no variation of D_{Li} with degree of intercalation), it can be shown that the concentration profile assumes a parabolic shape. This shape and steepness is retained as a function of time, but moves parallel towards higher concentrations [6], [7].

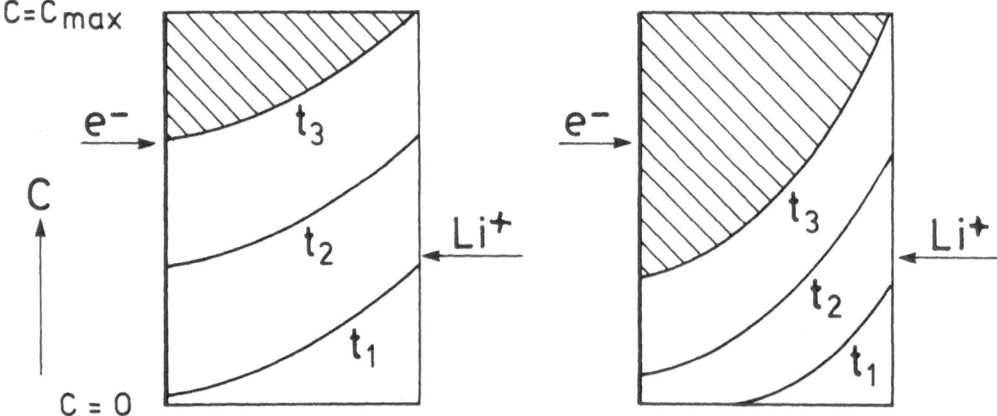

Figure 5. Concentration profiles of Li^+ ions in an insertion electrode for successive discharge times t_1, t_2 and t_3. At t_3, $C = C_{max}$ (See text).

Figure 5 shows the evolution of the concentration profile for two different current densities. As the concentration at the interface reaches C_{max}, the voltage drops, and the discharge stops. The non-utilized capacity is proportional to the hatched area in the figure. The capacity loss depends on the current density, on D_{Li} and on the thickness of the electrode, d_i. The discharge current in the right part of the figure is double that on the left, and the loss of utilization correspondingly larger.

The utilization depends on the ratio between the time constant for diffusion of the electrode, $t_i = d_i^2/D_{Li}$, and the stoichiometric discharge time T_D. This ratio is called the load factor, $L = t_i/T_D$ and is proportional to the current density. The relation between the load factor and the utilization depends on the electrode geometry. For planar electrodes, it can be approximated by two simple relations for low and high current drains, respectively:

$$U = 1 - \frac{L}{3} \quad (L<1.5) \quad ; \quad U = \frac{\pi}{4L} \quad (L>1.5) \tag{4}$$

These relations are depicted in figure 6. When doing discharge experiments, the utilization of the electrode should be kept above $\sim 50\%$, if the results are to be useful for comparing with realistic uses. The relations can be used to estimate the maximum thickness it is possible to use when the diffusion coefficient is known.

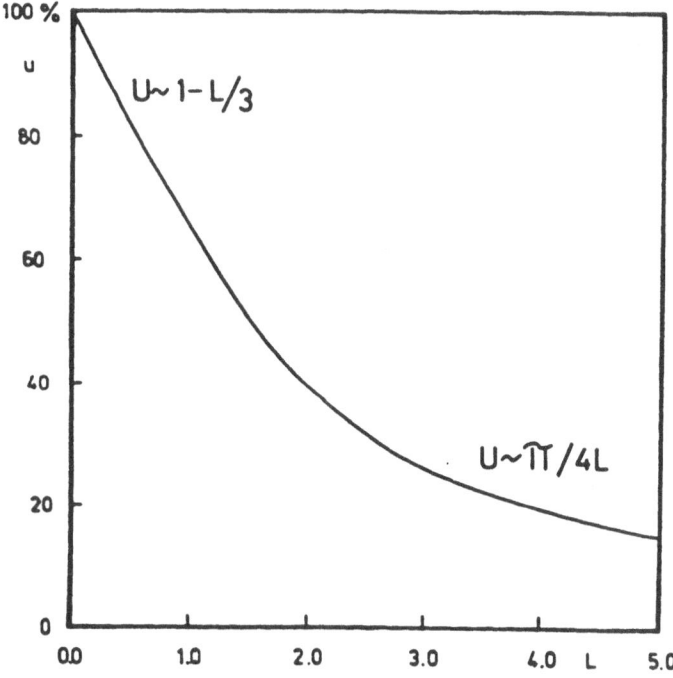

Figure 6. Utilization as a function of the load factor for a planar insertion electrode.

For pure TiS_2, $D_{Li} = 10^{-10}$ cm^2/s at 25 °C yields a maximum thickness of about 10 μm ($T_D = 10^4$ s). This unrealistically low value (even if considered

11

to be used only for purposes of experiment) illustrates that the motion of lithium in the bulk insertion compounds is rarely fast enough to be of practical use. This leads to the necessity of adding liquid, polymeric or solid electrolyte to the positive electrode, forming <u>porous and composite</u> electrodes, respectively. An ideal composite electrode consists of two contiguous, interconnected networks— of electrolyte and insertion compound particles. The ions are transported mainly in the electrolyte phase, the electrons in the insertion phase, and the cell reaction takes place along the large common interfacial area. Under simplifying conditions (e.g. linear *emf*-curve), it has been shown that the ionic transport can be characterized by an effective Li^+ diffusion coefficient D_c:

$$D_c = \frac{k\sigma_c}{FvC_{max}} \tag{5}$$

k is the slope of the *emf*-curve with respect to degree of intercalation (for TiS_2, k is ~ 0.5 V/stoichiometric equivalent), σ_c is the conductivity of the composite electrode as a whole and v is the volume fraction of insertion compound. v will usually be not too far from 0.5 since it is necessary to be above the percolation limit for both electronic and ionic motion. σ_c is equal to the conductivity of the <u>electrolyte phase</u> multiplied by a factor that depends on the detailed geometry of the mixture in a complex way. Reasonable estimates of this tortuosity factor is 0.2-0.3. In this way, the effective diffusion coefficient can be raised by several orders of magnitude compared to the pure insertion compound value D_{Li} [8].

All the equations developed above can now be used for composite electrodes by substituting D_c for D_{Li}. The time constant of the electrode is now $t_c = d_c^2/D_c$. The magnitude of D_c can be measured for instance by applying a current pulse and analyzing the voltage as a function of the square root of time [9], or by estimating the time constant from cyclic voltammograms at different scan rates [10].

OVERVIEW

The following table summarizes the conclusions. In using it, the characteristic times for the actual or planned geometry and current density are calculated and compared to the stoichiometric discharge time. If a time constant, $t \prec T_D$, the corresponding process can be ignored. A process with $t \succ T_D$ will contribute strongly to the overvoltage and may be the main limiting factor. The two characteristic depletion times t_d and t_* will be limiting if they are <u>smaller</u> than T_D. If only one process is limiting, the current/time relation can be calculated from the relevant equation. In addition to the relations in the table, other factors such as Ohmic voltage losses in the electrodes and electrolyte, charge transfer resistances, formation of blocking layers etc., may be of importance.

Current Density vs Time Relations for Ionic Transport Processes in Solid State Cells		
Process	**Characteristic time**	**Comment**
Discharge of cathode	$$T_D = \frac{F d_c v C_{max}}{I}$$	Stoichiometric discharge time
Lithium selfdiffusion	$$t_{Li} = \frac{d_{Li}^2}{D_{self}}$$	Anode time constant
Lithium depletion at interface	$$t_d = \frac{\pi D_{self} F^2 C_{Li}^2}{4 I^2}$$	Depletion occurs only for $$I > \frac{\sqrt{\pi} F D_{self} C_{Li}}{2 d_{Li}}$$
Salt diffusion in electrolyte	$$t_e = \frac{d_e^2}{D_s}$$	Electrolyte time constant
Salt depletion in electrolyte	$$t_* = \frac{\pi R T C_e \sigma t_+}{2(1-t_+) I^2}$$	Depletion occurs only for $$I > \frac{4 F C_e D_+}{d_e} = \frac{4 R T}{F} \frac{\sigma t_+}{d_e}$$
Lithium motion in composite cathode	$$t_c = \frac{d_c^2}{D_c} = \frac{v F d_c^2 C_{max}}{k \sigma_c}$$	Composite cathode time constant. Utilization is determined by: $$L = \frac{t_c}{T_D} = \frac{d_c I}{F v C_{max} D_c}$$

List of symbols

C_{Li}, C_e and C_{max}: Lithium concentration in anode, electrolyte and insertion compound (mol/cm^3).

d_{Li}, d_e and d_c: Thickness of anode, electrolyte and cathode (cm).

D_{self}, D_s, D_c and D_+: Self-, salt-, composite- and cation- diffusion coefficients (cm^2/s).

F, R: The Faraday (96485 C/mol) and the gas constant (8.3145 J/K·mol).
I: Current density (A/cm^2).
k: Slope of *emf*-curve of insertion compound (V).
L: Load factor of cathode.
t_+: Cation transport number.
T_D, t_{Li}, t_d, t_e, t_* and t_c: Characteristic times (s).
v: Volume fraction of insertion compound in composite cathode.
σ and σ_c: Total ionic conductivity of electrolyte and composite cathode (Ω^{-1}cm^{-1}).

REFERENCES

[1]   Skaarup, S., West, K., Yde-Andersen, Y. and Koksbang, R. (1990). In: *Recent Advances in Fast Ion Conductors, Materials and Devices*, eds. Chowdari, B.V.R., Liu, Q.-G. and Chen, L.-Q., (World Scientific Publishing), pp. 83-95.

[2]   Skaarup, S. (1988), *Proceedings of the International Seminar on Solid State Ionic Devices, Singapore*, eds. Chowdari, B.V.R. and Radhakrishna, S., (World Scientific Publishing), pp. 35-53.

[3]   West, K., Zachau-Christiansen, B., Østergård, M.J.L. and Jacobsen, T. (1987). *Journal of Power Sources*, **20**, pp. 165-172.

[4]   Jow, T.R. and Liang, C.C. (1983). *Solid State Ionics*, **9&10**, pp. 695-698.

[5]   Lodding, A., Mundy, J.N. and Ott, A. (1970). *Phys. Stat. Sol.*, **38**, p. 559.

[6]   Atlung, S., West, K. and Jacobsen, T. (1979). *J. Electrochem. Soc.*, **126**, pp. 1311-1321.

[7]   Atlung, S. (1985). In: *Solid State Batteries*, eds. Sequeira, C.A.C. and Hooper, A., NATO ASI Series E **101**, (Martinus Nijhoff Publishers) pp. 129-161.

[8]   Knutz, B. and Skaarup, S. (1986). *Solid State Ionics* **18&19**, pp. 783-787.

[9]   Weppner, W. and Huggins, R.A. (1977). *J. Solid State Chem.*, **22**, p. 297.

[10]   West, K., Zachau-Christiansen, B. and Jacobsen, T. (1983). *Electrochimica Acta*, **28**, pp. 1829-1833.

Materials for Solid State Batteries

R.V.G.K. Sarma and S. Radhakrishna

Institute for Advanced Studies, University of Malaya
59100 Kuala Lumpur, Malaysia

Solid State batteries have been prepared from a wide
range of electrolyte materials. Lithium, Silver and
copper electrolytes are used in the preparation of
microbatteries. This article is intended to provide
guidelines for the choice of material in the
preparation of batteries. The current trend in the
choice of mateials is discussed.

INTRODUCTION

Solid state batteries have attracted a great deal of
attention in the past decade because of their technological
importance. These batteries are made from specially prepared high
conducting solid electrolyte materials in which the relatively
fast motion of the ions is useful in several applications. Well
over 250 papers are being published annually and an International
conference is being held every two years attracting a large number
of scientists and an increasing number of participants from the
Industry. Besides, a regional workshop is being organized in Asia
once in every two years and all this massive literature is well
documented [1–6]. However, we feel that understanding of the
battery performance based on the choice of materials will give the
researcher a better insight into the parameters involved in
choosing suitable materials.

Fast ion conducting solids suitably prepared from glasses,
polymers and composite materials have been tried for optimum
performance. Several materials in their Polycrystalline, glassy,
and thin film forms have been used as electrolyte materials in the
preparation of solid state batteries. Electrolytes based on
Silver, Copper, Lithium and Sodium have been tried in the
fabrication of solid state batteries. Several methods have been
used to increase the electrical conductivity and to get better
thermal stability of these materials, which in turn improves the
battery performance characteristics. It is useful to understand
the criteria for choosing the materials for electrodes and
electrolytes, the advantages and disadvantages of different types
of solid electrolytes over liquid electrolytes and the advantages

15

associated with different forms of electrolytes used, before we go in for a description of the actual materials.

For instance, AgI taken in its β-AgI form shows fast ionic conduction only above 147°C. However, when AgI is mixed with compounds like RbI, enhancement in ionic conductivity is found at ambient temperature. Thus anodic or cathodic modification of AgI matrix gives raise to fast ion conduction at room temperature making it a suitable material in device applications. Very often, materials taken in glassy form have several advantages over its crystalline counterparts.

An Electrochemical cell consists of an anode, an electrolyte and a cathode. A cell basically provides dc current across the load of the external circuit and ionic current in the internal circuit. An electrolyte is used as a medium to pass the ionic current as well as a separator between the electrodes of the cell. Thus an electrochemical cell is a device which converts chemical energy into electrical energy. Depending upon whether the cell can be reused or not these batteries can be divided into primary and secondary batteries. In a primary battery the cell reactants are not created, and after use the cell has to be discarded. However, in a secondary battery the reactants can be reformed by using an external current source. This charge-discharge process is called a cycle and typical solid state cells are available which can last over hundreds of cycles. Because of the inefficiency of the reversal process the depth of the discharge that can be obtained each time depends on the nature of the reaction, the materials used and the geometry of the cell.

CRITERIA FOR CHOOSING ELECTROLYTE AND ELECTRODE MATERIALS

The electrolyte forms the most important component of the solid state battery. The ionic conductivity of the electrolyte should be as high as possible at ambient temperatures, so that the internal resistance of the cell does not change the operating voltage of the cell. The electronic conductivity of the materials should be as low as possible in comparison with the ionic conductivity of the material, in order to avoid the self discharge of the cell. The electrolyte should be fairly stable under the ambient conditions such as temperature, pressure and humidity to facilitate mass production. Ideally the electrolyte should show ionic transport number as close to unity as possible. The electrolyte should be compatible physically and chemically with the electrodes.

In contrast to a liquid electrolyte system, the solid electrolytes have non zero transport numbers. This results in the formation of the reaction product at the electrode/electrolyte interface leading to lower current density or lowering of the efficiency of the cell.

In the past fifteen years mass production of industrial products with their own sources of power has led to a new growth in research on new materials for solid state power sources. Solid state systems have several advantages like :

* simple design

* resistance to shocks and vibration

* inherently robust,spill proof and usually non corrosive

* suitable for miniaturization

* ability to form thin films

Some special problems are encountered with the solid state batteries inspite of the various advantages listed above. Solid/solid interface complicates the interfacial charge transport process, whereas liquid/solid interface ensure full contact during the discharge process. Moreover, the mechanical changes during the cell operation induces stress in the cell which results in the failure of the cell. During the operation of the cell the molar volume of the products generally differs from that of the reactants, thus causing expansion or contraction of the cell. Such stresses will lead to the breakage of the cell if the electrodes are unable to accommodate these changes. This is a major problem in based on vitreous electrolytes.

Slow mass diffusion across the electrode/electrolyte interface also impedes the cell performance. Formation of low conducting reaction product also results in the lowering of the cell voltage. Considerable amount of electronic conduction results in short shelf life especially in copper batteries. Ionic conduction by more than one species is a disadvantage in the case of polymer batteries.

In contrast to the electrolytes, electrodes should have high electronic conductivity and very low ionic conductivity. The electronic conductivity of the cathodes materials which are usually intercalation compounds , is not so high enough to be used as electrodes. These materials require the addition of some form of carbon. For example, vanadium oxides have low electronic conductivity values, compared to Titanium disulphide which can be used without any addition of carbon. Electrodes should be fairly stable and should not react with the electrolytes. The cathode should be highly electro negative and the anode should be highly electro positive.

BATTERY PERFORMANCE PARAMETERS

The *open circuit voltage (OCV)* of a solid state battery is one of the major parameters to be considered to check the performance of the battery. This is the terminal voltage measured under infinite load across the battery leads. Ideally this should be the emf of the cell reaction calculated from the thermodynamic data. But OCV of a cell generally differs from the emf because of the following reasons:

1. The actual cell reaction is not the same as was anticipated.

2. Increase in the internal resistance of the cell due to the reaction between electrode and current collectors

3. Internal discharge may occur because of the internal electronic conduction, bad construction of the cell and IR losses.

The open circuit voltage depends on the choice of anode and cathode couple. For a fixed anode and cathode combination efforts should be made to get higher current density.

The source of the IR drop is the internal resistance of the cell. The polarization losses at moderately high current densities is due to the ohmic loss. During design of the battery the following points should therefore be taken into consideration.

* To obtain electrolyte with high ionic conduction.

* To reduce electrode/electrolyte interfacial resistance.

* To optimize cell geometry to get uniform current density.

The *Power density* or *Energy density* of the battery are important parameters. These can be calculated based on either unit volume or unit mass and are expressed interms of W/Kg or W/L and energy density in units are W-hr/Kg or W-hr/L, where 1 W-hr = 3.6KJ. The calculations can be done by taking into consideration the current collectors, encapsulate etc., along with electrodes and electrolytes or not.

Current Density is expressed in terms of A/cm^2. For solid state batteries the area is usually taken as the cross sectional

18

area but not true area of contact. In solid state batteries the actual area of contact differs from the cross sectional area. The rate at which the recharging is possible is a important parameter in secondary batteries which is usually expressed interms of C/n, where C is the charge passed and n is the number of hours the recharge proceeds

Rechargeable battery systems utilize a intercalation cathode. During the cell reaction the cations are inserted into the intercalation material (for example TiS_2) forming a complex with the cathode. This process results in volume expansion, which might result in the breakage of the cell. Thus these materials need to be embedded in a flexible matrix (for example polymer like PEO), which absorbs the changes arising because of the cell reaction. Moreover, the rechargeable cathodes must be reversible to many cycles of charge/discharge without undergoing any significant irreversible changes.

VARIOUS FORMS OF THE ELECTROLYTE

The materials used in solid state batteries can be taken in various forms like polycrystalline, amorphous and thin films. Glassy materials have many advantages compared to crystalline counterparts like

* absence of grain boundaries

* ease of thin film formation

* wide range of composition selection thereby fine control of properties

These fast ion conducting glassy materials generally consist of a glass former (B_2O_3, SiO_2, P_2O_5, B_2S_3 etc.,), glass modifier (Ag_2O,Li_2O etc.,) and a doping salt. These glass modifiers interact strongly with the macromolecular chain formed by the network former, resulting in the breakage of the oxygen or sulphur bonds linking the two former cations. There is an increase in the number of non bridging oxygens with increase in content of the modifier, this increases the carrier concentration and thereby the ionic conductivity. Fig.1 shows two dimensional schematic representation of a glass and glass former with a glass modifier.

Spectroscopic studies on the ternary system revealed no modification in the vibrational frequency of the macromolecular chain, which shows that the addition of doping salt increases

19

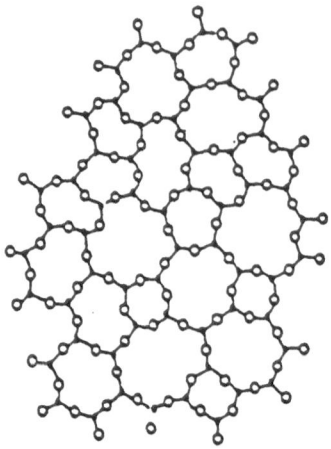

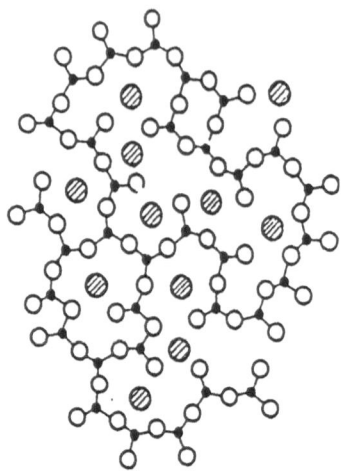

Fig. 1(a). Schematic representation of ordinary glass.
(B_2O_3, SiO_2, P_2O_5 etc.,)

Fig. 1(b). Representation
of modification of glass
structure by a glass modifier

(Ag_2O, Li_2O, Na_2O etc.,)

carrier concentration considerably without changing the network
structure. Addition of a second glass former to the network
structure is found to enhance the ionic conductivity [7].
Attempts to add a second doping salt resulted in decrease in the
conductivity [8].

The progress in microelectronics demanded miniaturization in
power sources. The progress in thin film technology made these
miniaturization of power sources possible. Various preparation
techniques like thermal evaporation, electrodeposition, flash
evaporation and sputtering have been used in the preparation of
the thin film electrolytes and electrodes for solid state
batteries [9]

The development of better solid state battery systems for
device applications depends on the preparation and
characterization of new electrode and electrolyte materials and
have been reviewed a number of times. Various experimental
techniques have been used in the characterization of solid
electrolyte materials [10]. Important techniques include

X – ray diffraction	Amorphous nature
Impedance Spectroscopy	Electrical Characterization
DTA/DSC	Thermal Characterization / Glass transition temperature
IR/Raman	Vibrational frequencies of structural units present
XPS(ESCA)	Surface Analysis
EDAX	Elemental Analysis
SEM	Surface Morphology
EXAFS	Local Structure and neighbourhood environment
NMR	Chemical shift, relaxation times

Measurement of diffusion coefficient and transport number also forms some of the techniques used in the characterization of materials.

The description of materials has been described in two sections.

1. Materials for low energy applications.

2. Materials for high energy applications.

MATERIALS FOR LOW ENERGY APPLICATIONS

LITHIUM SYSTEMS

Lithium solid state batteries with LiI as an electrolyte have been tried, which was not a successful attempt due to the poor conductivity of LiI ($10^{-7} \Omega^{-1} cm^{-1}$ at RT) and non compatibility with the electrodes. But the system I_2/Poly(2–vinylpyridine), P2VP, has been of commercial importance (used in heart pacemakers) in which LiI was formed *in situ* and the cell contained no electrolyte at the assembly stage [11].

The solid state micro batteries based on lithium systems embedded in a flexible polymer matrix are nearing commercial stage. Lithium batteries based on glassy electrolytes, ceramic and sulphate systems are also available. We are trying to give an overall picture of the systems, since more exhaustive review will appear in this volume. Lithium polymer batteries consists of a lithium anode, a nickel current collector on which electrolyte and cathode are coated. The electrolyte is generally a metal salt dissolved in a polymer matrix (for example PEO). The cathode is a intercalation compound like TiS_2, V_6O_{13} mixed with the electrolyte and acetylene black. The thickness of the electrolyte and electrodes of the order of microns.

Polymer electrolytes are interesting because of their special mechanical properties, ease of fabrication in thin film form, formation of good electrode/electrolyte interface contact. These polymer electrolyte comprises of a polymer such as Polyethylene Oxide (PEO), Polypropylene Oxide (PPO), Poly Vinyl Alcohol (PVA) complexed with a metal salt such as $LiClO_4$, $LiCF_3SO_3$, NaI and others. Most of the Li^+ and Na^+ polymer electrolytes are reviewed [12].

Composite polymer blends are the ones in which low molecular weight liquid polymer such as poly(ehtylene glycol) was blended with high molecular weight poly(ethelene oxide) in acetonitrile and was used as a polymer matrix for dissolving lithium salt [13]. Solid state cells of the configuration LI/LiNAGE/V_6O_{13},(PEO)$_8$LiCF$_3$SO$_3$,C (LiNAGE = Lithium Non Aqueous Gel Polymer Electrolytes) were fabricated, which gave an OCV of 3.3-3.6 volts and current upto 1mA. Fig.2 denotes the discharge characteristics of the above cell at room temperature (22°C).

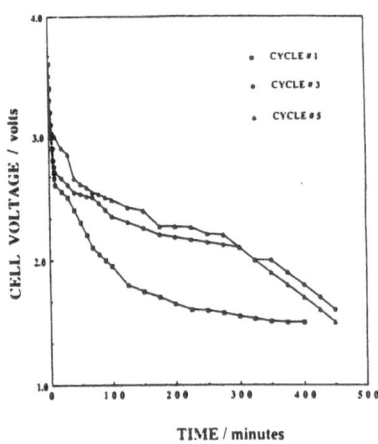

TIME / minutes

Fig. 2. Discharge characterstics of the cell configuration
Li/LiNAGE/V_6O_{13} at 22°C during first, third and fifth cycles.

22

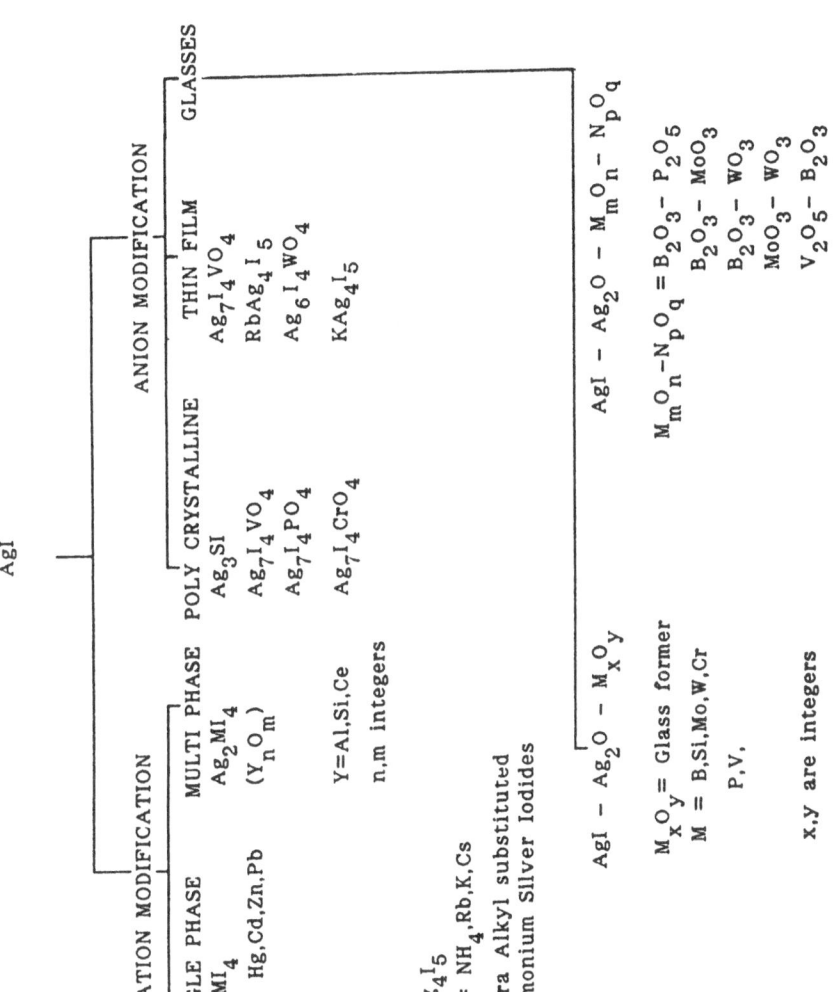

Fig. 3. Schematic representation of cationic and anionic modification of β-AgI

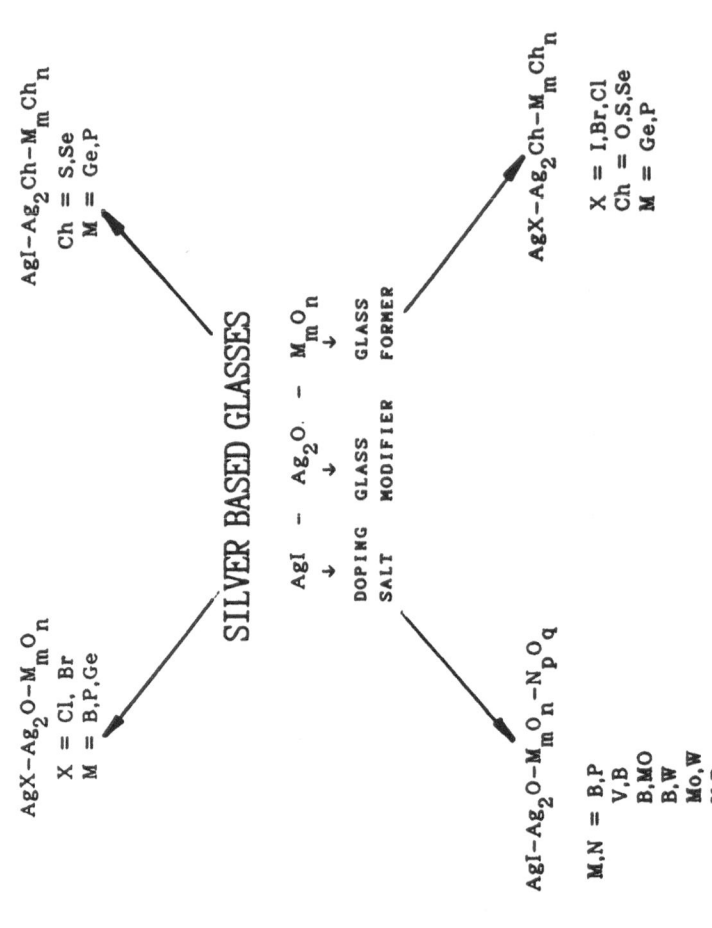

SILVER BASED GLASSES

AgI – Ag$_2$O. – M$_m$O$_n$
↓   ↓   ↓
DOPING GLASS GLASS
SALT MODIFIER FORMER

AgI–Ag$_2$Ch–M$_m$Ch$_n$
Ch = S,Se
M = Ge,P

AgX–Ag$_2$Ch–M$_m$Ch$_n$
X = I,Br,Cl
Ch = O,S,Se
M = Ge,P

AgX–Ag$_2$O–M$_m$O$_n$
X = Cl, Br
M = B,P,Ge

AgI–Ag$_2$O–M$_m$O$_n$–N$_p$O$_q$

M,N = B,P
V,B
B,MO
B,W
Mo,W
V,P

Fig.4. Different types of glasses with the modification in the AgI – Ag$_2$O – M$_m$O$_n$

SILVER SYSTEMS

Among the solids exhibiting high ionic conductivity, silver ion conductors are important in many respects. They possess highest conductivity that has been observed and they have highly disordered cationic distribution with low activation energies. Foremost among these is AgI, which exhibits fast ionic conduction above 147°C only. There has been a considerable effort in the past two decades to find good silver ion conductors which can be used at room temperature. Cationic and anionic modification of AgI has been attempted to bring down the transition temperature and to increase ionic conductivity. Promising among these are $RbAg_4I_5$, Ag_3SI, $Ag_7I_4VO_4$ etc., Fig.3 and Fig.4 denotes various compounds possible with anionic and cationic modification of AgI.

A typical silver based cell will give an open circuit voltage of 0.67 V and current densities of the order of 100 $\mu A/cm^2$. There has been continuous effort in improving the structure of the cathode material inorder to get better battery performance. Iodine has been used as cathode in earlier stages of battery development. Due to high diffusion coefficient, volatile nature and low electronic conductivity, different methods were tried to improve the efficiency of cathode material. Modifications such as I_2+Graphite, TMAI, I_2-phenothiazine, I_2-pyrene ,I_2+perylene and different types of cathode have been tried [14]. Different types of electrolytes like $RbAg_4I_5$, Ag_3SI, $NH_4Ag_4I_5$, KAg_4I_5[15] and silver based quaternary electrolytes have been used with different cathodes [16-18]. Fig.5 denotes the discharge characteristics of the silver borotungstate glass at different current densities with TMAI as cathode [17].

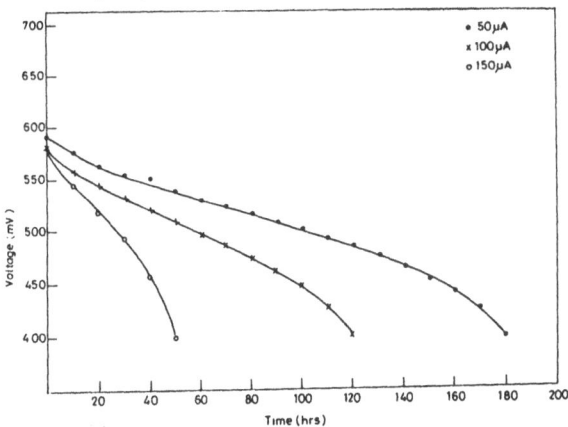

Fig. 5.   Discharge characteristics of a typical silver cell

COPPER SYSTEMS

Copper systems attracted much attention because of their low cost and can be easily handled unlike lithium ion conductors. These systems are based on the motion of Cu^+ ion. Major drawback of these systems being significant electronic contribution which leads to shorter shelf life.

Less amount of the work has been done in these systems in comparison with the silver systems. Most studied electrolyte being $Rb_4Cu_{16}I_7Cl_{13}$ and others are Cu_2CdI_4, Cu_2HgI_4, $RbCu_3Cl_4$, $Cu_3RbI_2Cl_3$, $CuI-CuCl$ etc., Copper ion conductors have been recently reviewed [19]. Polyiodides and I_2-perylene have been used as cathode materials in some of the primary batteries. Cathodes like Cu_2Se/Se and Cu_2Te/Te have also been investigated. The intercalation cathode TiS_2 which can be used in rechargeable systems has been investigated. The cell $Cu_{0.8}TiS_2$ / $4HMCH_3Br$ − $CuBr$ / $Cu_{0.4}TiS_2$ has shown best performance after being subjected to several charge-discharge cycles. Kanno etal have studied the cells of the configuration $Cu_4Mo_6S_8$ / $Rb_4Cu_{16}I_{6.8}Cl_{13.2}$ / Intercalation compound [20]. The metal dichalcogenides like TiS_2 and NbS_2 , copper thiomolybdate have also been considered.

Recently a solid state rechargeable battery using a paper form copper ion conductor $Cu_{16}Rb_4I_7Cl_{13}$ as electrolyte and copper chevrel compound $Cu_2Mo_7S_{7.8}$ as cathode has been reported [21]. They mixed Styrene-ethylene-butadiene-ethylene copolymer rubber with electrolyte and electrode to make the flexible battery and to minimize the leak current. They have reported a voltage of 0.55 volts and a current as high as $10mA/cm^2$ and charge-discharge cycle life of over 1000 times. Fig.6 denotes the construction of the paper thin battery. Thus electrolyte properties can be modified by the addition of plasticizers.

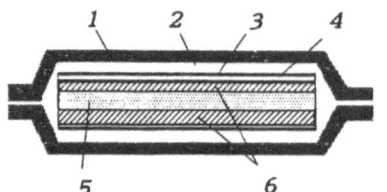

1. Aluminium Foil
2. Polypropylene
3. Copper foil lead
4. Carbon adhesive film
5. Solid electrolyte sheet
6. Chevrel electrode sheet

Fig.6. Construction of the paper thin battery [21]

Addition of insulating particles (Al_2O_3,SiO_2) to the poor ionic conductors like LiI, β-AgI as resulted in enhancement of conductivity. These are described as composite electrolytes of Dispersed Solid Electrolyte Systems (DSES). The material LiI(Al_2O_3) has been used as a electrolyte with Li/TiS$_2$ as electrodes, resulted in a current of the order of μA for about 4 years. These composite electrolytes were recently reviewed [22,23].

MATERIALS FOR HIGH ENERGY APPLICATIONS

Solid electrolytes have been used for high capacity applications. In many power station load levelling applications Na/S and Na/FeCl$_2$ batteries have been used while β-alumina based batteries have been used for converting thermal energy into electrical energy. These are referred to as AMTEC (Alkali Metal Thermal Energy Conversion) batteries. Sodium ion conducting batteries have been used in the past for vehicle propulsion. It is expected that power of the order of 1000 KW can be developed by such batteries. The lithium polymer batteries have held the best promise for electric car applications. These batteries are reviewed else where in this volume.

REFERENCES

1. Solid State Ionics
    9/10, 18/19,29/30,40/41, North Holland Publishing company, Amsterdam

2. Chowdari B V R and Radhakrishna S
    "Materials for Solid State Batteries", World Scientific, Singapore, 1986

3. Chowdari B V R and Radhakrishna S
    "Solid State Ionic Devices" World Scientific, Singapore, 1988

4. Chowdari B V R, Qingguo Liu and Liquan Chen
    "Recent Advances and Fast Ion Conducting Materials and Devices" World Scientific, Singapore, 1990

5. Takahashi T
    "High Conducting Solid Ion Conductors, Recent Trends and Applications" World Scientific, Singapore, 1990

6. Lasker A L and Chandra S
    "Superionic Solids and Solid Electrolytes : Recent Trends" Academic Press, 1990

7. Magistris A, Chiodelli G and Duclot M
    Solid State Ionics 9/10 (1983) 611

8. Chowdari B V R and Radhakrishnan K
    Solid State Ionics 24 (1987) 81

9. Ribes M in ref. 2 p.no.135,147

10. Linford R G and Hackwood S
    Chemical Reviews 81 (1981) 395

11. Greatbatch w, Lee J H, Mathias W, Eldrige M, Moser J R and Schneider A A
    IEEE Trans. Bio–Med Eng. 18 (1971) 317

12. Vincent C A
    Prog. Solid St. Chem.17 (1987) 145

13. Prasad P S S, Munzhi M Z A, Owens B B and Smyrl W H
    Solid State Ionics 40/41 (1990) 959

14. Linford R G in ref.3 p.no.527

15. Linford R G in ref 5 p.no.564

16. Sarma R V G K and Radhakrishna S
    J Power Sources 32 (1990) 327

17. Sarma R V G K and Radhakrishna S
    Solid State Ionics 40/41 (1990) 483

18. Prasad P S S and Radhakrishna S
    J Power Sources 5 (1989) 287

19. Viswanadh A K and Radhakrishna S in ref 5, p.no.280

20. Kanno R, Takeda Y, Oda Y, Ikeda H and Yamamoto O
    Solid State Ionics 18/19 (1986) 1068

21. Sotomura T, Kondo S and Iwaki T in ref 4 p.no. 325

22. Wagner J B Jr. in ref 5 p.no. 146

23. Maier J
    Mat. Chem. Phys. 17 (1987) 485

Polymer Batteries

Roger G. Linford

Head of the School of Applied Physical Sciences, Leicester Polytechnic
P.O. Box 143, Leicester LE1 9BH, England

Key issues in polymer electrolyte studies are identified and applications of these materials in polymer batteries and other devices are discussed. Typical techniques used for characterisation of polymer batteries and the materials they they contain are described and performance parameters are indicated.

WHAT ARE POLYMER BATTERIES?

All batteries, including polymer batteries, involve two electrodes, an anode and a cathode, separated by an electrolyte. They produce power by a redox reaction between anode and cathode species, oxidation taking place at the anode and reduction at the cathode. The ions involved in the cell reaction travel through the electrolyte whereas the electrons take a different pathway, through an external circuit. By doing so, the electrons can do external external work, providing power for heaters or motors. This is why batteries can be described as transducers, in which the chemical work associated with the cell reaction is converted into electrical work.

Batteries can be either primary or secondary power sources. The difference is quite simple. If the product of the cell reaction can be decomposed into the original reactants simply by applying a sufficient voltage to the battery for a long enough period of time, then the battery is called a secondary battery and is rechargeable. If the cell reaction is not reversible in this way, the battery is called a primary battery. Of course, as will be seen

later, most secondary batteries are not fully reversible.
The amount of charge (i.e. number of coulombs) obtainable
from the battery when it is first discharged usually cannot
be fully replaced during the recharging process, so that
less charge is obtained on the second discharge, and so on.
The charge/discharge process is normally called a cycle,
and the reduction of charge obtainable in second and
subsequent cycles is referred to as a loss of capacity.
(The term 'capacity' should not be confused with
capacitance which is a different type of parameter.) The
dependence of capacity loss on the number of
charge/discharge cycles is an important performance
parameter for a battery, and will be discussed later.

The word 'battery' originally meant several electrochemical
'cells' connected together either in series (to give a
voltage equal of the voltage for cell multiplied by the
number of cells) or in parallel (to give the same voltage
as for an individual cell, but a much larger current).
Nowadays, because the word 'cell' can have many different
meanings, electrochemical power sources are usually called
batteries. Sometimes, however, a battery during discharge
is called a galvanic cell, and a secondary battery under
recharge is called an electrolytic cell. For secondary
batteries the polarity of the electrodes, + or −, stays the
same whether the battery is under charge or discharge
conditions but, since reversing the cell reaction involves
changing an oxidation process into a reduction process and
vice versa, the positions of anode and cathode are
exchanged. This is shown in figure 1 and table 1.

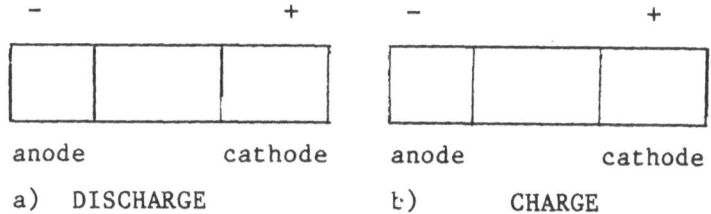

a) DISCHARGE   b)   CHARGE

Figure 1: Three-layer rechargeable batteries,
           consisting of left hand electrode/
           electrolyte/right hand electrode.

31

	charge	discharge
anode	+	−
cathode	−	+

Table 1. Polarity and cathode electrodes during charge or discharge processes.

So far we have been considering what a battery is. Now let us focus on what is special about a polymer battery. It is a solid state battery in which the electrolyte is a polymer electrolyte and the positive electrode (sometimes called the positive plate following the usage in the lead-acid battery field) may also contain flexible, plastic or elastic polymer material. The negative electrode is usually a metal foil. From now on, except when specifically referring to the recharging process, we will follow normal usage in the polymer battery field by calling the negative electrode the anode, and using the name 'catholyte' for the positive electrode. This is what is shown in figure 1a, except that we are now using catholyte in place of cathode. The reason is interesting, and arises from a special feature of the cell reactions used in polymer batteries (and indeed in certain other battery systems as well).

Polymer batteries, in practice, involve a reaction in which metal ions are inserted into a cathode material which is usually a layer lattice. The ions go between the layers and this process is called intercalation. During the intercalation process, the cathode material swells, and therefore the solid material surrounding the particles of cathode material has to be flexible enough to stay in contact even when the particles change size and shape. Consequently, a solid state battery involving an intercalation cathode material requires the particles of cathode material to be embedded in a plastic matrix, which is usually made from the polymer electrolyte material. In addition, fully intercalated cathode material is usually a poor electronic conductor and it is often necessary to add some carbon powder so that electrons can enter and leave the cathode particles on their way to and from the current collector. The region to the right hand side of a polymer battery therefore contains three components: particles of an intercalation material, normally called the 'active cathode'; particles of an electronic conductor such as carbon; and a surrounding material consisting of polymer electrolyte, which acts as an ionic conductor. These three components together form the catholyte compartment.

We can now consider what is meant by a polymer electrolyte. Being an electrolyte, it must conduct ions but not electrons (or electron holes), whereas an electrode must conduct both ionically and electronically. A more thorough description of electrochemical transport is given elsewhere [1, 2].

Most polymer materials on their own are insulators and conduct neither electrons nor ions. Certain polymers act as 'immobile solvents' for ionic salts and these solid solution materials are called polymer electrolytes; they are ionic conductors and electronic insulators. For readers who are new to the polymer field, a short description will now be given.

The term polymer describes many natural and synthetic materials. A broad definition will include any material which consists of large molecules. These macromolecules consist of many units, each being the size of a conventional small molecule. These units are linked to each other by conventional chemical bonds. This definition allows us to divide the properties of such materials into two categories:

i. Those properties which are independent of molecular size. They therefore depend on the nature of the atoms present and their bonding. They are entirely predictable without any need for a knowledge of polymer science.

ii. Those properties which are created or affected by the large molecular size.

The latter are our concern here. Most of them centre around the visco-elastic, or semi-solid, behaviour of polymers. However, the consequencies of this behaviour lead to a wide variety of structures and properties.

The polymeric materials to be considered here are principally those based upon chain structures. In the liquid state the chains are randomly coiled and, even when apparently solid, at least part of the material is amorphous, i.e. the randomness persists. This therefore excludes from our scope substances which are entirely crystalline, such as crystalline proteins, and also inorganic polymeric network structures such as zeolites which have symmetrically arranged units with bonding in three dimensions. The latter category includes many materials which are useful as solid electrolytes and intercalation electrodes, but these are not classed as polymeric for the purposes of this paper.

In a polymer chain, the individual units are often identical to each other, although not necessarily so. They are also usually divalent, having one linkage at each end. A chain of divalent units is called a linear polymer, where linear does not imply that the chain is straight; indeed the usual bond angle is 109°. The units at the ends of any polymer chain must be somewhat different from those in the centre; but such differences can usually be ignored unless specific chemical groupings, called end-groups, have been introduced to create particular chemical effects. The average number of units in the polymer chain is known as the degree of polymerisation. The length of the chains is usually measured and reported as the average relative molar mass (rmm), previously called molecular weight, which will be proportional to the degree of polymerisation and to the relative molar mass of the units.

Degrees of polymerisation may range from 10 to 10^6 with relative molar mass values being about 50 times these figures. They tend to be used in theoretical studies, being derived from a knowledge of the particular polymerisation process, or calculated from the relative molar mass. The molar mass can be determined by several experimental techniques, some (e.g. osmometry) being absolute whereas others (e.g. viscometry) are empirical. The results are averages. For some techniques the average is the conventional arithmetic mean, and the result is known as the number-average rmm. Other techniques yield a higher weight-average rmm. The difference between these arises because polymerisation methods often lead to a very wide spread of molecular sizes, covering several orders of magnitude.

Shorter chains may be referred to as oligomers. The
individual units are sometimes called monomer units, but
the term structural repeat unit (sru) is now preferred as
the units often differ from the parent monomer. An older
term for a structural repeat unit is a mer. The nature of
the end group is likely to be more significant for shorter
chains.

An occasional tri- or tetra-valent unit in a polymer chain
will produced a branched polymer. The presence of many
such units will lead to networks. These can also be
created by crosslinking linear polymers. Linear oligomers
with reactive end-groups may be crosslinked by added tri-
or tetra-functional reagents. Longer chains require
reactive sites, such as the double bonds in natural rubber,
or units added during copolymerisation. Polymers without
apparent reactivity can often be crosslinked by high-energy
gamma rays or neutron radiation.

Polymer electrolytes are thin solid films consisting of
ionic salts dissolved in an appropriate polymer. The
commonest polymer is poly(ethylene oxide), PEO, and the
most usual salts are $LiClO_4$ or $LiCF_3SO_3$ (lithium triflate).
Polymer electrolytes based on PEO are normally described as
$PEO_n:MX_z$ where n is the oxygen-to-cation ratio, $[O]:[M^{z+}]$,
M is the cation species of valence z, and X is a suitable
anion. Mechanically they behave like solids but the
internal structure, and consequently the conductivity
behaviour, closely resembles that expected of the liquid
state. The key features of these materials will be
discussed in the next section. It was shown by Berthier
et al [3] in 1983 from nmr studies that ionic conductivity
takes place in the amorphous regions.

An unusual feature of most polymer electrolytes when
compared with many other types of solid electrolyte, is
that not only cations (which migrate to the cathode during
galvanic discharge) but also anions (which go to the anode)
are mobile. This causes real problems in polymer
batteries, especially those that are designed to be
rechargeable. Why this is so can be clearly understood by
considering the type of cell reaction for which these
batteries were designed. A typical polymer battery,
designed for commercial application, is as shown in figure
2.

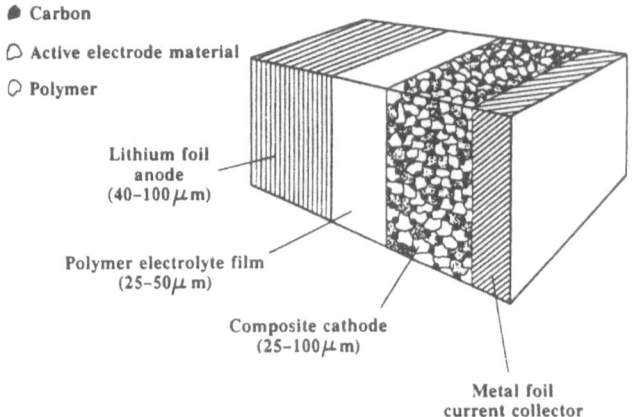

● Carbon
◗ Active electrode material
○ Polymer

Lithium foil
anode
$(40-100\,\mu m)$

Polymer electrolyte film
$(25-50\,\mu m)$

Composite cathode
$(25-100\,\mu m)$

Metal foil
current collector

FIG. 2. Solid-state cell configuration.

The designed cell reaction is made up of two half-cell reactions.

Oxidation at the anode: $\qquad\qquad xLi = xLi^+ + xe^-$ (1a)

Reduction at the cathode: $xLi^+ + V_6O_{13} + xe^- = Li_xV_6O_{13}$ (1b)

to give the overall reaction $\qquad V_6O_{13} + xLi = Li_xV_6O_{13}$ (2)

The Li^+ ions travel across the electrolyte and the electrons leave the anode through a wire, travel through the 'load' (heater, motor, lamp, etc.), and then proceed along a wire to a current collector (typically Ni foil) to which the catholyte is attached. It is comparatively easy to insert (and subsequently to remove on recharge) 8 Li^+ ions per formula unit of the V_6O_{13} active cathode material, i.e. $0 \leq x \leq 8$.

The anions are also mobile in many polymer electrolytes, examples of which include $PEO_n:LiClO_4$ and $PEO_n:LiCF_3SO_3$, $4 \leq n \leq 100$. Such anions move towards the anode and can react with Li^+ cations at the interface to form a thin layer of poorly conducting lithium salt.

$\qquad Li^+ + X^- = LiX \qquad\qquad\qquad \ldots$ (1c)

This increases the cell resistance, decreases the number of

36

Li$^+$ ions and reduces the capacity of the battery, as it is difficult to redissolve the ions on recharge. This anion migration process is limited by the fact that the cathode does not provide a source of anions to replenish those displaced in the electrolyte. Consequently depletion of anions from the right hand side of the electrolyte and increase in their concentration on the left hand side means that a concentration gradient, opposing the electric field-induced migration, is set up and the battery becomes concentration polarized with respect to anions.

It is useful to measure the proportion of ionic charge carried by positively charged and negatively charged species. These important performance parameters are called the cationic transference number τ_+ and the anionic transference number τ_- respectively. They are defined as:

$$\tau_+ = \frac{\sigma_+}{\sigma_i} \quad \text{and} \quad \tau_- = \frac{\sigma_-}{\sigma_i} = 1 - \tau_+ \quad \dots \quad (3)$$

Where $\sigma_i \, (= \sigma_+ + \sigma_-)$ is the total ionic conductivity and σ_+ and σ_- are the cationic and anionic contributions respectively.

The cationic transference number is the sum of the transport numbers (t) of each individual cation species and likewise for τ_- and anionic species. It might be thought that, in for example PEO$_n$:LiClO$_4$, the only cation species is Li$^+$, but this is not so. Typical values of n are in the range 8 to 30, and at these very high concentrations (2.5M to 0.7M), ion pairing and clustering are common Such species as $[Li_2ClO_4]^+$ and $[Li(ClO_4)_2]^-$ can arise, which each have their own transport numbers and these contribute to the total cationic and anionic transference numbers respectively [4]. Thus

$$\tau_+ = \Sigma(t_+)_j \quad \text{and} \quad \tau_- = \Sigma(t_-)_i \quad \dots \quad (4)$$

for each of the j species present. For PEO$_n$:LiClO$_4$ for example,

$$\tau_+ = t_{(Li+)} + t_{(Li_2ClO_4^+)} \quad \dots \quad (5)$$

Neutral ion pairs can also contribute to the ionic charge transport process [5] within polymer batteries! This must initially seem very surprising since ionic species need to be charged whereas ion pairs are not. The way the ion-pair conduction process works can best be understood by considering a test cell consisting of two identical metal electrodes in contact with a polymer electrolyte whose conductivity is being examined. Let us assume that, as in the particular case shown in fig.3, anions X^- may migrate under an electric field and neutral ion pairs M^+X^- may diffuse, but that cations M^+ are immobile. (An example in which the cation is in fact immobile is the divalent polymer electrolyte $PEO_{16}:CaI_2$ [6]). Let us look at what happens when a battery is connected to this test cell. A cation near the negatively charged electrode (which is the cathode in this situation which corresponds to the electrolytic, recharge configuration described earlier) deposits or 'plates out' on the electrode surface and is reduced to M. This leaves an isolated anion which can migrate across the cell under the electric field gradient. At the right hand electrode, it can pick up a cation to form an ion pair, which can then diffuse under a concentration gradient back to the left hand electrode. The cation can then plate out and the whole process can be repeated a very large number of times, transferring material from the right hand to the left hand electrode. Although the cation cannot migrate on its own, it can move when accompanied by an anion, like a prisoner being allowed to travel along a street but only under escort. This process is illustrated in figure 3.

In summary, the special features of polymer batteries are:-

1. both anions and cations are usually mobile in the electrolyte;

2. because intercalation cathode materials are used, the catholyte compartment also incorporates polymer electrolyte;

3. the battery configuration involves thin layers in intimate contact, so that the batteries are flexible as well as being compact, robust and spill-proof;

4. there are usually problems of declining capacity with successive charge/discharge cycles in secondary polymer batteries.

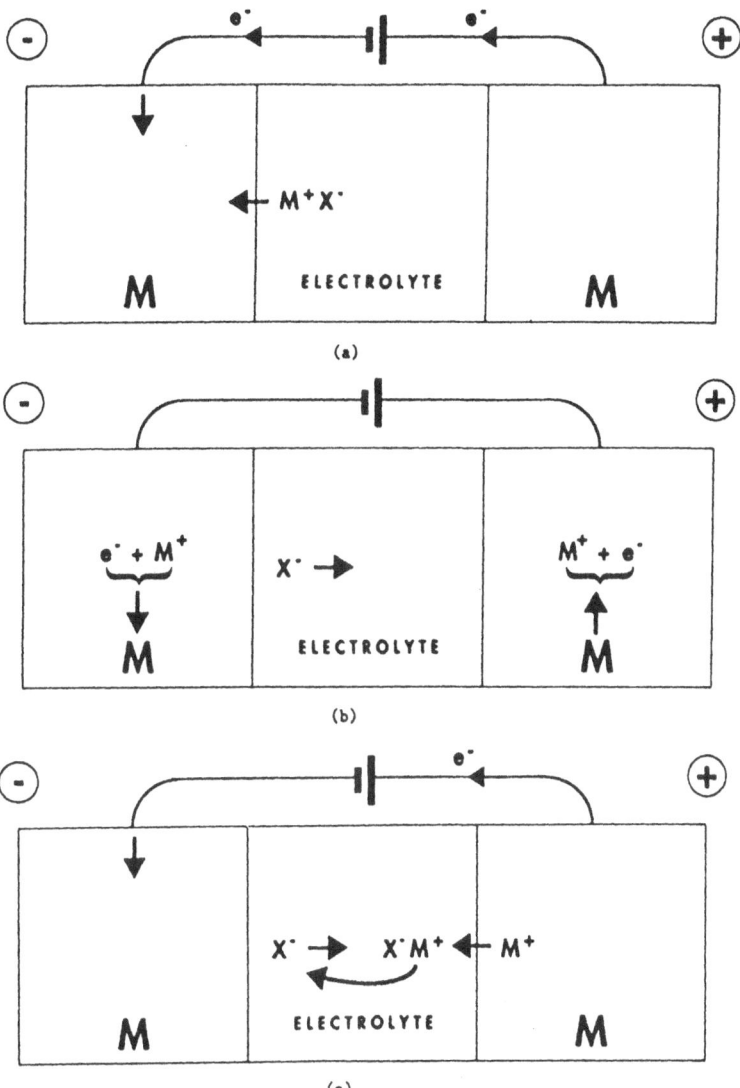

Fig.3. A test cell in which neutral ion pairs M^+X^- may diffuse and anions, X^-, may migrate under the influence of an applied field; cations, M^+, are treated as immobile within the bulk of the electrolyte. Stage 1: (a) Removal of the cation component of an ion pair by migration to the interface of the negative electrode, leaving an unpaired anion at the left-hand side of the electrolyte; Stage 2: (b) Migration of anion from left to right under the influence of both an electric field and a concentration gradient, and migration of the new cation from the right-hand electrode, leading to the formation of an ion pair near the right-hand interface; followed by diffusion of the ion pair from right to left under a concentration gradient. The net result is to transfer species M across the cell, despite the immobility of M^+.

What are Polymer Electrolytes?

Polymer electrolytes are thin solid films consisting of ionic salts 'dissolved' in an appropriate polymer; mechanically they behave like solids but the internal structure, and consequently the conductivity behaviour closely resembles that expected of the liquid state. The polymer acts as an 'immobile solvent', a term introduced by Michel Armand in 1983 [7]. The salt only remains in solution if an appropriate polymer is used. Such polymers contain hetero-atoms such as O or N; the most common example is poly(ethylene oxide), PEO, (typically of molecular weight of 4,000,000), for which the repeat unit is $-CH_2-CH_2-O-$.

Polymer electrolytes based on PEO are normally written as $PEO_n:MX_z$ where n is the oxygen-to-cation ratio, i.e. $[O]:[M^{z+}]$; M is the cation species, usually Li although a wide range of other monovalent (Na, K, Rb), divalent (Hg, Ca, Co, Ni, Cu, Zn, Cd, Hg, Pb) and even trivalent (Al, Nd) species have been studied; X is a suitable anion, usually one that can be described as 'soft' such as CF_3SO_3 (triflate), ClO_4 (perchlorate), CNS (thiocyanate), BF_4 (tetrafluoroborate), AsF_6 (hexfluoroarsenate), I (iodide), although other halides such as Br (bromide), Cl (chloride) and even, with certain cations, F (fluoride) have been used; z is the cation valence.

The PEO repeat unit is the same unit as found in crown ethers, which are used as phase-transfer agents for promoting the solubility of ionic salts in low polarity organic solvents. In the case of crown ethers, the lone pairs of electrons on the oxygen atoms co-ordinate with the metal cations and the methylene groups interact with hydrocarbon or similar solvents. The crown ethers can be thought of as amphiphilic molecules, the oxygen-containing regions being hydrophilic and the $-CH_2-CH_2-$ portions being hydrophobic.

The situation in polymer electrolytes based on PEO is similar to that in crown ethers; the cations appear to co-ordinate with the lone pairs on the oxygens. For both PEO based polymer electrolytes and for crown ethers, the anions are held loosely by Coulombic attraction in not precisely-known locations in the vicinity of the cations. In most polymer electrolytes, both cation and anion conductivity is possible.

The prime difficulty in developing an effective polymer electrolyte is to achieve a successful balance of ion-polymer interactions. These must be strong enough to keep the salt in solution, and yet weak enough to permit the ions to move.

If the cation-polymer interaction is too strong, then clearly the cations will have a very low mobility. This is not the only problem that results from over-strong cation-polymer interactions. Typically, a given cation is linked at any one instant of time, to four or more oxygens, which often are located on more than one polymer chain; thus, transient ionic cross-links are formed which greatly restrict the local freedom of motion of the polymer chains. Torrell et al in Sweden [8] have shown that ionic motion depends on the ability of the polymer chains to flex, and so the rigidity imposed by the transient cross-links (induced by the cations) also impedes the mobility of the anions. In addition, many cations interact with PEO in such a way that crystalline, high-melting salt-polymer complexes are formed. This obviously reduces the number of ionic carriers available for conduction.

Conductivity directly depends on the product of the number of carriers (N) and their mobility (μ), and it is clear that over-strong cation-polymer interactions reduce both N and μ.

When the cation-polymer interaction is insufficient, the salt deposits out on the surface of the film. This is a particular problem with PEO_n:RbX electrolytes. Often, this type of phase separation (sometimes erroneously called 'salting-out') occurs at elevated temperatures only, as would be expected for systems exhibiting lower critical solution temperature (LCST) behaviour.

In typical polymer electrolytes of the form PEO_n:MX_z, n is between 6 and 100. These values correspond to very concentrated solutions, far outside the range where Debye-Huckel or Debye-Huckel-Onsager behaviour is applicable. The nearest analogy is to molten salt behaviour, but even the classical theoretical approach of Fuoss cannot be applied, and one has instead to use the dynamic percolation theory developed by Ratner [9]. The practical consequence of such high concentrations is that ion-pairing is prevalent. This is a major problem that reduces the number of available carriers, and hence the potential conductivity, although Bruce and Vincent [10] and also Ingram [11] have proposed intriguing conduction mechanisms involving the migration of ion pairs under a concentration gradient, as discussed above.

Besides ion-pairing, another major problem is mixed morphology. In addition to the crystalline polymer-salt complexes referred to above, certain polymers including PEO have a propensity to form low-melting, salt free crystalline regions. These have the effect of increasing the actual salt concentration in the conducting amorphous regions to a higher value than would be anticipated from the overall oxygen-salt ratio, n, i.e. the value of n in the conducting region is lower than expected since some of the polymer oxygens are not co-ordinated to cations.

The crystalline material is spherulitic, i.e. it consists of thin crystalline sheets called lamellae or fibrils, radiating out in three dimensions from a central core, like the woollen strands in the 'bobble' or 'pom-pom' on a knitted ski-hat. Amorphous material is trapped between the lamellae so that the spherulites are always a mixture of crystalline and amorphous material. At equilibrium, all of the polymer film becomes spherulitic as the spherulites grow, initially with a circular cross-section until they touch, and subsequently so that they form straight, inter-spherulite boundaries similar to grain boundaries. Consequently, at equilibrium, the conducting amorphous region is in intimate contact with insulating but thin crystalline material. The conducting pathway is therefore somewhat tortuous. Furthermore, the dielectric behaviour within the spherulite resembles that of a leaky capacitor, which can be modelled in an equivalent circuit sense by a constant-phase-element (cpe). Such elements produce complex impedance plots which are characterised by depressed semi-circles and tilted spikes, exactly as found in practice for polymer electrolytes. Advantageously, however, the close juxtaposition of conducting (amorphous) and insulating (crystalline) regions resembles the situation in 'composite electrolytes' [12]. In these materials, insulating powdered substances such as alumina are incorporated into poorly conducting salts such as LiI, and conductivity enhancements of several orders of magnitude are observed. Various explanations have been advanced, that of Dudney [13] which involves plastic deformation at the interface, probably being the most plausible.

Work at Leicester and elsewhere suggests that the morphology can be controlled by careful attention to the preparation conditions and thermal history of the polymer [14-17]. The equilibrium situation described above is not always reached, as the kinetics of polymer crystallisation are slow. We have observed significant crystallisation in $PEO_8:LiCF_3SO_3$ within hours and the practical limit of crystallisation (about 60%) is reached in a few days, whereas for the similar system $PEO_8:LiClO_4$, complete crystallisation can take months. In some systems such as $PEO_8:ZnI_2$, the timescale, depending on preparation conditions, can take years. This increase in degree of crystallinity with time has an adverse effect on conductivity and means that, for practical applications, mixed morphology films have a shelf-life problem.

Many approaches have been used to suppress or control crystallinity. These include:-

i.   the use of non-crystallisable polymers such as poly(propylene oxide), PPO;

ii. the use of plasticisers such as ethylene carbonate (EC), propylene carbonate (PC), low molecular weight poly(ethylene glycol) PPG which is hydroxy-end-capped, or its methyl-end-capped analogue PPM;

iii. gentle irradiation cross-linking;

iv. the use of a second, structure-forming salt;

v.   the incorporation of 'fillers'.

Crystallinity is not entirely disadvantageous. Many workers, including Owens [18], argue that the desirable mechanical characteristics provided by a small degree of crystallisation more than outweigh the disadvantages of a slight loss in conductivity.

Compared with both liquid, and with other solid electrolytes, polymer electrolytes have very poor conductivities, typically 10^{-4} to 10^{-7} S cm^{-1} at room temperature. This is not an overriding problem however, as the performance parameter relevant to most applications is in fact the conductance, not the conductivity. Since the films can be produced with thicknesses of 10 to 100 µm, the film resistances become comparable to other materials. For example, a compacted disk made from powdered $Cu_{16}Rb_4I_{13}Cl_7$, which has the highest conductivity of any solid electrolyte, $\sigma = 0.47$ S cm^{-1} at 25°C, is difficult to produce in a form in which the conductance is more than a factor of 10 better than for a polymer electrolyte. The same is true for many glassy electrolytes [19]. None of these alternative solid electrolytes offer the flexibility of configurations that can be obtained for polymer electrolytes.

Attempts have been made to produce electrolytes with a cationic transference number of unity. In order to suppress anion conduction, poly-electrolytes with a negatively charged polymer backbone have been made [20-22] but such an approach appears to reduce chain mobility and reduce the cation conductivity to an unacceptable level.

Historically, polymer-salt complexes were first developed for divalent salts [23] in the mid-1960's. The 'salts' used were in fact covalent and non-conducting and the primary interest in these materials was from the point of view of polymer scientist who was interested in their structural behaviour.

The next benchmark date was 1973 when Wright developed PEO_n:LiX systems [24]. Again, his interest was from the perspective of a polymer scientist addressing structural questions.

The first realisation that useful electrolyte materials could be obtained was by Armand in 1978 [25]. He measured the conductivities of a range of polymer-salt systems and suggested guidelines, further developed in later work [26-28], about the potential suitability of various salts for battery applications. His studies developed from his PhD thesis investigation which involved the study of intercalation cathodes for battery applications. During the intercalation or de-intercalation process, significant volume changes (up to 15% for V_6O_{13} for example) are observed and consequently as discussed earlier, a flexible and deformable electrolyte is needed to maintain electrode-electrolyte contact. This necessitates the use either of a liquid electrolyte (non-aqueous in the case of Li which reacts with water and even with moisture vapour) or a plastic solid electrolyte and consequently he extended the work of Wright towards battery applications.

What are the Alternatives to PEO?

Until recently, there was considerable interest in alternative polymers such as:-

a) polyphosphazenes. These are essentially comb polymers consisting of a PN backbone, to which are attached side chains consisting of 2-12 repeat units of the $-CH_2-CH_2-O-$ type. Various acronyms, especially MEEP, have been used by Shriver and his co-workers to describe these materials [29]. They are amorphous which is good for conductivity, but have undesirable mechanical characteristics, being liquid-like and sticky.

b) PEI - poly(ethylene imine), used with NaI by Chiang [30];

c) PES - poly(ethylene succinate), used by Shriver [31];

d) PEA - poly(ethylene adipate), used by Armstrong's group [32];

e) mixed ionic-electronic conducting polymers, consisting of an electronically conducting polypyrrole backbone with ionically conducting CH_2-CH_2-O- side chains. These were developed by Owen [33].

A different approach to the problem of maintaining the solvation ability of PEO without suffering its attendant disadvantages of low room temperature conductivity and high crystallisation is to use a network polymer [4]. Typically these use polysiloxane junctions, each linked to four other such junctions by $-CH_2CH_2-O-$ units, like a fishermans net in which the knots are siloxane units and the rope lattice is like PEO.

This system was developed by Watanabe [34]. An overriding problem for divalent polymer electrolytes is that many divalent salts catalyse other reactions within siloxanes.

These network polymers represent a class of gel polymer electrolytes in which a pre-formed polymer film is swollen by infusion of a salt dissolved in a liquid solvent. Low salt concentrations and phase separation can both be problems in such systems, although it is known that for Li systems, concentrations as high as one $LiClO_4$ to eight PEO repeat units can be achieved. The network polymer concept was developed for battery use; the pre-formed films are rather thick, however, which reduces the conductance.

For use in electrochromic devices, there is interest in the USA in laser-type materials. These include polymer/neodymium salt electrolytes. There is a further interest, in smart window films using a variety of optically active configurations [35].

For sensors applications, a range of divalent and/or potentially redox active materials based on $PEO_n:MX_2$ (M = Co, Ni, Cu, Zn etc.) are being developed.

What Techniques are Used to Study Polymer Electrolytes?

This topic has been recently addressed in detail [36] and only a summary will be given here.

Conductivity and transport number measurements are clearly important. The special features pertaining to polymer electrolytes have been analysed [5].

Structure-conductivity relationships, of vital importance in the solid state electrochemistry field [37] need a complementary set of structural and morphological methods. These include:-

1. differential scanning calorimetry, DSC and other thermal techniques [38] which give indications of the degree of crystallinity;

2. variable temperature polarising microscopy (VTPM), Scanning Electron Microscopy and other microscopy-based techniques [39];

3. X-ray powder diffraction, to characterise the crystalline regions [40];

4. extended X-ray absorption fine structure (EXAFS) which provides local structural information in both the amorphous and crystalline regions [14, 17, 41].

What Techniques are Used to Study Polymer Batteries?

Again these have been recently reviewed [42] and will only be summarised here. The first performance parameter that is measured for a new developmental solid state battery is its open circuit voltage, ocv. This is not identical to the emf, which is the voltage expected from the supposed cell reaction, and which can be easily calculated from thermodynamic data. The ocv is the voltage actually measured under infinite resistive load and may differ from the emf for several reasons, including:

a) the actual cell reaction may not be the one that was anticipated;

b) the working cell may be in series with a corrosion or similar cell involving reaction between the electrodes and the current collectors;

c) self-discharge may occur because of significant internal electronic conduction through the electrolyte or, if the cell is badly designed and/or constructed, through the encapsulant. This lowers the ocv by the product of the current flowing and the ionic self-discharge resistance, i.e. by an internal IR drop.

Under load, the measured voltage will be lower than the ocv, both because of an IR drop, and because the battery components may be modified by the passage of charge. An extreme, but common, example of the latter occurs in batteries involving ion-insertion electrodes.

Power and energy densities. These can be calculated on either a volumetric (per unit volume) or a gravimetric (per unit mass) basis, and in each case either the collectors, encapsulants etc., can be included (i.e. 'gross') or the densities can be calculated on the basis of the electrodes and electrolyte only (i.e. 'net'). The units for power density are typically W/kg or W/L and for energy density

47

are W-hr/kg or W-hr/L, where 1 W-hr = 3.6 kJ; the distinction between net and gross is not always made clear.

Currrent density, expressed as A/cm^2 is also important.

For solid state batteries, the area is usually taken as the nominal cross sectional area rather than the true area of contact. The latter can be considerably smaller than the former for relatively rigid materials whereas for other systems such as liquid electrolytes in contact with porous electrodes, the true area of contact can be very much larger than the apparent cross sectional area.

The battery efficiency is a useful parameter, normally calculated as the charge actually passed during the discharge lifetime, t_L, of the cell, divided by the charge theoretically available from the limiting electrode (i.e. mzF/M where m is mass, z is stoichiometric charge number, F is the Faraday constant, 96487 C/mol, and M is the relative molar mass).

For secondary batteries, the rate at which recharge is possible is technologically important. This parameter is normally quoted in terms of C/n where C is the charge passed either in the preceeding or in the initial discharge half-cycle (the distinction is not always made explicit) and n is the number of hours over which recharge proceeds. Most rechargeable room temperature solid state systems cannot be recharged as rapidly as say a lead-acid accumulator or a NiCd battery and typical recharge rates are C/30.

For a solid electrolyte the relevant performance parameter for conduction is ionic conductance rather than conductivity. As already mentioned, even materials such as LiI in which the conductivity is as low as 10^{-7} S cm^{-1}, can be used in a thin film configuration that yields a conductance of 10^{-4} S. This conductance value corresponds to current densities in the range 1-100 μA cm^{-2} for typical battery voltages of 0.4-4 V and cross-sectional areas of $\sim$ 4 cm^2.

Acknowledgements

The British Council and the Conference organisers are thanked for travel and subsistence support for this meeting. Duracell (UK) Ltd., SERC and the former UK National Advisory Board are thanked for financial suppaort of the research programmes at Leicester on which parts of this paper are based. Helpful comments on this manuscript from Dr.R.J.Latham are gratefully acknowledged.

References

1. Latham, R.J. and Linford, R.G. (1987) in Electrochemical Science and Technology of Polymers 1, edited Linford, R.G., Elsevier Applied Science, London, pp.1-21.

2. Linford, R.G. and Hackwood, S. (1981) Chemical Reviews, 81, p.327.

3. Berthier, C., Gorecki, W., Minier, M., Armand, M.B., Chabagno, J.M. and Rigauol, P. (1983) Solid State Ionics, 11, p.91.

4. Vincent, C.A. (1990) in Electrochemical Science and Technology of Polymers 2, edited Linford, R.G., Elsevier Applied Science, London, pp.47-96.

5. Linford, R.G. (1990) in Electrochemical Science and Technology of Polymers 2, edited Linford, R.G., Elsevier Applied Science, London, pp.281-318.

6. Farrington, G.C. and Linford, R.G. (1989) in Polymer Electrolyte Reviews 2, edited MacCallum, J.R. and Vincent, C.A., Elsevier Applied Science, London, pp.255-284.

7. Armand, M.B. (1987) presented at the 6th International Conference on Solid State Ionics, Garmisch-Partenkirchen, paper A4-6 (unpublished).

8. Kakihana, M., Schantz, S., Torrell, L.M. and Morjesson, L. (1988) Mat. Res. Soc. Symp. Proc., 135, p.3651.

9. Druger, S.D., Nitzan, A. and Ratner, M.A. (1983), Solid State Ionics, 9/10, p.115.

10. Bruce, D.G. and Vincent, C.A. (1989), Faraday Discuss. Chem. Soc., 88, p.43.

11. Cameron, G.G., Ingram, M.D. and Harview, J.L. (1989), Faraday Discuss. Chem. Soc., 88, p.55.

12. Liang, C.C., Joshi, A.V. and Hamilton, N.E. (1978), J. Appl. Electrochem., 8, p.445.

13. Dudney, N.J. (1988), Solid State Ionics, 28-30, p.1065.

14. Cole, M., Sheldon, M.H., Glasse, M.D., Latham, R.J. and Linford, R.G. (1989), Applied Physics A, 49, p.249.

15. Bruce, P.G., Krok, F., Evans, J. and Vincent, C.A. (1988), Brit. Polymer J., 20, p.193.

16. Glasse, M.D., Linford, R.G. and Schlindwein, W.S. (1989), in Second International Symposium on Polymer Electrolytes, edited by Scrosati, B., Elsevier Applied Science, London, p.203.

17. Latham, R.J., Linford, R.G., Pynenburg, R. and Schlindwein, W.S. (1991), Mat. Res. Soc. Symp. Proc., 210, p.273.

18. Munshi, M.Z.A. and Owens, B.B. (1987), Appl. Phys. Commun., 6, p.279.

19. Akridge, J.R. and Vourlis, H. (1986), Solid State Ionics, 18-19, p.1082.

20. Dominquez, L. and Meyer, W.H. (1988), Solid State Ionics, 28-30, p.941.

21. Maier, J., Mater, Chem. Phys. (1987), 20, p.317.

22. Poulsen, F.W., J. Power Sources (1987), 20, p.317.

23. Blumberg, A.A., Pollack, S.S. and Hoere, C.A.J. (1964), J. Polym. Sci. A., 2, p.2499.

24. Fenton, D.E., Parker, J.M. and Wright, P.V. (1973), Polymer, 14, p.589.

25. Armand, M.B., Chabagno, J.M. and Duclot, M. (1978), Second Int. Conf. on Solid Electrolytes, St.Andrews, paper 65 (unpublished).

26. James, D.B., Wetton, R.E. and Brown, D.S. (1979),
    Polymer, 20, p.187.

27. Wissbrun, K.F. and Hannon, M.J. (1975), J. Poly. Sci.
    Polym. Phys. Ed., 13, p.223.

28. Fontanella, J.J., Wintersgill, M.C., Calame, J.P. and
    Andeen, C.G. (1985), J. Polym. Sci. Polym. Phys. Ed.,
    23, p.113.

29. Blonsky, P.M. Shriver, D.F., Austin, P.E. and Allcock,
    H.R. (1986), Solid State Ionics, 18/19, p.258.

30. Chiang, C.K., Davis, G.T., Hardin, C.A. and Takahashi,
    T. (1986), Solid State Ionics, 18/19, p.300.

31. Dupon, R., Papke, B.L., Ratner, M.A. and Shriver, D.F.
    (1984), J. Electrochem. Soc., 131, p.586.

32. Armstrong, R.D. and Clarke, M.D. (1984), Electrochim.
    Acta, 29, p.1443.

33. Minett, M.G. and Owen, J.R. (1988), Solid State Ionics,
    28-30, p.1192.

34. Watanabe, M., Nagano, S., Sanul, K. and Ogata, N.
    (1986), Polymer J., 18, p.809.

35. Deroo, D. (1990) in Second International Symposium on
    Polymer Electrolytes, edited Scrosati, B., Elsevier
    Applied Science, London, p.433.

36. Linford, R.G. (1988) in Proceedings of the
    International Seminar on Solid State Ionic Devices,
    edited Chowdari, B.V.R. and Radhakrishna, S., World
    Scientific Press, Singapore, p.551.

37. Transport-Structure Relations in Fast Ion and Mixed
    Conductors, edited Poulsen, F.W., Anderson, N.K.,
    Clausen, K., Skaarup, S. and Sorenson, T.T. (1985),
    Riso National Laboratory, Roskilde, Denmark.

38. McGhie, A.R. (1990) in Electrochemical Science and
    Technology of Polymers 2, edited Linford, R.G.,
    Elsevier Applied Science, London, p.201.

39. Cope, B.C. and Glasse, M.D. (1990) in Electrochemical Science and Technology of Polymers 2, edited Linford, R.G., Elsevier Applied Science, London, p.233.

40. Wendsjo, A., and Yang, H. (1990) in Second International Symposium of Polymer Electrolytes, edited by Scrosati, B., Elsevier Applied Science, London, p.225.

41. Latham, R.J., Linford, R.G. and Schlindwein, W.S. (1989), Faraday Discuss. Chem. Soc., 88, p.103.

42. Linford, R.G. (1988) in Proceedings of the International Seminar on Solid State Ionic Devices, edited Chowdari, B.V.R. and Radhakrishna, S., World Scientific Press, Singapore, p.527.

Dynamical Properties of Fast Ion Conducting Borate Glasses

M. Balkanski[*1], R.F. Wallis,[1] J. Deppe[1] and M. Massot[2]

[1]Department of Physics, University of California, Irvine, CA 92717, USA
[2]Laboratoire de Physiques des Solides, Université Pierre et Marie Curie
4, Place Jussieu-75252 Paris Cedex 05, France

Experimental and theoretical work which leads to an improved understanding of the physics of ion diffusion in lithium doped borate glasses is presented. The matrix vibrations of undoped, i.e., B_2O_3, and doped, i.e. B_2O_3-xLi_2O, borate glasses are discussed in connection with Raman scattering and infrared reflectivity measurements. Addition of the modifier Li_2O is shown to cause a transformation of three- to four-coordinated boron atoms. The introduction of dopants, e.g. Li_2SO_4 or LiX (with X = F, Cl, Br, I) and their effect on the ionic conductivity are discussed.

1.   INTRODUCTION

Fast ion conducting glasses have attracted considerable interest in recent years in view of their use as separators in solid state batteries. A solid state cell is formed of three components: an ion source, usually Li; a dielectric insulator but good ion conductor such as the lithium borate glass, for example, which acts as a separator between the anode or ion source and the cathode or electron exchanger; and an insertion compound acting as an electron exchanger. The requirements of the separator are to be shape adaptable in order to assure good contact with the ion source and the exchanger and to have a good ionic conductivity.

Another important characteristic is that the separator should be as perfect a dielectric as possible in order to prevent electronic leakage and assure a long shelf lifetime of the batteries. The borate glasses are one of the best materials among those known today to respond to all these requirements.

The boron oxide glass is a very good insulator and constitutes the glass former. Upon the addition of a modifier such as Li_2O, for example, the structure of the glass is significantly changed. In addition, some fraction of the Li_2O dissociates releasing free lithium ions and thus considerably increasing the ion conductivity of the glass. A further increase of the conductivity is achieved by adding a dopant containing the mobile ion, Li in this case. The effect of the dopant upon the concentration of free ions available for the conduction process has been investigated recently [1]. Spectroscopic investigations on the structure of borate glasses with different modifier concentrations and different dopants have also been extensively developed and a summary is given in a recent publication [2].

The structural modifications conditioning the fast ion conduction in borate glasses are inferred mainly from spectroscopic data which reflects the dynamical properties of the glass matrix. It was therefore important to develop theoretical studies on the vibrations of the glass in parallel with the experimental investigations. We shall present here simultaneously the theoretical and experimental results on the vibrational properties of the borate glass with the aim of reaching some conclusions on the conditions for increasing the ionic conductivity of the glass.

2.   VIBRATIONS OF BORON OXIDE GLASS

The structure of the boron oxide glass has been studied by X-ray [3], Raman [4], infrared [5], nuclear magnetic resonance [6] and neutron diffraction [7] measurements. According to these observations, boron oxide glass is formed of an infinite network of boroxol groups B_3O_3. The six atoms inside the group form a planar ring and the B-O-B angle is 120°. These boroxol hexagons are interconnected by an oxygen bridge -0-, and the external B-O-B angle θ, less accurately determined, is about 130°. The relative geometry of two successive boroxol planes is not well known.

One of the characteristic features of the vibrational spectra of v-B_2O_3 observed by Raman scattering [2] and shown in Fig. 1 is the presence of a sharp peak at 800 cm^{-1} which is strong and well polarized [8]. Usually the vibrational spectra of amorphous solids are expected to

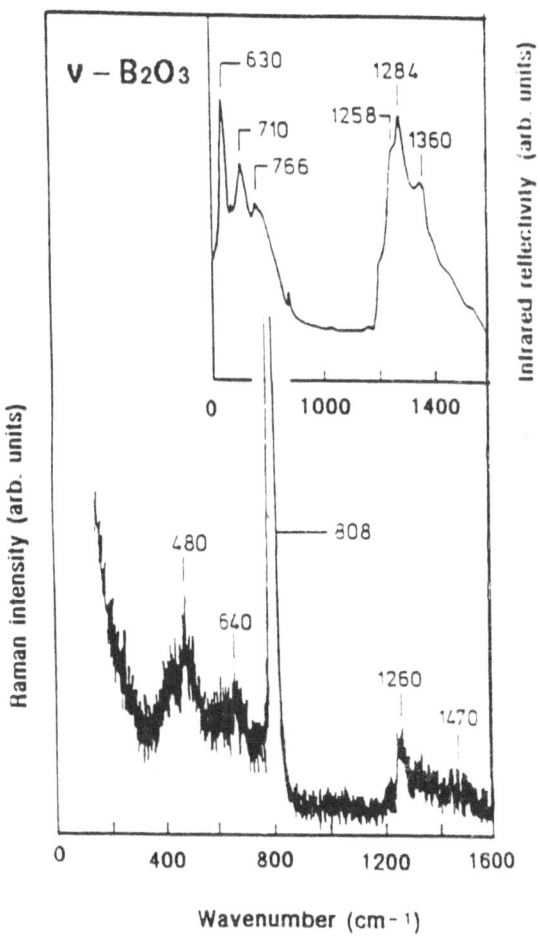

Fig. 1. Raman and infrared data for pure B_2O_3. The sharp peak at 808 cm^{-1} is attributed to the breathing mode of the six-membered boroxol ring.

consist of broad bands (see for example ref. 9). This is because the localized molecular mode responsible for a sharp peak couples with all other atoms and broadens when a network is formed. For example, Galeener et al. [8] have calculated vibrations of v-B_2O_3 using the Bethe lattice approximation (described in detail below) and obtained only broad bands due to extended modes.

There have been a few attempts to understand theoretically the possible existence of sharp features in the vibrational spectra. Kristiansen and Krogh-Moe [10] calculated the vibrational frequencies of the boroxol molecule with atoms of variable mass attached to the outer oxygen atoms and found that the mode at 800 cm^{-1} does not vary much when the mass is changed. Galeener et al. [11] explain the decoupling of rings from the rest of the network by a somewhat fortuitous cancellation of central and noncentral forces.

Galeener and Thorpe [12] have also discussed the introduction of 6-fold boroxol rings into a continuous random network model of v-B_2O_3. They conclude that the 800 cm^{-1} mode consists of a breathing motion of the intra-ring oxygen, and therefore does not couple to the outside network.

Kanehisa and Elliott [13] have demonstrated this decoupling explicitly using a cluster Bethe-lattice approximation.

3.   EFFECT OF THE ADDITION OF MODIFIER Li_2O to v-B_2O_3: THE SYSTEM B_2O_3-xLi_2O

3.1   Experimental Results

The Raman spectra of the binary glass B_2O_3-xLi_2O are shown in Fig. 2. At low Li_2O concentration, x = 0.1, the 800 cm^{-1} band corresponding to the breathing oxygen motion in the boroxol ring, the ν_2 mode trigonal deformation of the ring shown in Fig. 3, is still narrow and strong. A new band appears at 780 cm^{-1} which corresponds to the trigonal deformation of six-membered borate rings with one or two BO_4 units [2] as shown in Fig. 4 for the tetraborate, triborate or diborate configuration. The band at 500 cm^{-1} corresponds to the breathing mode ν_3 (Fig. 3) where the boron and the bridging oxygen atoms move in phase.

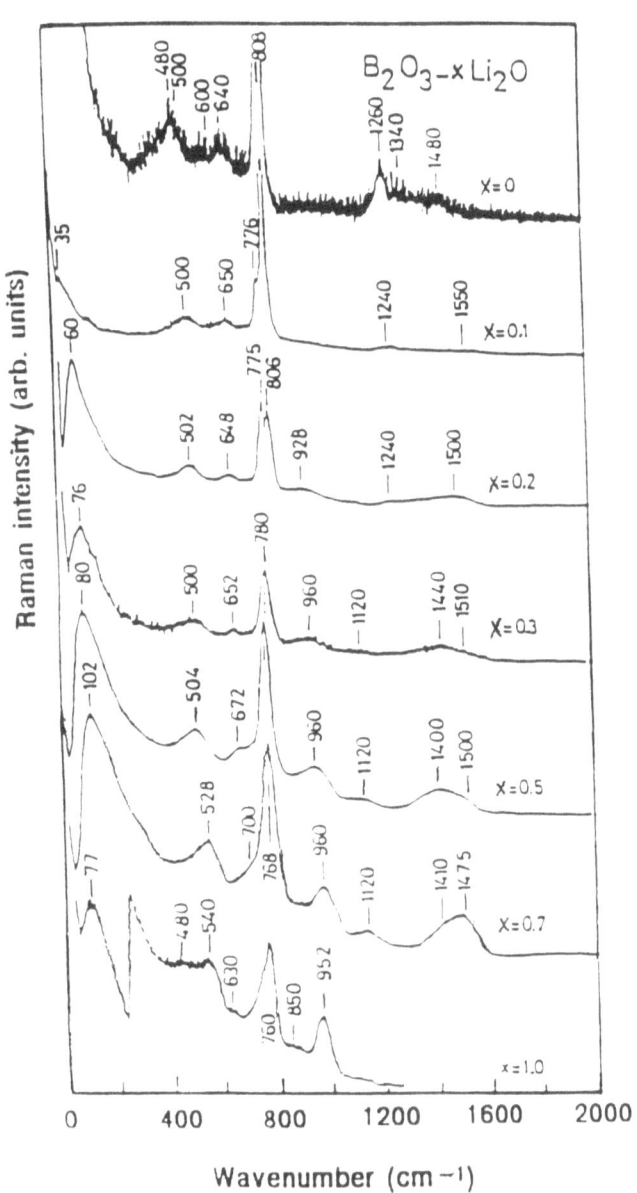

Fig. 2. Raman spectra of the binary glass system B_2O_3-xLi_2O.

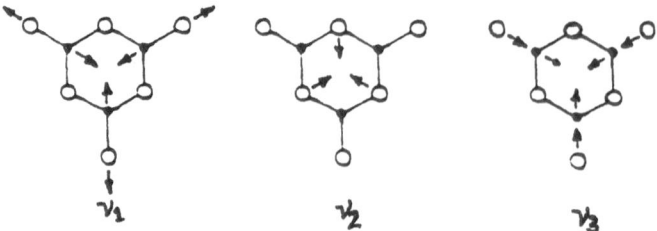

Fig. 3. The three symmetrical vibrations of the boroxol rings.

The mode ν_1, at 1240 cm^{-1}, is associated with the symmetric vibration of the same atoms. The band between 1400 and 1550 cm^{-1} is associated with the stretching vibrations of the B-O bonds in BO$_3$ units. The three regions associated with the stretching vibrations of the B-O bonds which link the boron-oxygen network will be modified upon addition of Li$_2$O, which leads to the transformation of trihedrally coordinated boron, B^{III}, to tetrahedrally coordinated boron, B^{IV}, according to the reaction

$$
\begin{array}{c}
^{-}O \\ \diagdown \\ \diagup \\ _{-}O
\end{array}
B\!-\!O\!-\!B
\begin{array}{c}
O^{-} \\ \diagup \\ \diagdown \\ O_{-}
\end{array}
+ \, Li_2O
\quad
\substack{\rightarrow \\ \leftarrow}
\quad
\left[
\begin{array}{c}
^{-}O \\ \diagdown \\ \diagup \\ _{-}O
\end{array}
B\!-\!O\!-\!B
\begin{array}{c}
O^{-} \\ \diagup \\ \diagdown \\ O_{-}
\end{array}
-O
\right]^{2-}
+ 2Li^{+}
$$

For low Li$_2$O concentration, the abundance of B^{IV} increases, (with increasing modifier concentration); as a consequence, the intensity of the line at 780 cm^{-1} increases while that at 806 cm^{-1} decreases and finally vanishes at x = 0.3. For x larger than 0.5, the frequency of the B^{IV} band decreases until it reaches 760 cm^{-1} in the metaborate glass x = 1.0.

At high Li$_2$O concentration (x > 0.5), the formation of non-bridging oxygen atoms is inferred from the relative intensity increase of the two bands at 960 and 1480 cm^{-1}.

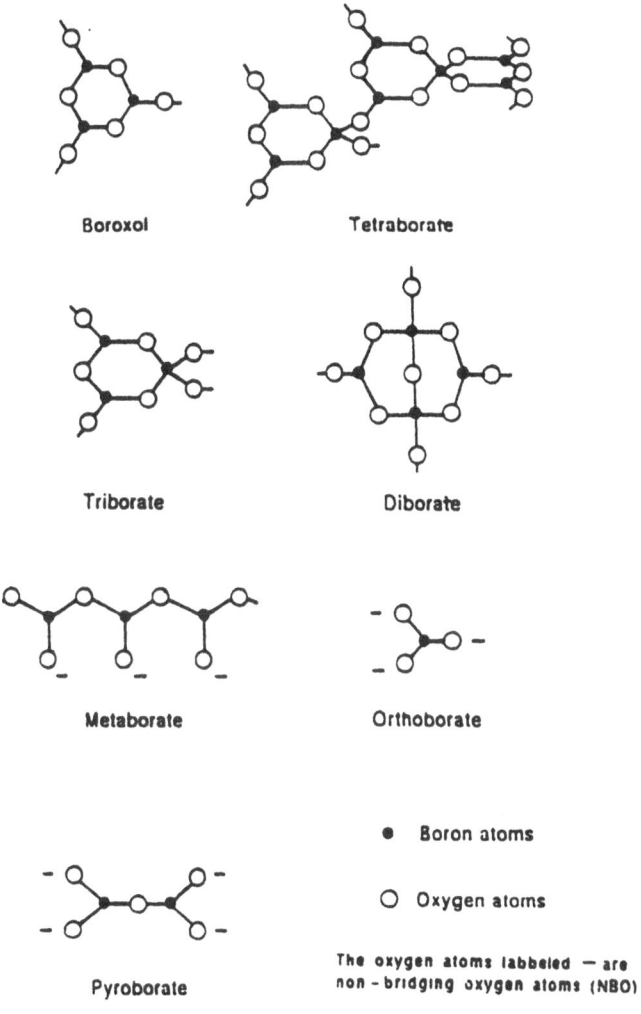

Boroxol

Tetraborate

Triborate

Diborate

Metaborate

Orthoborate

● Boron atoms

○ Oxygen atoms

The oxygen atoms labbeled — are non - bridging oxygen atoms (NBO)

Pyroborate

Fig. 4. Some typical borate groups.

Two stages of modification can be distinguished when the concentration of the modifier increases. During the first stage, the tetrahedrally coordinated boron units increase and lead to an increase of the linkage of the boron-oxygen network up to $x = 0.5$. Above this concentration, formation of BO_3 units with non-bridging oxygen atoms "opens" the borate units and consequently the degree of linkage of the network is decreased.

3.2 Theory

With the vibrational frequencies of groups of atoms in a given configuration deduced from spectroscopy data, it is of interest to calculate the vibrational spectrum with some specific short-range configuration. Toward this end, we have employed the network-dynamics theory and lattice dynamics in order to effect an indirect determination of the glass structure. Central force network-dynamics seems an ideal method to investigate the relation between local environmental changes and the modes of vibration, since the short range order can be easily modified.

3.3 Methods Of Calculation

3.3.1 Network dynamics

The network-dynamics method, introduced by Sen and Thorpe [14], considers only nearest-neighbor central forces in the continuous random network under consideration: the other forces, for example, long-range Coulomb and angle bending forces, are much smaller in many cases.

Thorpe and Galeener [15] have put the central-force model of Sen and Thorpe in a more general framework using a Lagrangian formulation.

As discussed in the previous section, we have seen that the addition of Li_2O to B_2O_3 is accompanied by the appearance of groups containing fourfold-coordinated borons. In addition, the concentration of these structures increases proportionally with the concentration of the modifier Li_2O up to a critical value x_c. We shall consider a network containing both threefold- and fourfold-coordinated borons. Indeed, for the composition B_2O_3-$0.5Li_2O$, one can envision a network of equal numbers of threefold- and fourfold-coordinated borons. Here we consider one such network, with each threefold-coordinated boron surrounded by fourfold-coordinated borons and vice versa. The equations of motion and their solution are presented in a separate publication [16].

60

3.3.2 Lattice dynamics

The lattice-dynamics calculations were carried out as an independent check on the network-dynamics results. As such, they were done in the nearest-neighbor central-force approximation. No disagreement was found between the lattice-dynamics and network-dynamics calculations. The two methods are quite complementary in examining the effect of varying various parameters, such as force constants, intermolecular bridging angles, or configurations on the vibrational spectrum of the solid.

3.4   Results And Discussion

In Fig. 5, the results of network- and lattice-dynamics calculations [16] are presented for $\beta/\alpha = 0.8$ where α and β are the force constants for four- and three-coordinated boron atoms, respectively. Note that the general features in the spectra, such as the narrowing of the middle frequency gap as compared to the data of pure B_2O_3, are found in the network-dynamics results. The widening of the high-frequency gap, presumably due to the difficulty of the modes associated with BO_3 groups to hybridize with those of the stiffer BO_4 groups, is also found. The narrow peak at 808 cm^{-1} in the Raman data for $x = 0$ is no longer visible, indicating that there are no longer any rings without at least one fourfold-coordinated boron. The new peak in the vicinity of 780 cm^{-1}, corresponding to the trigonal deformation of six-membered rings containing one or two $\overline{BO_4}$ units [17] increases in intensity, as can be seen by comparing Figs. 1 and 2. Note also that as the bridging angle θ approaches 90°, the modes decouple into a low-frequency singlet and high-frequency triplets characteristic of the threefold- and fourfold-coordinated molecules.

It is interesting to note that such simple calculations should yield results representing well the trends in the spectral distribution of different vibrational modes in the binary glass system as a function of modifier concentration. This result suggests that the force constants used are reasonable, and again underscores the importance of local order in understanding the vibrational response of covalently bonded glass.

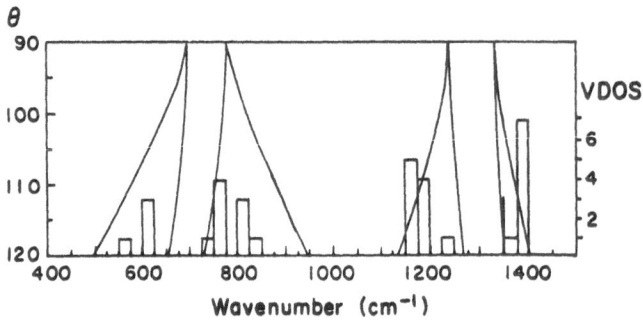

Fig. 5. Raman spectrum, network and lattice dynamics results for B_2O_3-$0.5Li_2O$. The ratio of the force constants of three-coordinated B-O bonds to that of four-coordinated ones is $\beta/\alpha = 0.8$. Histograms indicate the number of modes in the corresponding frequency band according to the lattice dynamics calculation for $\theta = 120°$.

4.   THE EFFECT OF DOPANTS

4.1   Structural Modifications Induced By The Dopant

Two types of dopants which introduce additional free Li ions and increase the ionic conductivity of borate glasses are examined: the lithium halides LiX (X = F, Cl, Br, I), and lithium sulfate Li_2SO_4. A general observation is that the dopants do not significantly modify the glass structure which simply localizes the anion leaving as the only mobile species the small lithium cation. The larger the anion, the stronger is its interaction with the host network.

In the ternary glass B_2O_3-$0.57Li_2O$-yLiX, in which the modifier concentration is such that the glass does not contain boroxol rings, the

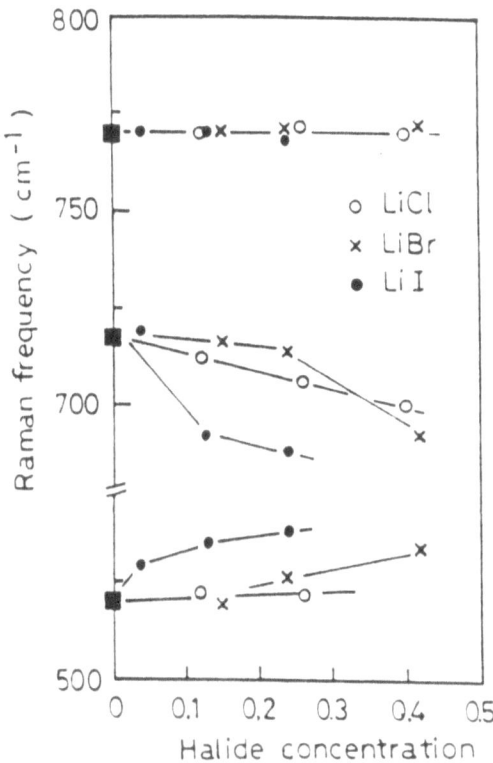

Fig. 6. The shift of the Raman frequencies versus the halide concentration.

modification due to the introduction of dopant salts is characterized in the Raman spectrum by the appearance of a shoulder at about 720 cm^{-1} on the 770 cm^{-1} band due to the vibration of four-coordinated boron atoms. For a given halide anion this new line shifts toward lower frequencies and its intensity increases with increasing dopant salt concentration, y, as shown in Fig. 6. At the same time the frequency of the band at 250 cm^{-1} due to the in-phase motion of the bridging oxygen and the boron atom increases. The stiffening of this mode, which is higher the larger the anion size, corresponds to a compression of the B-O-B band.

There is also an increase of intensity of the bands at 960 and 1480 cm^{-1} attributed to the vibrations of groups with non-bridging oxygen atoms. Sharpening of these bands and a shift toward higher frequencies is observed when going from LiCl to LiI with the same concentration.

These modifications of the boron-oxygen network of the ternary glasses have shown that in all cases the lithium halide addition results in an expansion of the network, the effect being most pronounced with LiI. These results agree with density and glass transition temperature measurements.

These observations are of prime importance for the understanding of the ion transport properties of the borate glasses. The BO_4 groups may be considered as negative ions with a large ionic radius which provide binding sites with small binding energy for the mobile lithium ions. At low modifier content the increase of ionic conductivity comes from the increase of the number of BO_4 units and of the number of free lithium ions. But for higher oxide content there is competition between two competing mechanisms: an increase in the number of free lithium ions provided by the dissociation of the dopant salt and conversion of the BO_4 units into BO_3 units with non-bridging negatively charged oxygen atoms which trap the lithium ions. In the range of large Li_2O modifier content the conductivity is less sensitive to the addition of the doping salt.

4.2   Effect Of The Dopant Li_2SO_4: B_2O_3-xLi_2O-yLi_2SO_4

Comparison of the Raman scattering spectra of the binary B_2O_3-0.7Li_2O and ternary B_2O_3-0.7Li_2O-0.42Li_2SO_4 systems, displayed in Fig. 7, shows that the spectrum of the Li_2SO_4 doped glass consists of the superposition of the spectra of B_2O_3-0.7Li_2O and Li_2SO_4. The peaks appearing at 456, 644, 1004 and 1100 cm^{-1} correspond respectively to ν_2, ν_4, ν_1 and ν_3 vibrations of the sulfate ion [18].

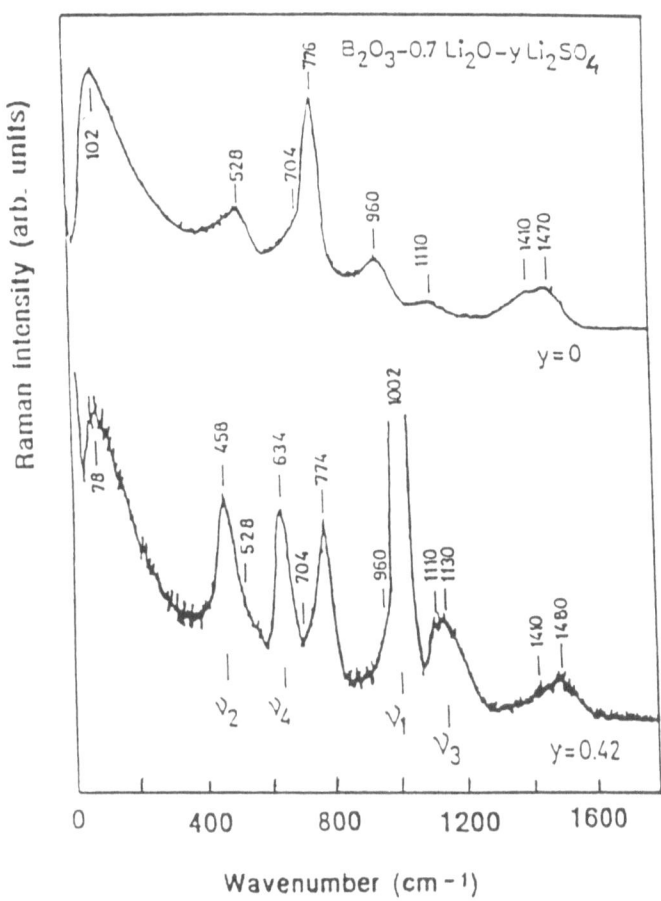

Fig. 7. Raman spectra of the ternary sulfoborate glasses
B_2O_3-0.7Li_2O-yLi_2SO_4 versus the lithium sulfate content.

5. CONCLUSIONS

The effect of the addition of the modifier Li_2O, was investigated experimentally by Raman and infrared spectroscopy and theoretically by network dynamics and lattice dynamics calculations. The modifier was found to increase the number of free lithium ions which act as charge carriers, and also to significantly modify the host structure through the transformation of trihedrally coordinated boron atoms into tetrahedrally coordinated boron. The concentration dependence shows that this second effect goes through a maximum for x = 0.4 and then decreases as a result of the formation of BO_3 groups with non-bridging oxygen atoms which carry a negative charge. Compared with the experimental data, the results of network and lattice dynamics calculations seems to give an adequate picture of the structure and vibrational response of lithium doped borate glasses.

Introduction of dopants in the form of alkali halide salts produces by dissociation an additional free lithium concentration and also a modification of the host configuration dependent on the size of the anion. Flourine enters in the network in substitutional positions forming BO_3F and BO_2F_2 units. Chlorine, bromine and iodine enter interstitial positions which produces breaking of B-O-B links and formation of BO_4 units. This modification of the boron oxygen network results in a less linked network facilitating the free ion diffusion. The effect is more pronounced for LiI and decreases with the anion size.

Doping with Li_2SO_4 leads to analogous effects. The effect of Li_2SO_4 on the Raman spectra can be understood as simply a super-position of the vibrations of the host network and that of the SO_4 groups. The modifications of the host network with preferential formation of diborate groups tends to produce an expansion of the network in order to accommodate the sulfate ions and facilitate the lithium mobility.

References

*Permanent address: Laboratoire de Physique des Solides, Université Pierre et Marie Curie, 4 place Jussieu, 75252 Paris Cedex 05, FRANCE.

1. M. Balkanski, R. F. Wallis, I. Darianian and J. Deppe, Mater. Sci. Engin. B1, 15 (1988).
2. M. Massot and M. Balkanski in Festschrift for Sir R. J. Elliott edited by J. A. Blackman, Oxford University Press.
3. R. L. Mozzi and B. E. Warren, J. Appl. Crystallogr., 3, 251 (1970).
4. J. Goubeau and H. Keller, Z. Anorg. Chem. 272, 303, (1953).
5. J. Krogh-Moe, Phys. Chem. Glasses, 6, 46 (1965).
6. G. E. Jellison, Jr., L. W. Panek, P. J. Bray and G. B. Rouse, Jr., J. Chem. Phys., 66, 802 (1977).
7. R. N. Sinclair, J. A. E. Desa, G. Etherington, P. A. V. Johnson and A. C. Wright, J. Non-Cryst. Solids, 42, 107 (1980).
8. F. L. Galeener, G. Lucovsky and J. C. Mikkelsen, Jr., Phys. Rev. B, 22, 3983 (1980).
9. M. H. Brodsky, in Light Scattering in Solids, ed. M. Cardona, Springer, Berlin, 1975, p. 205.
10. L. A. Kristiansen and J. Krogh-Moe, Phys. Chem. Glasses, 10, 96 (1968).
11. F. L. Galeener, R. A. Barrio, E. Martinez and R. J. Elliott, Phys. Rev. Lett., 53, 2429 (1984).
12. F. L. Galeener and M. F. Thorpe, Phys. Rev. B28, 5802 (1983).
13. M. A. Kanehisa and R. J. Elliott, Mater. Sci. Eng. B3, 163 (1989).
14. P. N. Sen and M. F. Thorpe, Phys. Rev. B22, 3078 (1980).
15. M. F. Thorpe and F. L. Galeener, Phys. Rev. B22, 3078 (1980).
16. J. Deppe, M. Balkanski and R. F. Wallis, Phys. Rev. B41, 7767 (1990).
17. T. W. Brill, Phillips Research Rep., Suppl. 2 (1976).
18. E. Cazanelli and R. Fresh, J. Chem. Phys. 79, 2615 (1983).

Layered Intercalation Compounds

M. Balkanski

Laboratoire de Physique des Solides, associé au CNRS
Université Pierre et Marie Curie, 4 Place Jussieu-75252 Paris Cedex, France

1. Introduction : Insertion Materials

One of the most striking and probably the most promissing applications of layered intercalation materials is their use as insertion electrodes in solid state batteries. The voltage as well as the capacity or the energy and power densities of a battery are entierly defined by the nature of the intercalant and the insertion material.

The electromotive force (EMF) of a solid state battery, connected to an external load, is given by the potential difference between the metal : ion source and the intercalation material characterized by their respective work functions. The EMF is equal to the quasi Fermi level difference of the ion source and the intercalation compound. Open circuit voltage of solid batteries with intercalation compound cathodes tested today range between one and four volts and the energy densities are between 500 and 1200 Whkg^{-1}.

Layered type solids represent a large and important group of materials ranging from graphite with a simple structure to oxide superconductors with multiple atomic planes within a unit cell.

A layered solid bears a strong anisotropy in most of its structural and physical properties. Such a solid usually displays on axial symmetry and its properties can be distinguished between those belonging to the basal plane and those perpendicular to it and therefore it is often called a "two dimensional" solid. Due to the weak interlayer bonding implying highly anisotropic electronic band structure and Fermi surface, the electronic, magnetic and optical properties shows strong anisotropy which is also reflected in the mechanical properties in the case of cleaving and in compressibility.

The weak interlayer forces of these materials offer the possibility of introducing foreign atoms or molecules between the layers. This is the process of intercalation. In this process intercalant species are inserted in the Van der Waals gap, which is generally accompanied by a charge transfer between the intercalant species and the host lattice. It is believed that the tendency for this charge transfer is the driving force for the intercalation reaction[1]. Intercalation process is the basis on which certain layered compounds are used as the cathode of solid state batteries. Its practical applications are likely to increase, particularly in microelectronics. At the same time intercalation is now being recognized as an important practical way of controlling oxygen content and introducing cation doping to oxide superconductors.

The simplest layered materials are graphite and BN in which each layer consists of a simple sheet of atoms. Generally layered compounds have layers each consisting of three atomic planes : one metal atom sheet sandwished between two non-metal atom sheets. These three atomic sheets are strongly bound by either a covalent or ionic bond, but the layers are only weakly bound with each other by Van der Waals interactions. Such structures are formed by compounds of transition metal dichalcogenides such as ZrS_2, NbS_2 and MoS_2. Numerous other variants with more complex structures such as those of the first row transition metal phosphorus trichalcogenides MPX_3 with M = Fi, Ni, Co and X = S, Se and group III chalcogenides GaSe, InSe, In_2Se_3. Some of the recently discovered oxide superconductors may also be regarded as of this class of materials.

In many cases a rigid band model is a useful first approximation for discribing the changes in electronic properties of the host material with intercalation. We shall see, nevertheless, that this model is not appliable to all of the layered intercalation materials. One may argue that the applicability of the rigid band model may be taken as a test for the properties most desirable in a good cathodic material[2] this needs yet to be more extensively documented.

We shall discuss lattice dynamics and electronic band structure modifications upon intercalation with the aim of better understanding of the intercalation process and establishing some quide lines for improving the performances of these materials in their most eminent applications.

2. Non-transition metal chalcogenides

2.1. *Structure*

Among the layered compounds of the III-VI group, indium selenide has been recently most widely studied and systematically used as intercalation cathode in Li solid state batteries.

Each layer of InSe consists of four close packed, covalently bounded, monoatomic sheets in the sequence Se-In-In-Se. Successive layers are separated by Van der Waals gaps. According to the way in which successive layers are piled, four different polytypes are possible. Bridgeman grown crystals are generally of type γ. For the discussion here we shall consider type γ-InSe whose structure is shown in figure 1.

2.2. *Electronic properties, pure γ-InSe*

The first band structure calculation for this family of materials was made for GaSe[3]. A calculation for InSe with a tight-binding model has been carried out in a two-dimensional approximation[4]. Extentions to the three-dimensional case have been developed using the pseudopotential method[5] and the tight-binding method[6]. Including the spin-orbit interaction in the tight-binding calculations[7] has a non-negligible effect on the InSe band structure.

The optical properties near the band edges of pure InSe have been investigated and compared with the calculated band structure[7]. Sharp excitonic peaks are observed in the absorption spectrum at low temperature corresponding to the three absorption thresholds as shown schematically in figure 2.

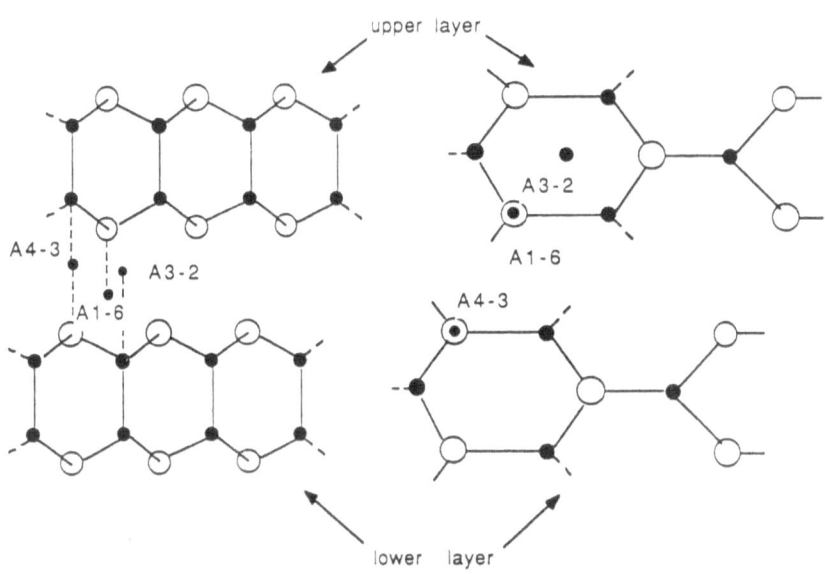

SIDE VIEW   TOP VIEW

upper layer

A4-3
A3-2
A1-6

A3-2

A1-6

A4-3

lower layer

Figure 1 :   Structure of γ-InSe.

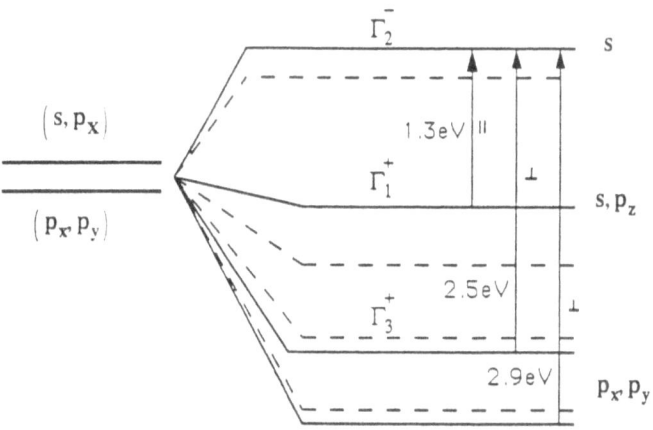

Figure 2 :   Schematic representation of the optical transition at k = 0 between the top of the
valence band (s.p_z) and the bottom of the conduction band (s) allowed in
polarization E ∥ c ; and the transition between spin-orbit split-off valence band
(p_x, p_y) allowed in the polarization E ⊥ c.

71

The first threshold, in the optical absorption spectrum, at 1.3 eV is related to the direct transition between the s, p_z valence band states and s conduction band states. The second and third thresholds at 2.5 eV and 2.9 eV are related to the transitions from the spin-orbit split-off valence bands to the s conduction band.

The electronic band structure of γ-InSe was recently calculated[8] on the basis of the tight-binding scheme taking into consideration the 4s and 4p atomic states of Se and 5s and 5p states of In. The calculated band structure is shown in figure 3.

2.2.2. Li intercalated γ-InSe

The effect of Li intercalation on the interband optical transitions is rather weak, but nevertheless, clearly observable. In figure 4 are shown the shifts in energy of the smallest direct band gap E_1 and the next-to-smallest band gap E_2 in γ-InSe as function of the lithium content[9] deduced from the position of the n = 1 excitonic peak.

The persistence of the excitonic transitions in highly intercalated InSe suggests that the Li 2s electrons form a low mobility impurity band or are efficiently trapped into localized states. In the lithium intercalated InSe a new photoluminescence peak appear at a photon energy somewhat less than that of the fundamental photoluminescence peak of pure InSe. This new peak is due to radiation emitted by the decay of excitons associated with an electron in a Li-2s impurity band and a hole at the top of the valence band.

The modifications of the electronic states suggested by this experimental finding are shown in figure 2 (dashed lines). Upon Li intercalation the conduction band shifts downward slighly whereas the top valence band shifts drastically. The second valence band shifts slightly upwards.

Li atoms intercalated in InSe occupy the sites of lowest potential energy in the Van der Waals gap between layers. Those sites are determined by comparing the total energy of different trial configurations calculated using the density functional theory within the local density approximation[10]. The most favorable sites are shown in figure 1. From these results it follows that the diffusion[11] of Li occurs between the layers, by successive jumps between the A3 and A1 sites, following a zig-zag path.

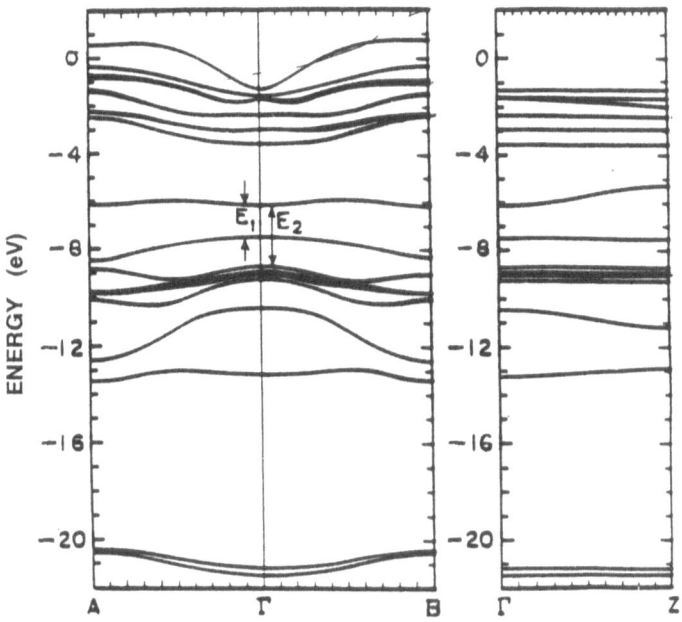

Figure 3 : Energy bands for pure γ-InSe (from ref. 8).

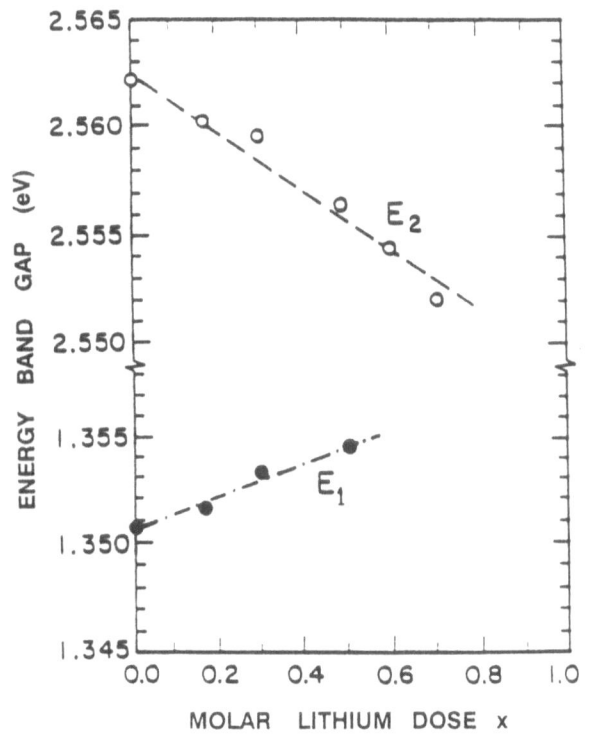

Figure 4 : Energy gap versus lithium content x in $Li_x InSe$ for the smallest and next to

smallest energy gaps E_1 and E_2 at 5 K (from ref. 9).

73

The electronic band structure of Li intercalated γ-InSe shown in figure 5 was calculated neglecting any change in the interlayer tight-binding parameters of InSe in view of the fact that lithium intercalation leads to only a small increase in the intralayer spacing. The overlap and interaction parameters for the interaction of a Li atom with the nearest Se and In atoms were estimated using tabulated atomic functions[12].

A band associated with the Li-2s state is seen to lie just below the conduction band edge and is separated from the latter by 0.03 eV at Γ. In table I are given the values of the energy differences E_1 and E_2 calculated for pure and Li intercalated γ-InSe which compare very well with the experimental results represented in figure 4.

The occurence of the Li-2s band in the fundamental gap of InSe provides a natural explanation for the new photoluminescence peak[13] produced by Li insertion shown in figure 6.

The semiconducting properties of lithium intercalated γ-InSe are consistent with the Li-2s band lying within the valence-conduction band gap.

2.3. *Lattice dynamics*

2.3.1. Host material, pure InSe

InSe, being partially ionic, long range coulomb interaction have to be considered along with the short range covalent-band contributions to the potential energy in the calculation of the dispersion relations. The simplest model taking both types of interactions into account is the rigid ion model, where the ion charges are approximated by point charges $\pm Ze$ centered at the nucleis.

Theoretical[14] dispersion curves for γ-InSe are shown in figure 7. Notice that in the direction of propagation perpendicular to the layers and the weak Van der Waals type coupling between layers, the dispersion curves are quite flat.

2.3.2. Effect of Li intercalation

The thoeretical dispersion curves for γ-InSe intercalated with lithium are shown in figure 8. The extra Li bands are very flat like the electronic energy bands due to the

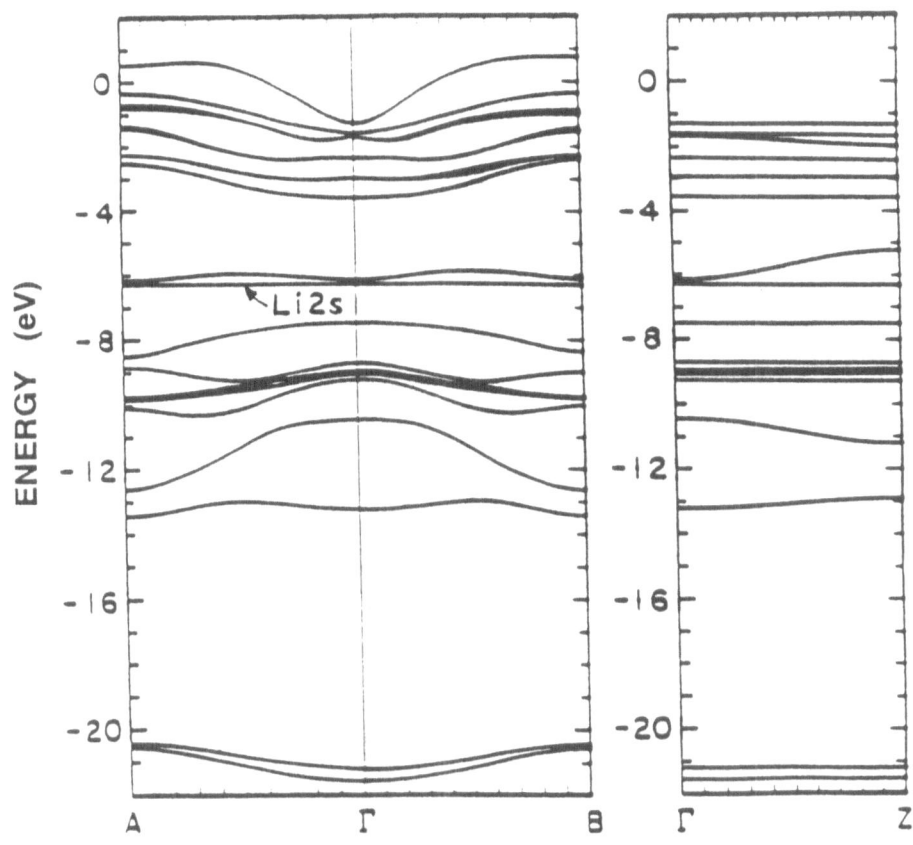

Figure 5 : Energy bands for lithium intercalated γ-InSe with composition $Li_{0.5}InSe$. The gap between the Li-2s band and the conduction band just above it is 0.036 eV at Γ, it is exaggerated in the drawing for clarity (from ref. 8).

Table I

	pure γ-InSe	Li-intercalated γ-InSe
E_1	1.3507 eV	1.3546 eV
E_2	2.5622 eV	2.5565 eV

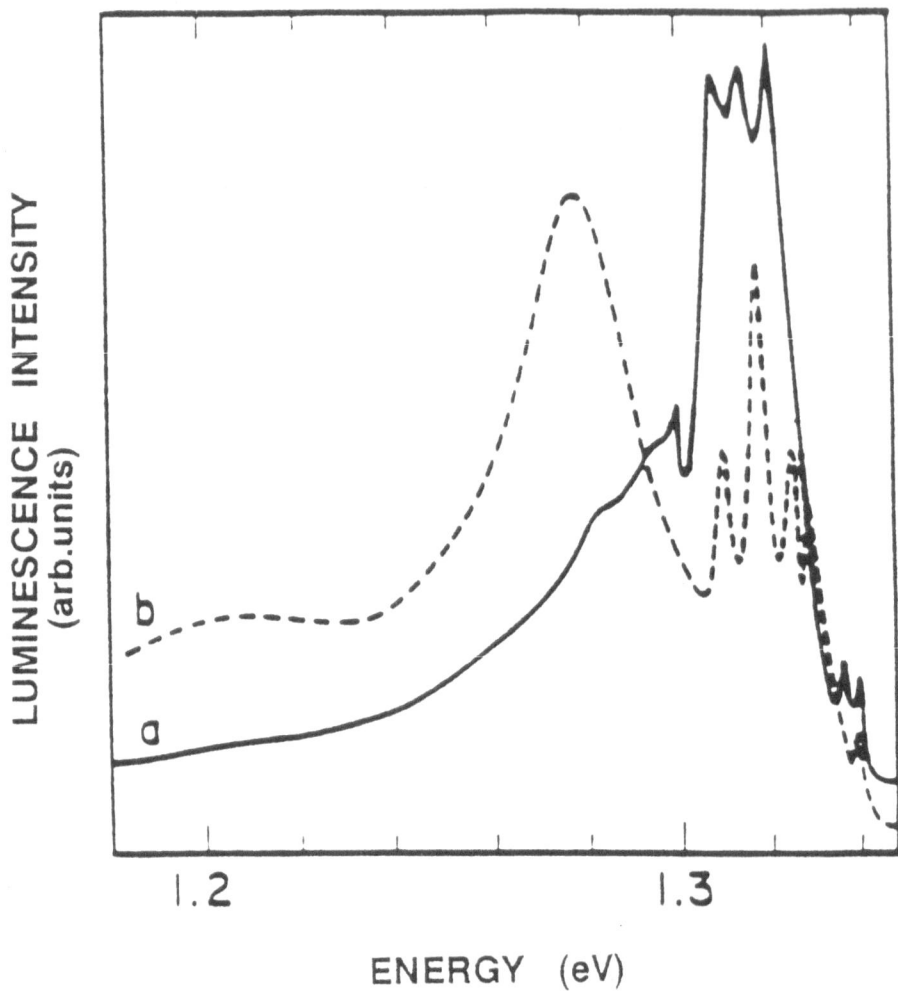

Figure 6: Photoluminescence spectra for γ-InSe before (a) and after (b) lithium
intercalation taken at 5 K under 1.916 eV excitation by a Kr⁺ laser (from ref.

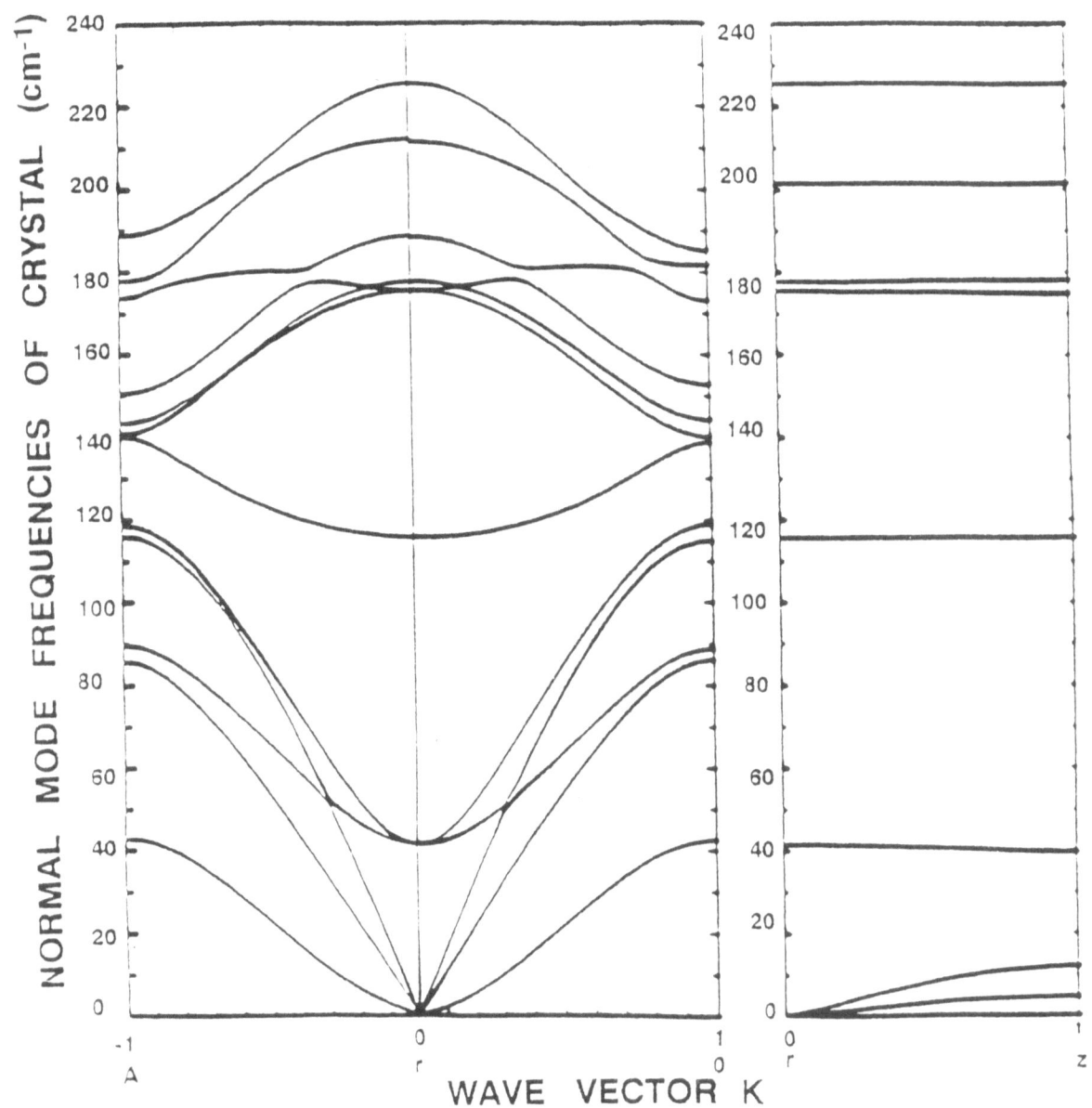

Figure 7 : Dispersion curves for the normal modes of γ-InSe (from ref. 14).

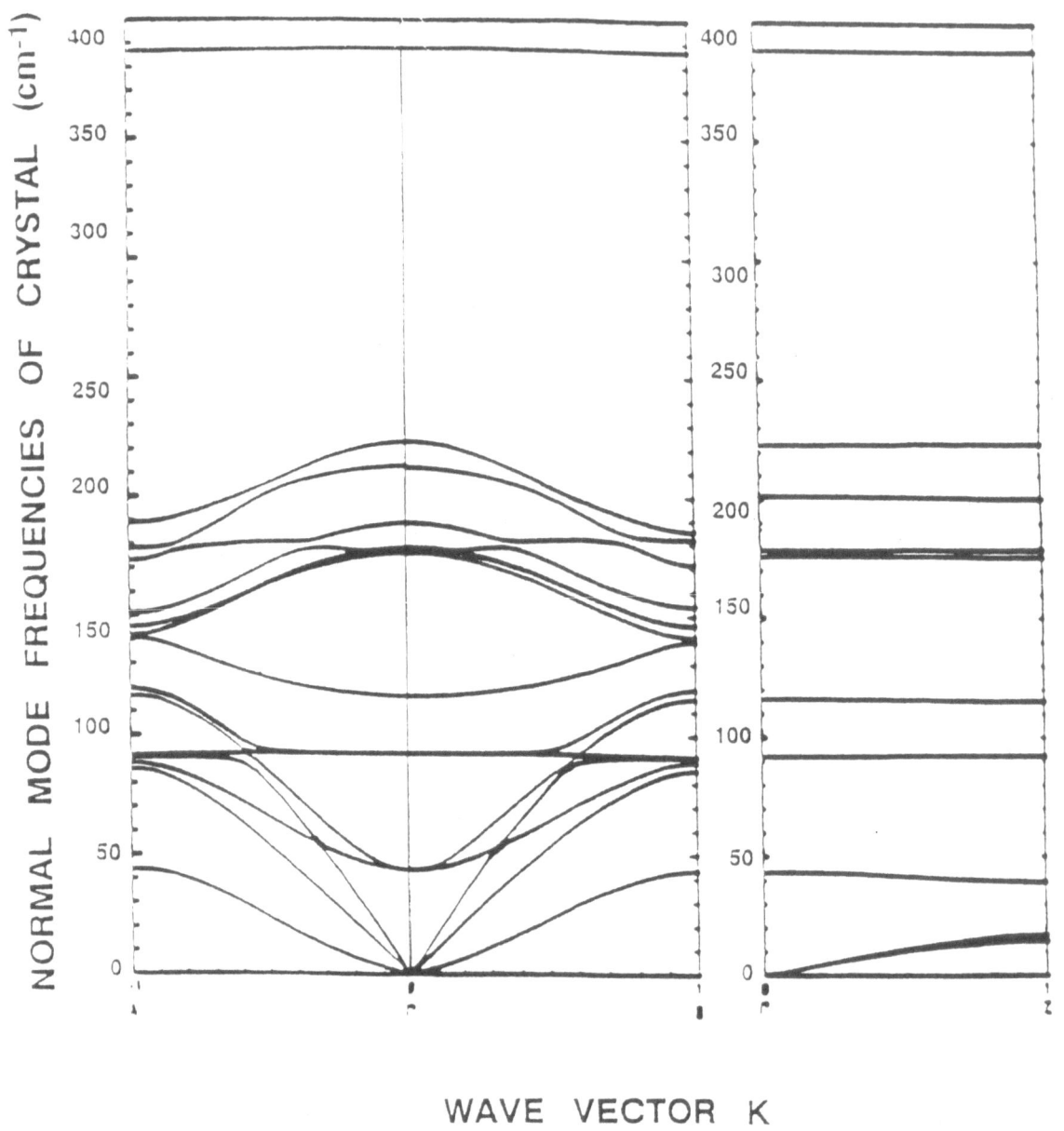

Figure 8 : Normal mode frequencies for γ-InSe intercalated with Li (from ref. 14).

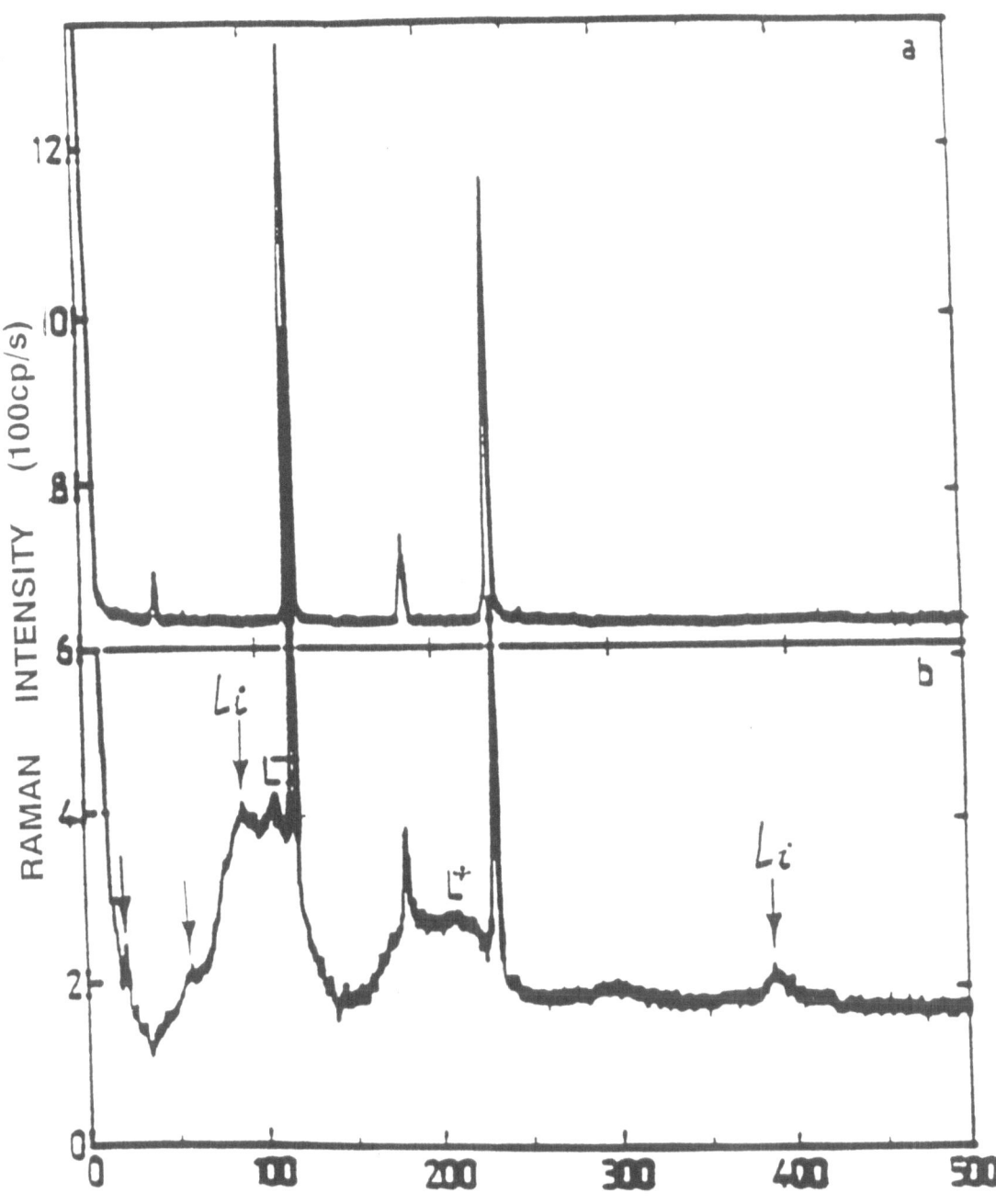

Figure 9 :   Raman spectra of pure InSe (a) and Li$_x$InSe (b) at 5 K, excited with a 514 nm

laser beam (from ref. 15).

smallness of the interaction of Li with the InSe. Dynamically, Li atoms behave as if they were decoupled from the host lattice, having only a very small interaction with the layers adjacent to the Van der Waals gap. This means also that the Li ions move easily through the lattice and are not greatly affected by the vibrations of the In and Se atoms.

Experimental[15] Raman spectra for pure γ-InSe and γ-InSe intercalated with Li are shown in figure 9. Extra modes due to the presence of Li are observed at 92 cm^{-1} and 388 cm^{-1} corresponding to Li vibration in the plane and perpendicular to the plane of the Van der Waals gap respectively.

3. Transition metal chalcogenides

3.1. *Structure*

Most of the transition metals from IV, V, VI and some from group VII and VIII of the periodic table can form layered structure dichalcogenides. Table II.

In the dichalcogenide family MX$_2$ each atomic plane forms a two-dimensional close packed structure, shown in figure 10a, and the planes are stacked along the c-axis separated by Van der Waals gaps. In an octahedrally coordinated structure, the sandwich layers are made up of atomic planes arranged in AbC, as shown in figure 10b, and the layers may be stacked in a variety of ways resulting in a large number of polytypes. The group VI metal compounds mainly adopt the trigonal prismatic coordination in which the sandwich layer is made up of atomic planes arranged in AbA, shown in figure 10c.

3.2. *Electronic properties*

3.2.1. Host materials

Group IV transition metal dichalcogenides are semiconductors with large band gaps, TiSe$_2$ is a semimetal due to overlapping valence and conduction bands. Transition metals from group V form dichalcogenides which are metal and superconductors, and those of groupe VI are semiconductors with a small gap. This trend is clearly pictured in the schematic density of energy states sequence of diagrams show in figure 11. The sequence is

Table II

Groups	Transition metal M					Chalcogen X
	IV	V	VI	VII	VIII	VI
						S
	Ti	(V)	(Cr)		Ni	Se
	Zr	Nb	Mo		Pd	Tc
	Hg	Ta	W	Re	Pt	

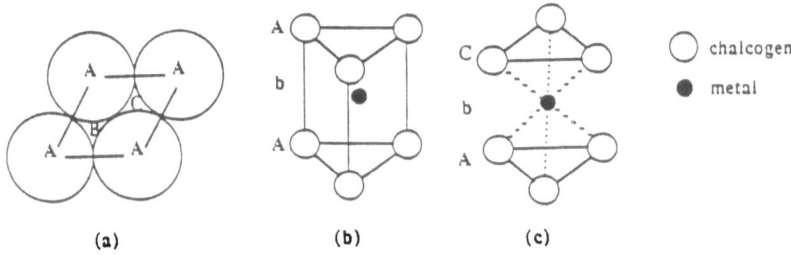

Figure 10 : (a) Basal plane view showing two-dimensional close-packed structure, (b) octahedrally coordinated unit and (c) trigonal prismatically coordinated unit of MX_2 (from ref. 2).

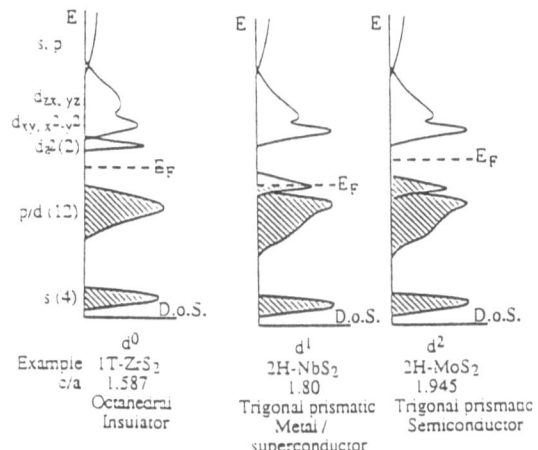

Figure 11 : Schematic density of energy level diagrams of groups IV, V, VI transition metal dichalcogenides. Number in parentheses next to the orbital notations gives the number of states in a given band (from ref. 2).

the result of increasing the number of valence electrons by one on each move to a higher group. The major difference between this three groups of compounds is the occupation of d_z^2 band which shift in relation to other bands with increase electron occupation and this is accompanied by a trigonal distortion. The main driving force for such a distortion is to minimise the total energy by lowering electron occupied conduction bands.

Results of detailed band structure calculations support this qualitative band scheme. One such calculation based on muffin-tin orbitals and the atomic sphere approximation which also include exchange-correlation corrections has been applied to 2H-TaS$_2$ and the results[16] is shown in figure 12.

3.2.2. Intercalation in transition metal chalcogenides

The same result can be achieved by adding electrons to the host compound by intercalation. Thus Na$_x$ZrS$_2$ with x = 1 behaves in many respects like NbS$_2$ and is also a superconductor. The transport properties are dominated by the occupation of the lowest conduction band, often referred to as the d_z^2 band. An additional feature is the tendency of the d electrons to become tightly bound due to the narrow d-band with the consequence of Anderson localization or the formation of charge density waves.

The outer atomic orbitals of transition metals and chalcogens which form the valence and conduction band structures are : Zr = d^2s^2, Nb = d^3s^2, Mo = d^4s^2 and S, Se = s^2p^4.

The electrons occupying the valence bands form strong chemical bonds which serve to underpin a common structure for all members of the dichalcogenide family, and the addition of further electrons to the d_z^2 conduction band produces relatively small perturbations to this part of the band structure. This is the basis of the rigid band approximation often used as a first approximation to describe effects on the electronic structure resulting from the intercalation.

Intercalant spacies are inserted into the Van der Waals gap and the process is generally accompanied by a charge transfer to the host layer and an expansion of the layer-layer distance.

Magnetic transition temperatures for iron and nickel phosphorus trichalcogenides.

Compounds	T_N(K)			II (J mol^{-1})
FePS$_3$	118 ± 1[1]	116[2]	120[3]	390 ± 10[1]
FePSe$_3$	106 ± 1[1]	112[2]		120 ± 20[1]
NiPS$_3$	151 ± 1[1]	155[2]		29 ± 1[1]

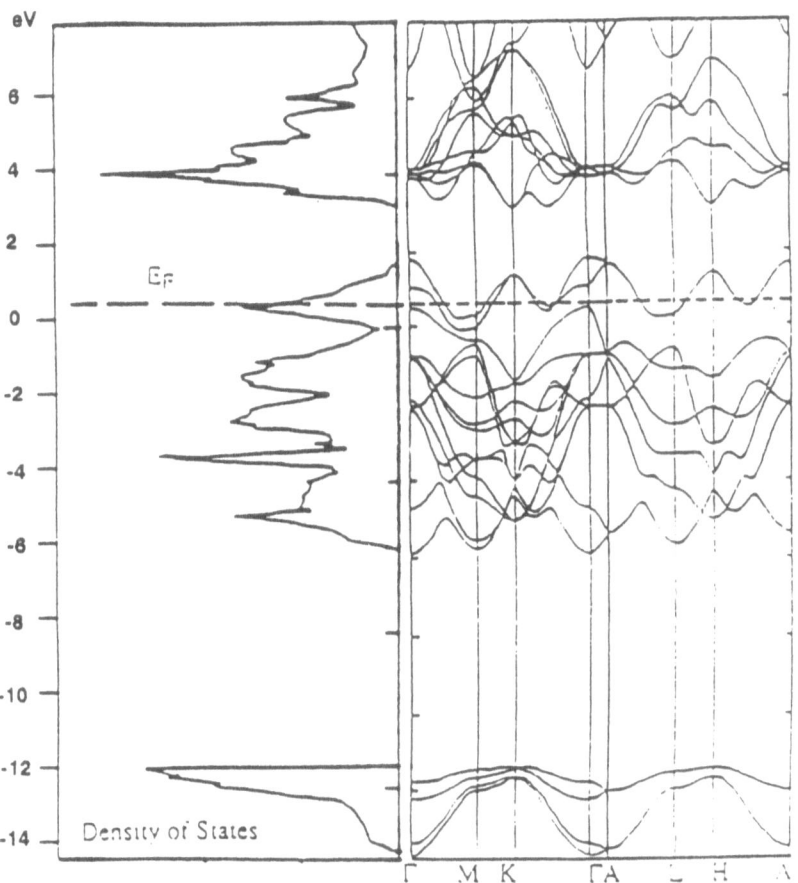

Figure 12 : Band structure and calculated density of states function of 2H-TaS$_2$ (from ref. 16).

83

It is belived that the tendency for charge transfer is the driving force for the intercalation reaction. Transition metal dichalcogenides are known to form intercalation complexes with electron donor species only. The charge transfer process may be used to "fine tune" the electronic properties of the host material. Control of band filling in this way is a feature unique to low dimensional material structures of this type, and provides an extra variable of great value. It is thus possible to achieve semiconductor to metal and metal to semiconductor transitions with intercalation.

Different categories of intercalant complexes are known, we shall mention here only the simplest and the most widely studied : simple metals.

Li, Na, K, Pb and Ag being highly electro-positive ions constitute good electron donors whose effect is to increase electron filling of the bands with minimum perturbation of the band structures.

Potassium intercalation in different transition metal dichalcogenides leads to a change of electron occupation, but also, such changes of electro-occupation may lead to structural changes and consequently to a complete breakdown of the rigid band approximation as demonstrated by the following examples[17]

HfS_2 (semiconductor) + xK $\rightarrow$ K_xHfS_2 (metal)

$NbSe_2$ (metal and superconductor $T_c = 7.4$ K) + xK $\rightarrow$ K_xNbSe_2 (semiconductor)

MoS_2 (semiconductor) + xK $\rightarrow$ K_xMoS_2 (metal x = 0.5 superconductor with $T_c = 6.5$ K)

Trigonal prismatic coordination $\rightarrow$ Octahedral coordination

3.3. *Lattice dynamics*

3.3.1. Host material MoS_2

A remarkable property of the layered materials is their high anysotropy. Phonon spectroscopy offers an excellent way of quantifying the degree of anisotropy not only by distinguishing inter and intralayer normal modes but also by determining the shear moduli in different directions.

The Raman spectrum of MoS_2 at room temperature is shown in figure 13a. The lowest frequency mode E_{2g}^2 at 33.5 cm^{-1} has been identified as the rigid layer mode while all higher frequency modes involve both intralayer as well as interlayer forces. A shear force constant of 152 $N.m^{-1}$ corresponding to interlayer forces and 2.67 $N.m^{-1}$ corresponding to intralayer forces have been deduced. The ratio of 57 between these two shear forces constants indicates the high degree of anysotropy of MoS_2.

3.3.2. Effect of intercalation

Recently Raman spectra of Li intercalated MoS_2 have been measured and a simple lattice dynamical model used to calculate the frequencies of the new modes appearing after Li intercalation[18]. In this model it has been assumed that host lattice parameters are unchanged after intercalation. Raman spectra are shown in figure 13b. After intercalation a new band is observed at 205 cm^{-1}. The two pristine intralayer modes are observed with small frequency shifts but at the low frequency side of each one a new peak appears. These split-off bands form two Davidev pairs due to the splitting of the intralayer modes caused by the interlayer interaction resulting in a reduced Brillouin zone and folded optical modes.

4. Transition metal phosphorous trichalcogenides MPX_3 (M = V, Mn, Fe, Co, Ni, Zn, X = S, Se)

4.1. Structure

The MPX_3 compounds have layered structures formed by compact stacking of type ABC or AB of S or Se sheets. Alternatively, the space between chalcogen atom sheets is eitherly empty—forming a Van der Waals gap—or filled with metallic and phosphorous ions. A projection along the b axes of a MPS_3 structure showing alternating slabs and Van der Waals gaps is represented in figure 14[19].

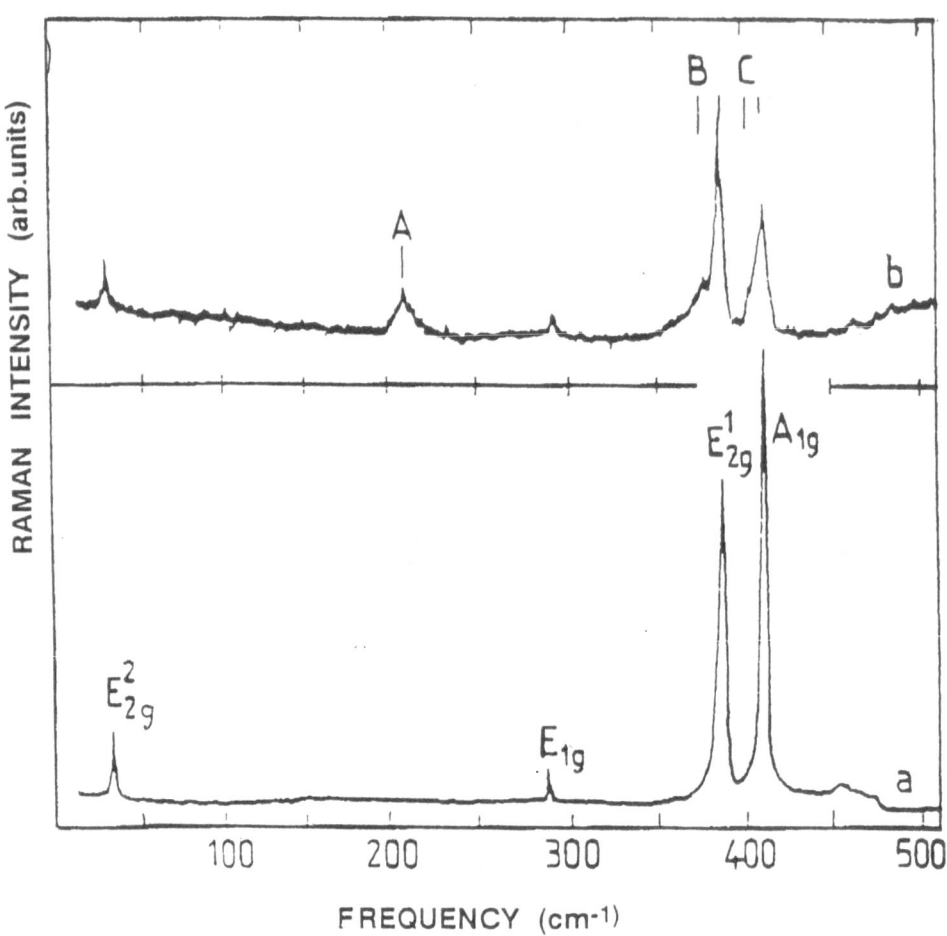

Figure 13 : Raman spectrum of MoS_2 at room temperature (a) and Li intercalated MoS_2 (b).

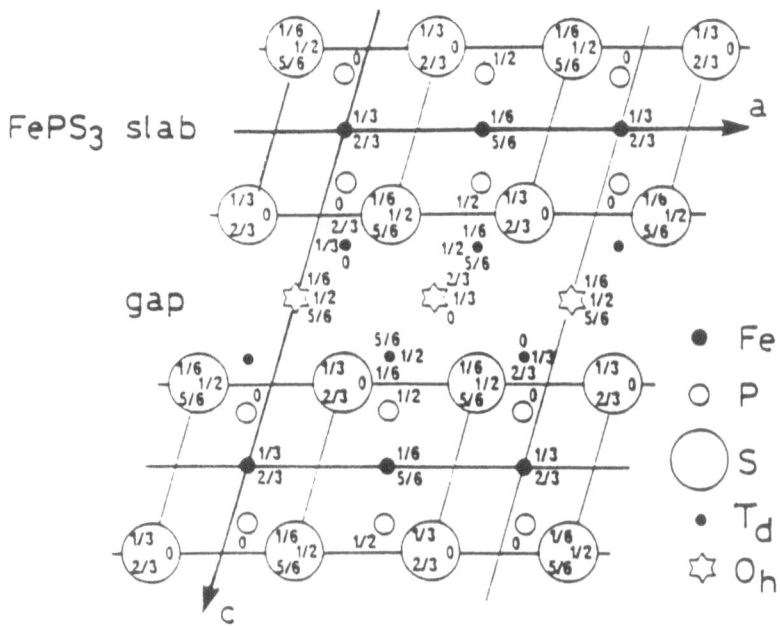

Figure 14 : Projection along the b axis of a MPS$_3$ structure showing alternating slabs and Van der Waals gaps (from ref. 19).

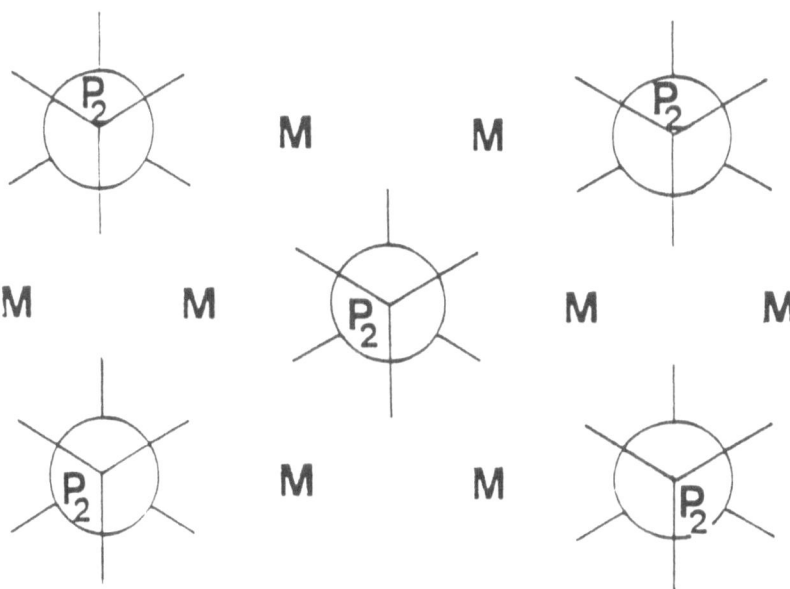

Figure 15 : Schematic representation of M^{2+} cations and (P$_2$S$_6$)$^{4-}$ anions for a MPX$_3$ structure (from ref. 19).

This family can be viewed as salts constituted by M^{2+} cations and $(P_2S_6)^{4-}$ anions as shown in figure 15[19]. An elementary cell of D_{3d} symmetry corresponding to this arrangement is shown in figure 16.

The MPX_3 compounds are antiferromagnetic with strong anisotropy of the magnetic properties due to their layered structure. The magnetic ions form nearest—neighbour ferromagnetic chains in the ab plane. Neighbouring chains are antiferromagnetically coupled. The interlayer coupling between nearest—neighbour chains differs from one compound to another. In $FePS_3$ the coupling is antiferromagnetic, forming a magnetic superstructure, doubled with respect to the primitive unit cell along the c-axis. The magnetic structures of $FePS_3$ are given in figure 17. The magnetic transition temperature of $FePS_3$ is 118 K.

4.2. *Electronic properties*

Following a model of weak interaction, used by Wilson and Yoffe[20] in the calculation of the band structure of transition metal dichalcogenides, Khymaco and Hughes[21] deduced the density of state distribution for $FePS_3$ given in figure 18.

The electronic band structure is that of a wide gap magnetic semiconductor. The valence band is formed of 3p levels of P and S states and the conduction band is derived from the phosphorous antibinding orbitals. The Fe 3d levels are localized in the forbidden energy gap. The electronic conductivity of this materials is very low. Intercalation would result in a magnetic phase change and could cause a considerable perturbation to the total energy. Considerable shift of the Néel temperature has been observed upon Li intercalation[22]. The electrons in MPX_3 are essentially not mobile and do not contribute to the conductivity of the intercalated compounds. This factor must ultimately pose a serious constraint on the usefulness of most of these materials as cathodes.

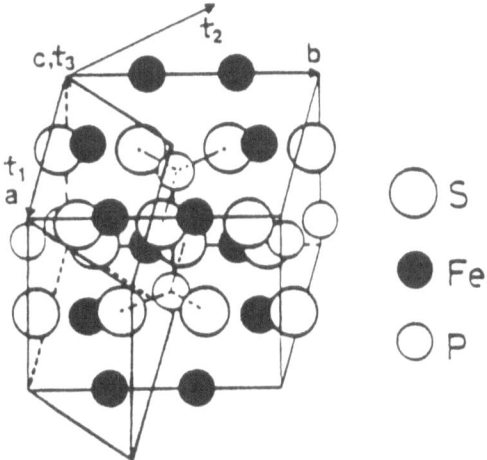

Figure 16 : Crystal structure of $FePS_3$. The thick lines denote the primitive unit cell and t_1, t_2 and t_3 are unit vectors of the primitive unit cell.

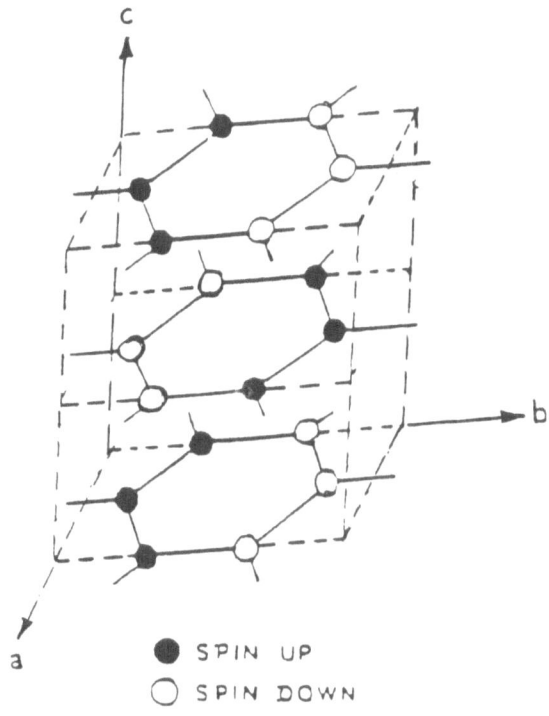

Figure 17 : Magnetic structure of $FePS_3$ presented by Fe atoms. Closed and open circles are up and down atomic spins, respectively (from ref. 19).

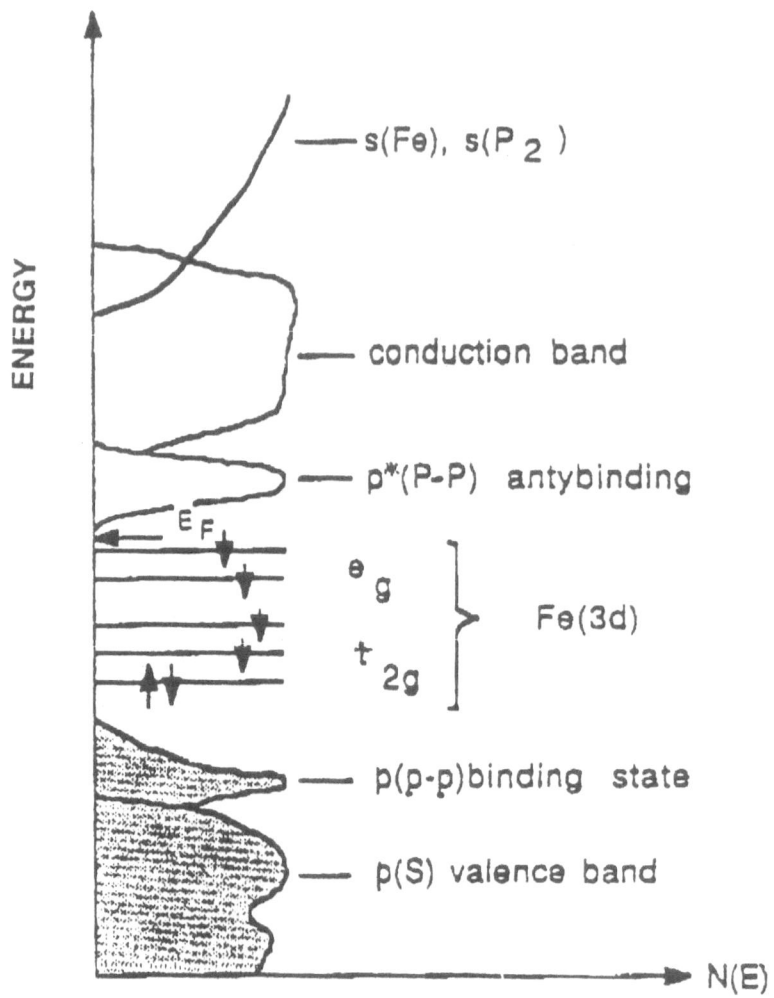

Figure 18 : Schematic electronic band structure of FePS$_3$ in the strong binding approximation (from ref. 21).

4.3. Lattice dynamics of MPX₃ compounds

4.3.1. Host material

The lattice dynamics of MPX_3 compounds has been approached in the framework of an axially symmetric force-constant model generated by short-range two-body potentials[23]. Such a simple model yields quite similar phonon dispersion curve for the different MPX_3 compounds. An example is given by the $FePS_3$ phonon dispersion curves along the symmetry direction of the Brillouin zone shown in figure 19.

Presently only Raman and IR data are available for comparison with the experiment. In figure 20 is shown the Raman spectrum of $FePS_3$ recorded at room temperature[24].

The identification of the Raman peaks is quite easy if one considers a simple pseudo-cell containing one P_2S_6 anion and two iron cations, shown in figure 15. The P_2S_6 molecule belongs to the D_{3d} symmetry group and six Raman-active modes are expected. The three polarized modes at 243, 376 and 466 cm^{-1} are assigned to A_{1g}-type modes of the P_2S_6 ion. The three polarized modes at 215, 275 and 573 cm^{-1} are assigned to E_g-type modes of the P_2S_6 ion. The modes in the low-frequency region, below 150 cm^{-1} are related to the cation motion.

4.3.2. Spin ordering induced phonon Raman scattering

The low temperature Raman spectrum shown in figure 21 is drastically different from that obtained at room temperature. Three new bands appear for $T \leq T_N$ at 88, 95 and 109 cm^{-1}. The intensity of the peak at 249 cm^{-1} is significantly enhanced and the intensity of the peak at 380 cm^{-1} changes only slightly.

The drastic difference between the spectra at room temperature shown in figure 20 and the spectra taken at liquid He temperature, i.e. below T_N shown in figure 21 can be accounted for by the Brillouin zone folding effect due to formation of a magnetic superstructure, i.e. it is attributed to inelastic scattering from the zone boundary phonons and the mid-zone phonons induced by the elastic "Bragg" magnetic scattering from the magnetic superstructure. For $FePS_3$ the magnetic unit cell along the c axis is double with respect to the crystallographic unit cell. The phonon branches are folded along the Δ

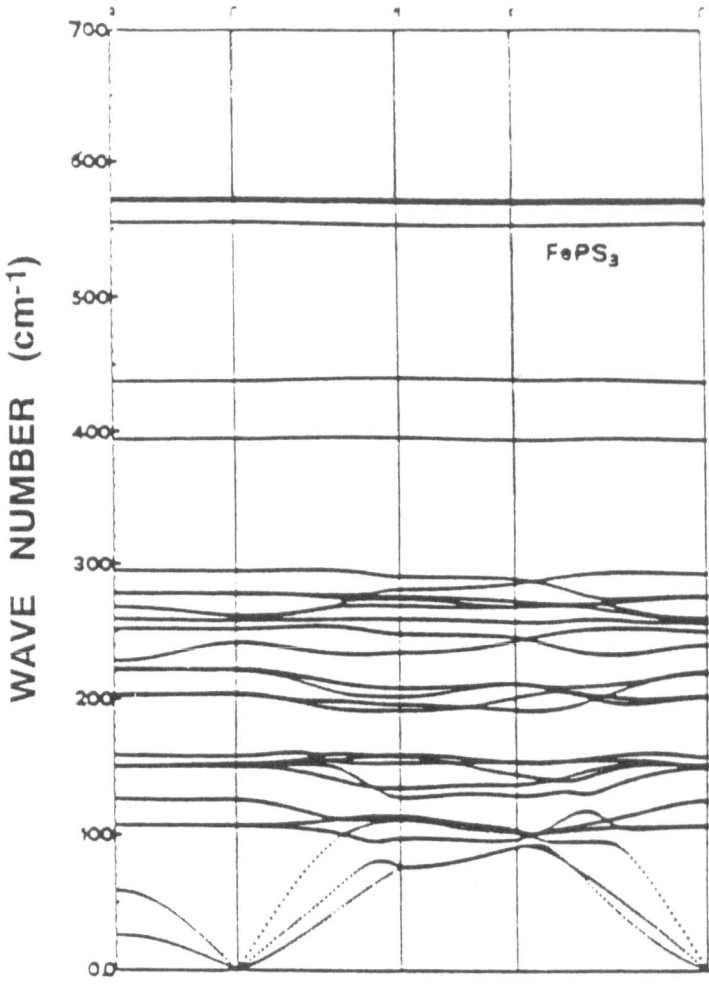

WAVE VECTOR

Figure 19 : Phonon dispersion curves of FePS$_3$ along the symmetry directions of the BZ. The dispersionless branches with energies above 300 cm^{-1} are produced by molecular vibrations of P$_2$S$_6$ group. The poor dispersion along ΓA direction and the upwards curvature of the in-plane TA branches are typical of layered compounds (from ref. 23).

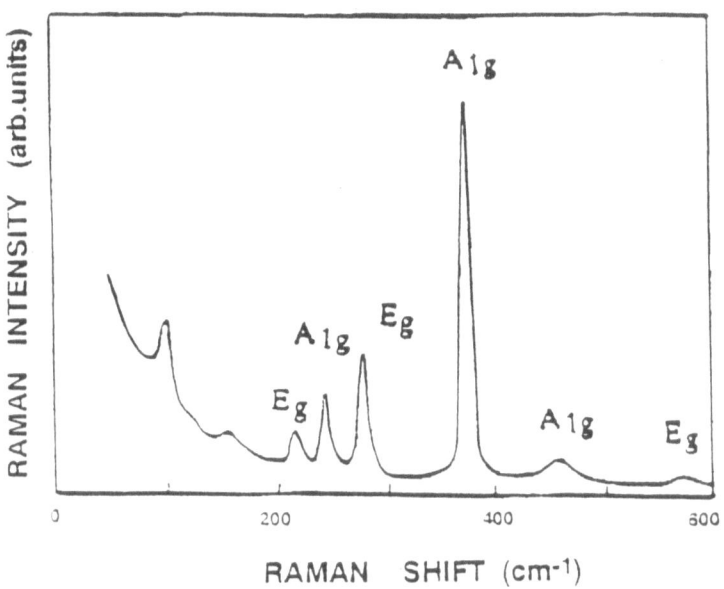

Figure 20 : Raman spectrum of FePS$_3$ recorded at room temperature excited by the 488.0
nm line of an Ar$^+$ laser (from ref. 24).

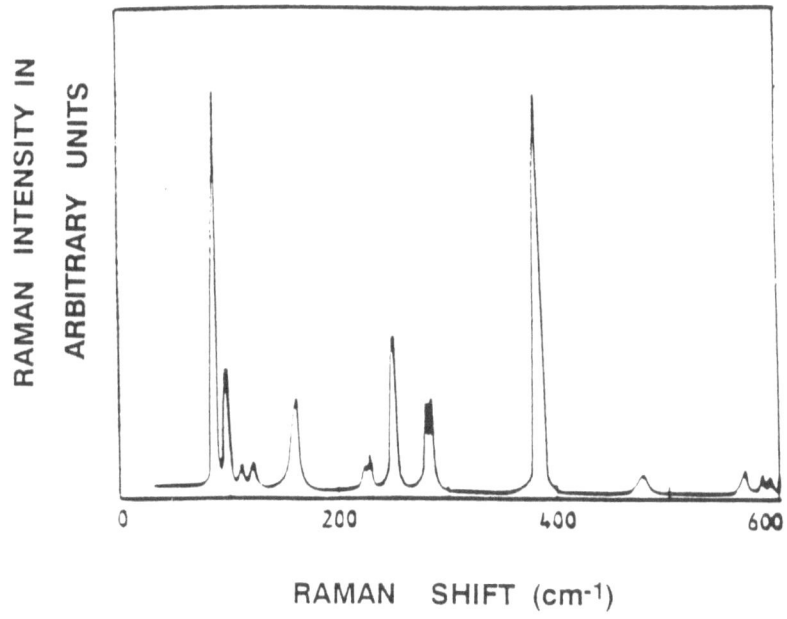

Figure 21 : Raman spectrum of FePS$_3$ recorded at liquid-helium temperature, excited by
388.0 nm line of an Ar$^+$ laser (from ref. 24).

direction. As a consequence modes from the A point of the Brillouin zone are folded into the Γ point and can therefore be observed in Raman spectra. The three new peaks, at 88, 95 and 109 cm^{-1}, not observed in the paramagnetic phase but appearing in the magnetically ordered phase below the Néel temperature correspond to folded acoustic branches.

Although below T_N magnetic superstructure is formed not only along the basal plane but also along the c-axis, the Raman peaks at 88.94 and 108 cm^{-1} correspond to zone-boundary phonons at $q = t_2/2$ in good agreement with the calculated frequencies (25) and non zone-boundary phonons are observed at $q = t_3/2$ and $q = t_2/2 + t_3/2$ where t_2 is a vector in the basal plan and t_3 is parallel to the c direction and perpendicular to the basal plane. This result suggests that the spin phonon interaction along the basal plane is much stronger than that perpendicular to the basal plane reflecting the two dimensionality of the spin-phonon system. This result should be correlated to the fact that the magnetostriction effect is observed only in the basal plane (26).

A phenomenological theory developed by Suzuki and Kamimura (27) gives the spin-dependent phonon Raman scattering in magnetic crystals in terms of the correlation function of the spin dependent polarizability tensors. The temperature dependence of the reduced integrated intensity is thus given by

$$I(T) = |R + M <S_0.S_1> / S^2|^2 \tag{1}$$

R is the spin independent term, $M <S_0.S_1>/S^2$ is the spin dependent term of the Raman tensor and $<S_0.S_1>/S^2$ is the reduced nearest neighbor spin correlation function.

The spin dependent Raman tensor has the same temperature dependence as the correlation function. This is true independently of the microscopic mechanism responsible for the scattering process.

The temperature dependence of the spin-dependent Raman intensity of the peaks at 88 and 161 cm^{-1} is shown in figure 22 compared with the square of the reduced magnetic moment[28]. Both functions are constant up to 0.6-0.7 T_N and then abruptly decrease approaching the transition temperature, in agreement with the 2D Ising character of FePS$_3$.

94

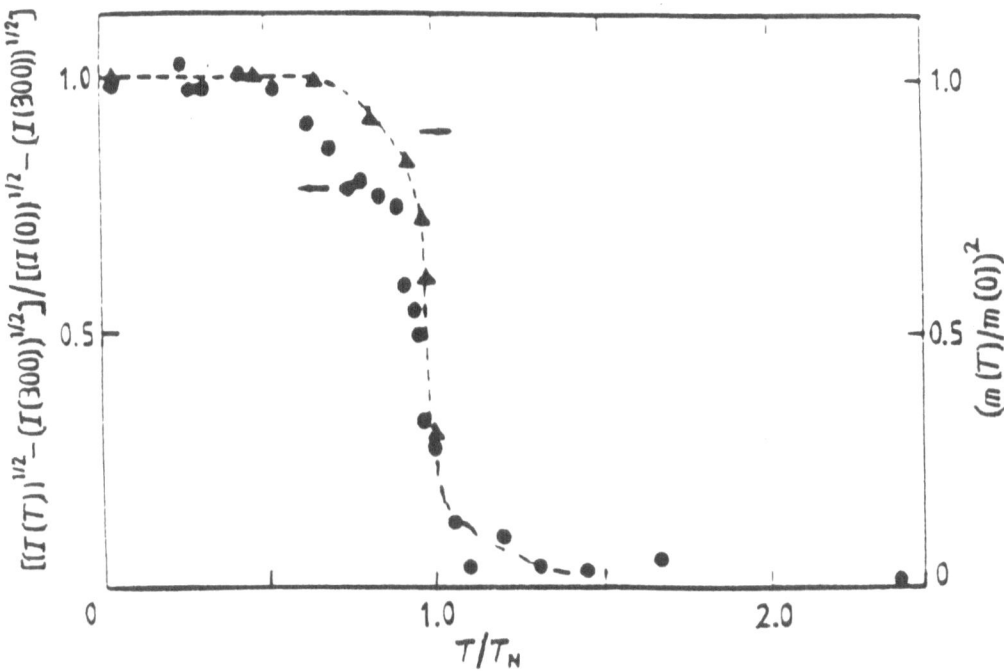

Figure 22 : Reduced nearest-neighbor spin correlation function obtained from the temperature dependence of the intensity of the phonon at 249 cm^{-1} (·) in the spectrum of FePS$_3$. The square of the reduced magnetisation m(T)/m(0) (Δ) is reported for comparison. The broken line is shown as a visual aid.

In addition to the effect of magnetic ordering on the phonon Raman scattering this type of material offers the possibility of investigating the effect of spin fluctuations on the Raman spectra of the 2D antiferromagnet $FePS_3$ and to study the crytical phenomena in the antiferromagnetic phase transition.

4.3.3. Spin disorder induced phonon Raman scattering

Taking into consideration that the spin system of $FePS_3$ is described by the Ising model, using mean field theory to calculate the magnetic susceptibility one can calculate the phonon Raman scattering induced by spin disorder[29]. The Raman intensity for the n-branch mode reflects the one phonon density of states.

In figure 23 is shown the temperature dependence of the low frequency Raman spectra and the calculated Raman spectra due to spin disorder induced Raman scattering from the TA phonon. The low frequency broad band with an assymmetric peak at about 100 cm^{-1} is manifestly due to the spin disorder induced Raman scattering from the TA phonon in which all atoms vibrate approximately along the z axis with $q = 0$ on the basal t_3 plane. The agreement between the theoretical and experimental spectra is quite good except for the low frequency region. The intensity of the low frequency spectrum, below 30 cm^{-1}, is weaker than the theoretical one above 148 K.

4.3.4. Magnetic critical scattering

The observation that the intensity of the low frequency spectrum is much weaker than the theoretical one leads to the conclusion that the square of the coupling constant C_o^2 has a strong frequency dependence in the low frequency region, which decreases with decreasing ω.

The plot of the induced intensity $I(\omega)/[n(\omega)$ 1] at 10 cm^{-1} which is approximatly proportional to C_o^2 at 10 cm^{-1} has a stronger temperature dependence than C_o^2 and shows a maximum at T_N which suggests that there exists another light scattering process in the low frequency region. The calculated curves for magnetic critical scattering in $FePS_3$ and the points obtained by substituting the calculated ω dependence of C_o^2 from the

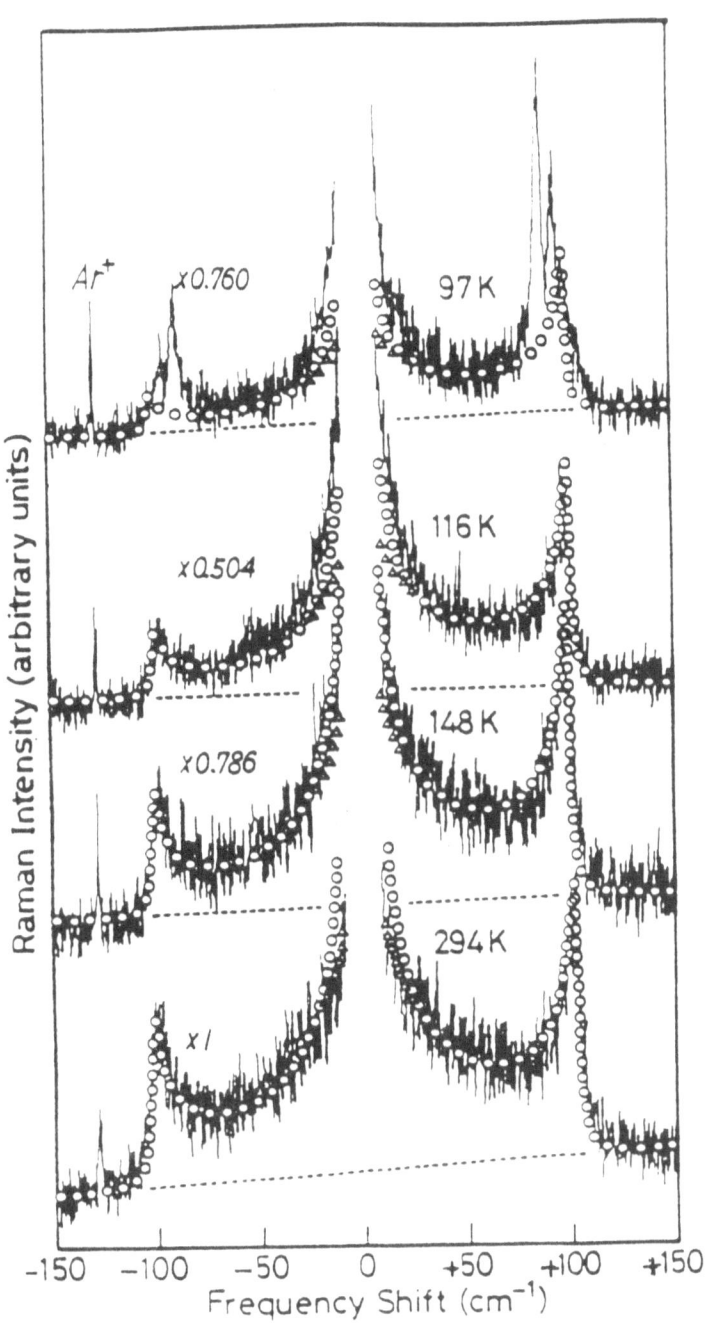

Figure 23 : Experimental Raman spectra and calculated (open circles) Raman scattering due to spin disorder induced scattering from the TA phonon (from ref. 29).

experimental Raman spectra are shown in figure 24. It has thus been shown that the central component is due to the magnetic critical scattering observed[29] for the first time, in addition to the spin disorder induced Raman scattering from the TA phonon.

The triangles in figure 23 denote the theoretical spin disorder induced phonon Raman spectra below 30 cm^{-1} when they are calculated taking into account the ω dependence of C_0^2. Near the Néel temperature the quasielastic scattering due to the magnetic critical scattering is enhanced.

4.3.5. One-magnon Raman scattering

In figure 25 which shows the temperature dependence of Stokes Raman spectra at low temperature in $FePS_3$, one observes a peak appearing at 122 cm^{-1} below T_N having its frequency shifted to the low frequency side as the temperature is increased. An infrared absorption peak having the same temperature dependence is also observed. No other phonon peaks showing such a large frequency shift are observed by Raman scattering or infrared absorption measurements. This peak is clearly assigned to one magnon transitions[29].

It should be notted that spin disorder induced phonon Raman spectrum disappears at the same temperature at which the one-magnon Raman scattering begin to emerge.

4.3.6. Effect of Li intercalation on lattice dynamics of MPX_3 compounds

All observations published up to now converge to the conclusion that the interlayer separation does not change upon Li intercalation and in addition there is no evidence for in-plane modification of the host material. This structural integrity leads one to consider, for the investigation of lattice dynamics, three perfect crystals with stoichiometric composition $Li_{0.5}FePS_3$, $LiFePS_3$ and $Li_{1.5}FePS_3$ by adding respectively one, two and three lithium atoms in the unit cell, with no modification of the pure phase Bravais lattice. Also the position of the host atoms in the unit cell have not been modified with intercalation.

The dynamics of $Li_{0.5}FePS_3$, $LiFePS_3$ and $Li_{1.5}FePS_3$ has been investigated within the same dynamical model[23] as used for the pure materials. The additionnal

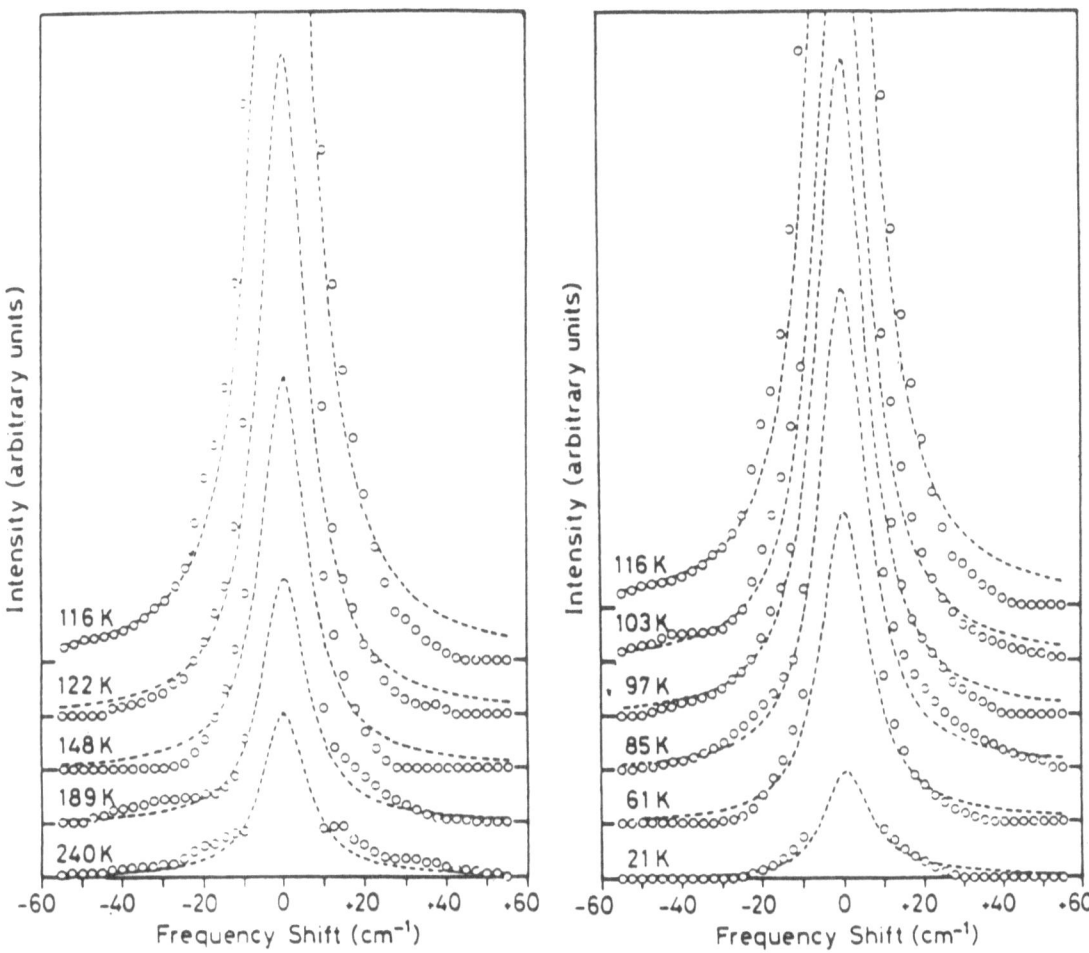

Figure 24 : Magnetic critical scattering in FePS$_3$. Open circles are obtained by substructing the theoretical curves calculated by taking into account the ω dependence of C_0^2 from the experimental Raman spectra. Dotted curves show the theoretical curves above T_N and below T_N.

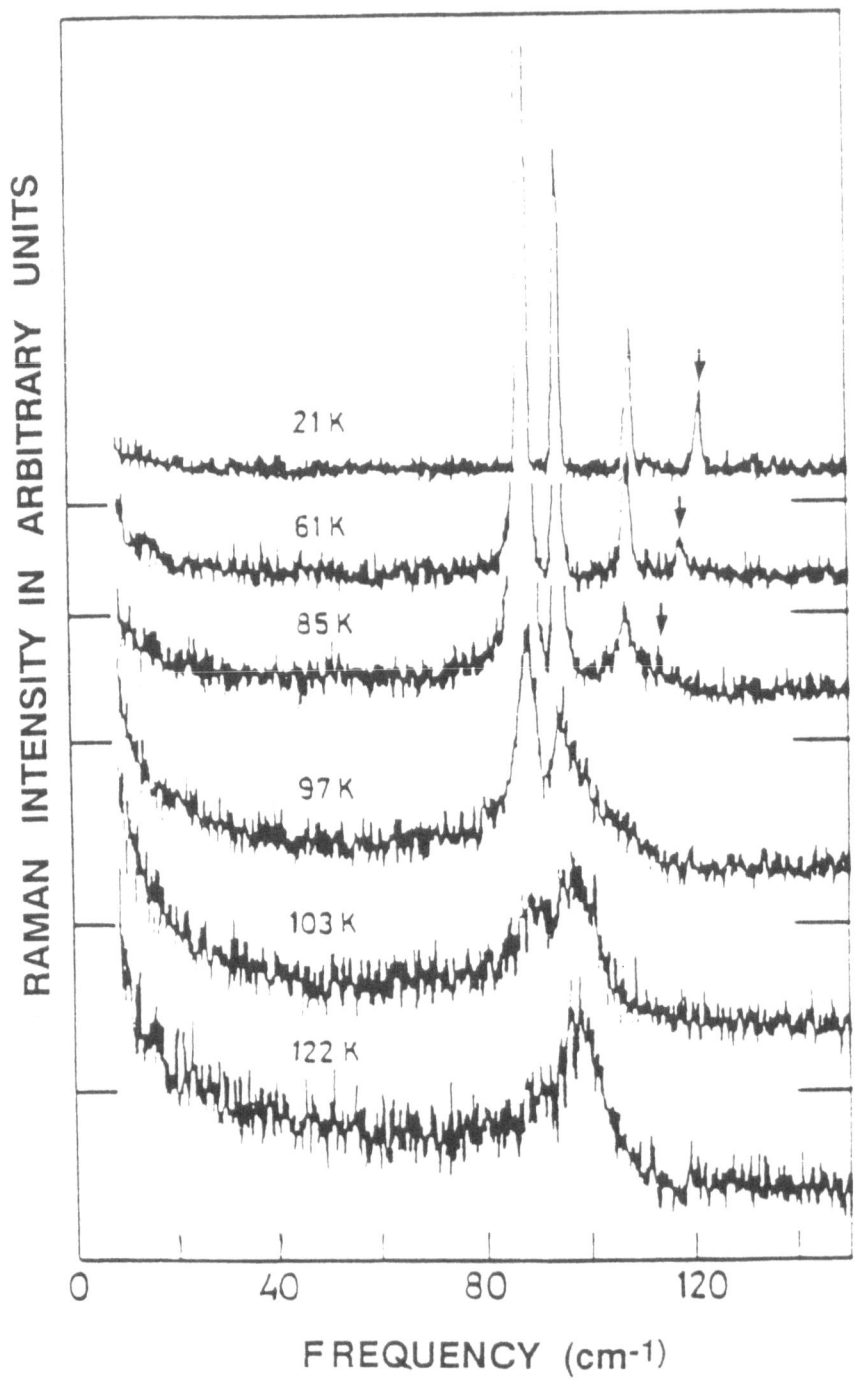

Figure 25 : Temperature dependence of stokes Raman spectrum at low temperature in FePS$_3$. Arrows show the one magnon Raman peaks.

interaction between Li and S ions concerns only nearest neighbors and is described by a two-body potentials as well. This very simple model should be sufficient to get a qualitatively satisfactory picture of the dynamical effects of the intercalation process. IR absorption spectrum[30] show that the most pronounced structures present in the spectra of the pure material do not sensibly change with intercalation. By increasing the Li content the low frequency shoulder in the band at 257 cm^{-1} becomes more and more prominent and new bands at 336 cm^{-1} and 384 cm^{-1} grow in intensity. Analysis suggest that the IR band at 336 cm^{-1} has to be attributed to the vibration of lithium cations in the chalcogen cage.

The calculated modes attributed to the experimental band at 336 cm^{-1} correspond to the in plane translations of Li ions, while the weak peak at 384 cm^{-1} has been associated with the translation of Li ions along the direction perpendicular to the layers. The three modes are not degenerate because the chalcogen cage has trigonal symmetry.

In $Li_{0.5}FePS_3$ there are three additionnal modes of ungerade symmetry with respect to the pure phase, while in $Li_{1.5}FePS_3$ there are six new ungerade modes and three new gerade modes. The latter are probably detectable by Raman spectroscopy. In $LiFePS_3$ the particular occupation of crystal sites by Li atoms removes the inversion symmetry and all the six new modes are both IR and Raman active.

The phonon dispersion curves of the three intercalated phases along the symmetry directions of the Brillouin zone are calculated and shown in figure 26.

The results shown here point toward a reduction of anisotropy in the intercalated phases as a consequence of a stiffer interaction between adjacent layers produced by the interposed Li planes.

One of the most interesting findings in the investigation of the dynamics of Li intercalated $FePS_3$ is the determination of modes highly sensitive to the degree of intercalation. The Raman mode at 242 cm^{-1} in the pure phase, for example, shifts to 252, 263 and 271 cm^{-1} in the $Li_{0.5}FePS_3$, $LiFePS_3$ and $Li_{1.5}FePS_3$ respectively. In this mode the two chalcogen planes move rigidly in antiphase along the z-direction, producing the largest modulation of the Van der Waals gap. The presence of Li hinders the deformation of the Van der Waals gap, inducing the aforementioned energy shift.

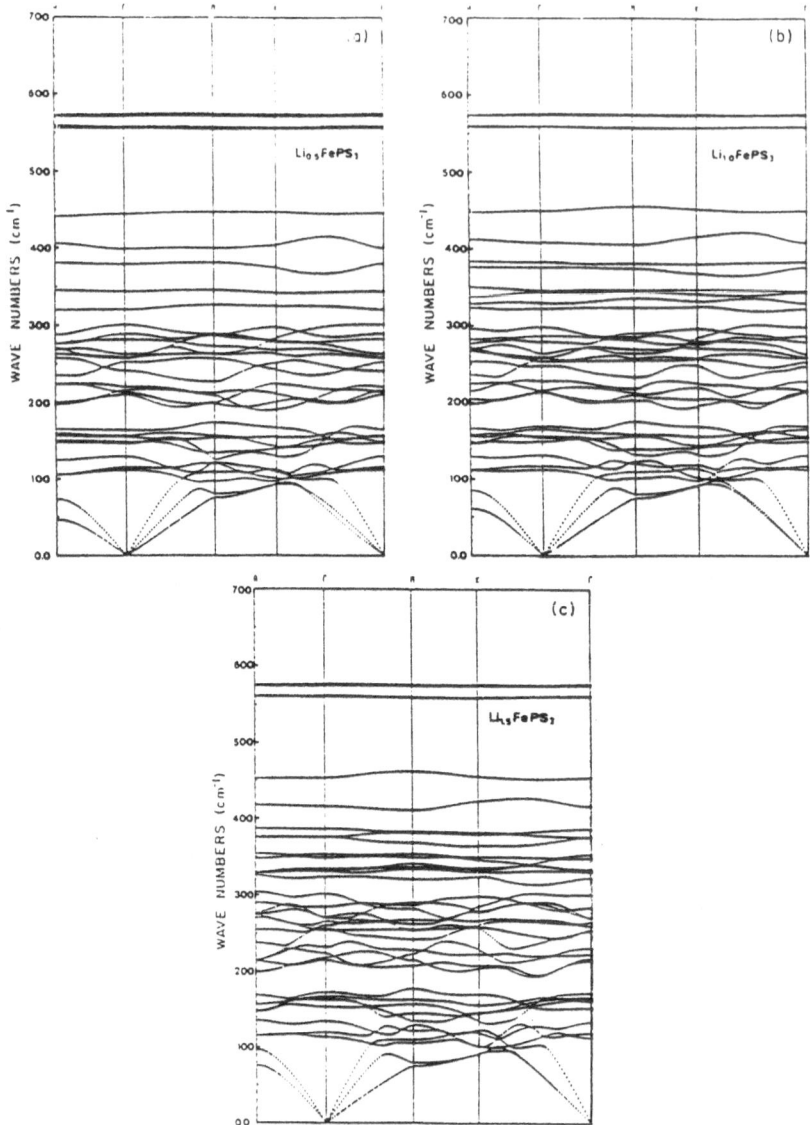

Figure 26 : (a) Phonon dispersion curves of $Li_{0.5}FePS_3$ along the symmetry directions of the Brillouin zone (BZ). The BZ of the three stoichiometric compositions studied is the same as in the pure phase. (b) Phonon dispersion curves of $LiFePS_3$, and (c) Phonon dispersion curves of $Li_{1.5}FePS_3$.

The determination of this mode in the Raman spectrum could provide a method for real time monitoring of the Li concentration in intercalated samples through the measurement of its frequency. Furthemore, exploring the sample surface with a fine focused laser beam would give the possibility of imaging the inhomogeneities of Li distribution.

5. Superconductivity in transition metal dichalcogenides

Some of the transition metal dichalcogenides which exhibit considerable anisotropy in their normal metallic state like $NbSe_2$, or having acquired the metallicity via intercalation like $K_{0.5}MoS_2$, are superconductors at low temprature.

An interesting question is whether superconductivity can persist in individual layers and whether this may be regarded as examples of two dimensional superconductivity. Correlated questions are what is the effect of interlayer interaction on the transition temperature T_c in high temprature superconductors and what will be the effect of intercalation, which, as we have seen, changes the interlayer interaction.

Evidence for believing individual layer superconductivity comes from a number of observations.

1) The transition temperature T_c appears to be insensitive to the changing separation between layers. For $2H-TaS_2$, for example, the transition temperature remains at around 4K irrespective of the size of the organic molecules used as intercalants which give a range of layer separation from 6 Å to 12 Å.

2) Intercalation with chromocene, which possesses strong local moments, does not affect T_c.

3) $NbSe_2$ crystals of unit cell thickness (two layers) remains superconducting with T_c of ~ 5K compared with 7.4 K in the thick specimen.

Many of the oxide superconductors have layered structure and share many of the behaviours described for the layered intercalation compounds. The anisotrophy of normal conductivity with current path parallel and perpendicular to the c-axis generally fall in the

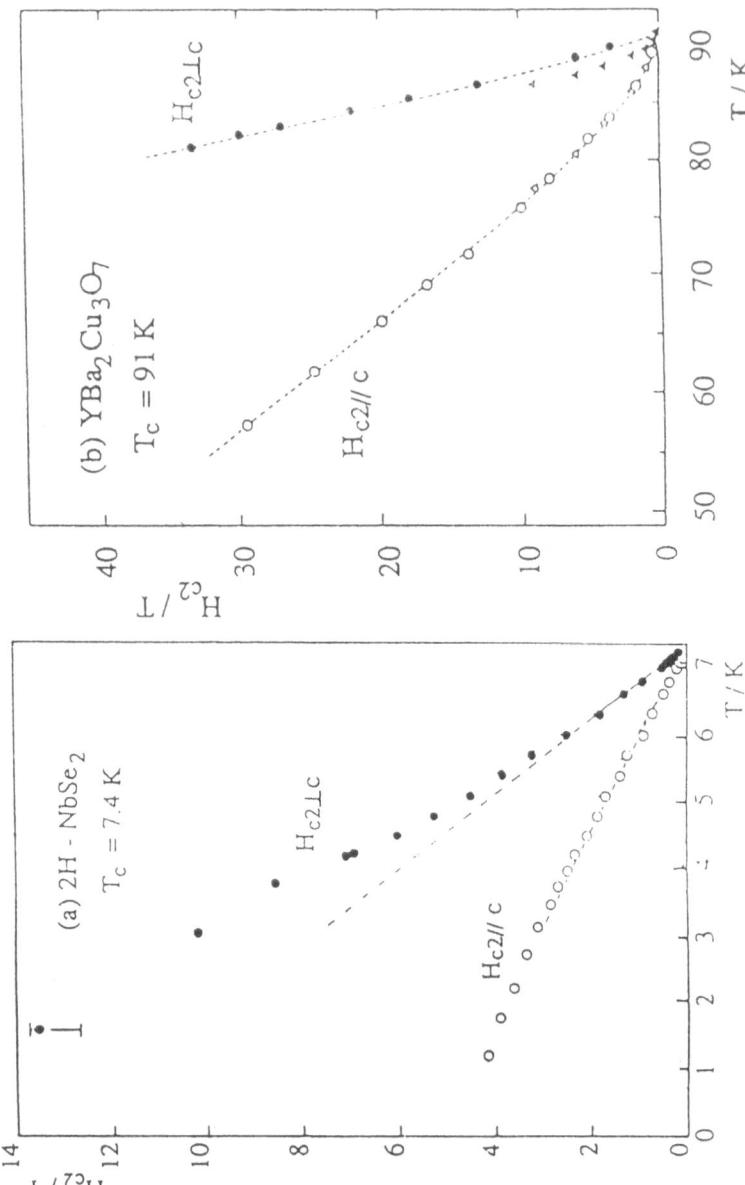

Figure 27 : Extreme anisotropy in H_{c2} with respect to field applied parallel or perpendicular to the c-axis (a) 2H-NbSe$_2$ (from ref. 31) and (b) YBa$_2$Cu$_3$O$_7$ (from ref. 32).

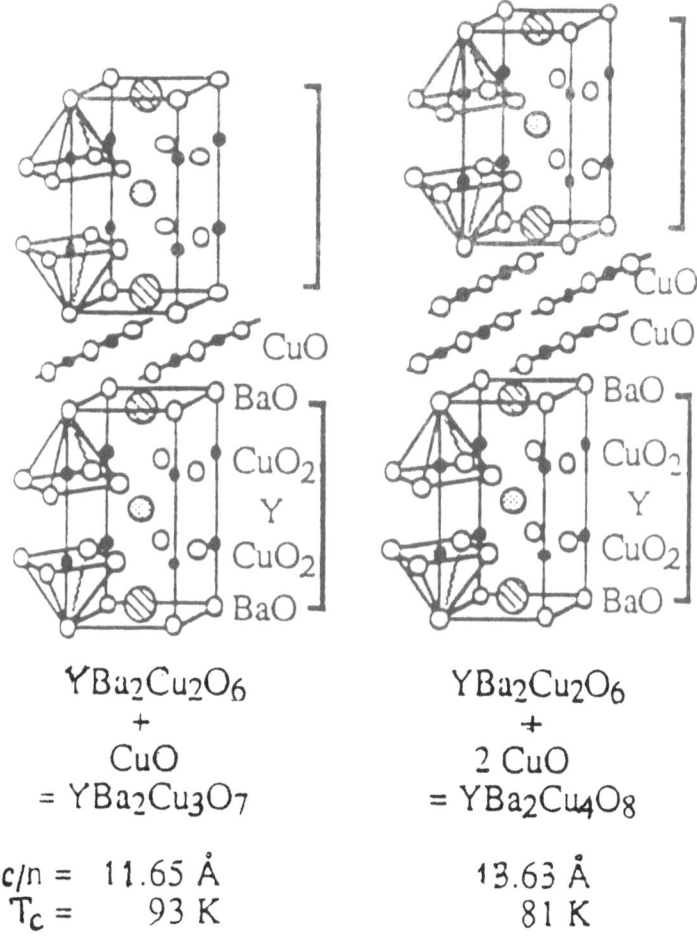

$$YBa_2Cu_2O_6$$
$$+$$
$$CuO$$
$$= YBa_2Cu_3O_7$$

$$YBa_2Cu_2O_6$$
$$+$$
$$2\,CuO$$
$$= YBa_2Cu_4O_8$$

$c/n = 11.65\ Å$ $13.63\ Å$
$T_c = \ \ \ 93\ K$ $81\ K$

Figure 28 : Structure of oxide superconductors which can be built up from the basic block of $YBa_2Cu_2O_6$ or $CaBa_2Cu_2O_6$ together with a variaty of insertion layers ; c/n represents the one formula unit sandwich thicknes in the c-direction (from ref. 17).

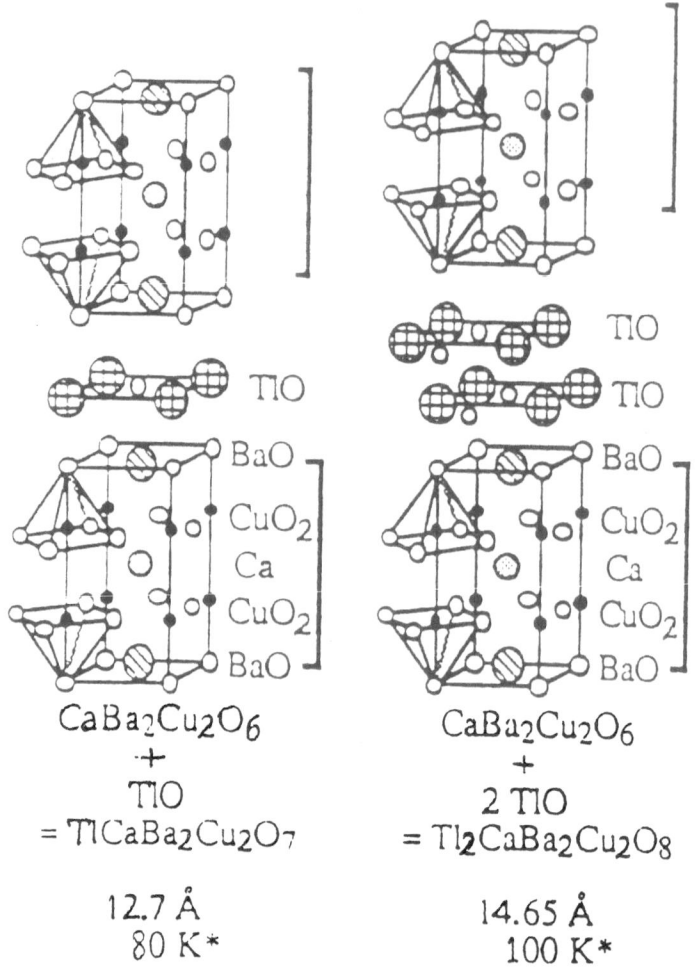

$$CaBa_2Cu_2O_6$$
$$+$$
$$TlO$$
$$= TlCaBa_2Cu_2O_7$$

12.7 Å
80 K*

$$CaBa_2Cu_2O_6$$
$$+$$
$$2 TlO$$
$$= Tl_2CaBa_2Cu_2O_8$$

14.65 Å
100 K*

Figure 29 : Structure of oxide superconductor built up by substitution of Y by Ba in the layers and introduction of one or two sheets of TiO in the interlayer space (from ref. 17).

range 30 to 100 at room temperature which is similar to NbSe$_2$. The anisotropy in the upper critical fields H$_{c2}$ for 2H-NbSe$_2$ is comparable to that of YBa$_2$Cu$_3$O$_7$ as shown in figure 27.

The high temperature oxide superconductors can be viewed[17] as layers of YBa$_2$Cu$_2$O$_6$ separated by one or two sets of chains of CuO as shown in figure 28.

Variations of this structures are possible by full or partial substitution of Y by Ca leading to the structure shown in figure 29 where the interlayer space is occupied by one or two sheets of TlO. The nature and number of intercalated sheets do not affect drastically the high temperature superconductivity : T$_c$ remains in the same temperature range with slight changes. It is well known that the majority of the electronic states at the Fermi level in the oxyde system concentrate in the CuO$_2$ layers. It is also belived that their spin exchange interactions and the specific corner sharing structure of CuO$_2$ are essential in determining the high T$_c$ in these materials. Two further factors are important in order to optimize the superconducting properties. One relates to the carrier density and the other to the ion sizes. Both factors have implications on the chemical stability of the compounds. In many compounds it is possible to increase T$_c$ by suitable substitution of a cation in the basic building layers, or by an appropriate choise of the insertion atomic sheets.

This point of view might in the future demonstrate to be very fruitful and stimulating.

References

1. Liang W.Y. in Solid State Ionics edited by M. Balkanski and C. Julien, North-Holland, 1989.

2. Liang W.Y., Electronic Properties of Intercalation Compounds in "Intercalation in Layered Materials", edited by M. Dusselhauss, Plenum Press (1986).

3. Schlüter M., Il Nuovo Cimento B 13 (2), 313 (1973).

4. Mc Canny J.V. and Murray R.B., J. Phys. C 10, 1211 (1977).

5. Bourdon A., Chevy A. and Besson J.M., Inst. Phys. Conf. Ser. N 43, 1371 (1979).

6. Doni E., Girlanda R., Grasso V., Bolzorotti A. and Piacentini M., Il Nuovo Cimento B 51 (1), 154 (1979).

7. Piacentini M., Doni E., Girlanda R., Grasso V. and Bolzorotti A., Il Nuovo Cimento B 54 (1), 269 (1979).

8. Gomes da Costa P., Balkanski M. and Wallis R.F., to be published in Phys. Rev.

9. Balkanski M., Julien C. and Jouanne M., J. Power Sources 20, 213 (1987).

10. Kunc K., Zeyher R. and Molinari E., Proc. 19th Int. Conf. Phys. Semiconductors, Warsaw, 1988, edited by W. Zawadski (Inst. of Physics, Polish Acad. Sci., Wraclaw, 1988, p. 1119).

11. Kunc K. and Zeyher R., Europhys. Lett. 7, 611 (1988).

12. Clementi E. and Roetti C., Atomic Data and Nuclear Data Tables 14, 177 (1974).

13. Julien C., Halzikroniotis E., Paraskevopoulos K.M., Chevy A. and Balkanski M., Solid State Ionics 18-19, 859 (1986).

14. Wallis R.F., Balkanski M. and Gomes da Costa P., to be published in R.C Leite Festshrift.

15. P.A. Burret, C. Julien, M. Jouanne and M. Balkanski, Mater. Sci. Eng. B 3, 39 (1989).

16. Guo G.Y. and Liang W.Y., J. Phys. C 19, 995 (1986).

17. Liang W.Y., Proc. NATO Adv. Research Workshop on Condensed Systems of Low Dimensionality, Mormoris, Turkey, April 23-27 (1990).

18. Julien C., Sekine T. and Balkanski M., to be published in Solid State Ionics.

19. Brek R., Intercalation in Layered Materials, ed. by M.S. Dresselhaus, Plenum Press (1986).

20. Wilson J.A. and Yoffe A.D., Adv. Phys. 18, 193 (1969).

21. Khumalo F.S. and Hughes H.P., Phys. Rev. B 23, 5375 (1981).

22. Sekine T., Jouanne M., Julien C. and Balkanski M., Mater. Sci. Eng. B 3, 91 (1989).

23. Benedek G., Benasconi M. and Miglio L., "Microionics - Solid State Integrable Microbatteries" ed. by M. Balkanski, Elsevier (1991).

24. Balkanski M., Jouanne M., Ouvrard G., Scagliotti M., J. Phys. C : Solid State Physics 20, 4397 (1987).

25. Bernasconi M. Marra G.L., Benedek G., Miglio L., Jouanne M. Julien C., Scagliotti M. and Balkanski M., Phys. Rev. B 38, 12089 (1988).

26. Jemberg P. Bjarman S., Wäppling R., J. Magn. Mater. 46, 178 (1984).

27. Suzuki N. and Kamimura H., Solid State Commun. 11, 1603 (1972).

28. Kurosawa K., Saito S. and Yamaguchi Y., J. Phys. Soc. Japan 52, 3919 (1983).

29. Sekine T., Jouanne M., Julien C. and Balkanski M., Phys. Rev. B 42, 8382 (1990).

30. Barj M., Sourisseau C., Ouvrard G. and Brek R., Solid State Ionics 11, 179 (1983).

31. Toyota N., Nakatonji H., Noto K., Hoshi A., Kobayashi N., Muto Y. and Onodera Y., J. Low Tamp. Phys. 25, 485 (1976).

32. Iye Y., Tamegai T., Sakakibara T., Goto T., Minra N., Takeya H. and Takei H. Physica C 153-155, 26 (1988).

33. Liang W.Y., Progress in High T_C Superconducting Materials, to be published in Mater. Sci. Eng. B (1991).

Thin Films of Solid Electrolyte and Their Applications

K. Radhakrishnan

Department of Physics, National University of Singapore, Kent Ridge, Singapore

Extensive studies have been concentrated on the development of solid state micropower sources for a variety applications from batteries to miniature electronic devices. Identificaion of a good solid electrolyte with excellent electrical properties and a compatible electrochemical couple is of paramount importance. This review describes the current status of solid electrolytes studied in thin film form and their applications. Emphasis is made on vitreous lithium based fast ion conductors, and examples from the investigations carried out in the recent past have been discussed in detail. Different preparation techniques currently used in the fabrication of microcells have also been briefed.

1. INTRODUCTION

The phenomenon of electrical conduction in solids as the result of the motion of ions has attracted scientific attention since the first experiments conducted on nonmetallic solids (Ag_2S, PbF) by Michael Faraday well over 150 years ago. However, the interest in high ionically conducting solids was not simulated until Yao and Kummer published their report on sodium beta alumina in 1965. Several studies that followed were primarily directed towards understanding the relationship between structure, composition and ionic conductivity. Now, it is well known that ionic conduction in solids is possible based on a variety of monovalent cations such as Li, Ag, Cu, Na, etc., or anions such as F . Bivalent cations such as Mg_2, Ca_2, etc. have also been known to provide ionic transport in solids. The solid materials which show high ionic conductivity can be crystalline, polycrystalline, or amorphous including glasses and polymer complexes. Another group of materials, basically composite solids, such as the dispersion of Al_2O_3 in LiI where conduction is believed to take place in the thin interfacial region surrounding the dispersed particles is also of interest. The technological importance of all these materials, variously known as fast ion conductors, suprionic conductors or solid electrolytes, is exemplified by their use in a growing number of applications such as solid state batteries, fuel cells, sensors, etc.

During the last two decades several solid electrolyte materials have been developed which exhibit high ionic conductivity. Nevertheless, it is generally accepted that conductivity values of these materials are still lower than that of the concentrated aqueous solutions at room temperature. However, for a solid electrolyte the relevant process parameter is conductance, ρ, rather than the conductivity, σ, since the electrolyte acts simply as a separator in the devices (between the electrodes) for ionic transport. As conductance ($\rho = \sigma A/L$, A is the area of cross section and L is the thickness) is inversely proportional to the thickness of the electrolyte, it can be increased significantly by decreasing the thickness. Thus, even a poor conductor with conductivity as low as $10^{-7}\Omega^{-1}cm^{-1}$ can be used in thin film form that yields a conductance of $10^{-4}\Omega^{-1}cm^{-1}$. Moreover, thin film cells offer specific advantages such as a good quality interface between the electrolyte and the electrodes. Thin films of solid electrolytes are useful in the development of microbatteries which can be integrated in miniaturized devices for low current applications such as LSI memories. For example, power requirements of the order of few hundred pW can be supplied during electrical failures by a microbattery which is integrated to the component. Other applications include electrochromic windows, display devices etc.

2. Thin Film Deposition Technology

There are number of deposition techniques available for thin film formation. Table 1 provides an exhaustive list of deposition methods currently used, especially in the field of 'microelectronics' [1]. Thus, the knowledge of microelectronics can be of immense help for successful fabrication and characterization of devices based on solid electrolytes.

For our purpose, the techniques used to prepare thin film can be broadly classified into two categories: (i) Chemical Vapor Deposition (CVD) and (ii) Physical Vapor Deposition (PVD). In the first category, starting materials engage in specific chemical reactions at the hot surface of the substrate to form thin layers of the desired material. Therma oxidation, liquid-phase chemical formation, anodizing, electroplating, chemical reduction plating, etc., are the examples of CVD.

PVD is exemplified by thermal evaporation, flash evaporation, electron beam evaporation, glow-discharge processes, etc. Glow discharge processes can be further divided into two sub-categories: sputtering and plasma deposition as listed in Table 1. Since the evaporative and sputtering techniques have been largely used in the preparation of solid state electrochemical cells, these methods will be discussed below in detail.

Evaporative Methods: Thermal evaporation is commonly referred to as vacuum evaporation. The process of film formation by this method involves the following three stages: (i) evaporation or sublimation of the charge to form vapor (ii) transfer of atoms or molecules from evaporation source to substrate and (iii) condensation of vapor on the substrate.

Evaporation or sublimation process requires the provision of sufficient energy to surmount the attractive intermolecular forces existing within the starting material Techniques used are indirect resistance heating, flash evaporation and electron beam heating.

Indirect resistance heating involves placing the evaporation material in a container made of refractory metals like Mo, Ta or W which can be in the form of boat, coil

Table 1 (from Ref 1)

Classification of Thin Film Deposition Technologies

Evaporative Methods

Vacuum Evaporation
    Conventional Vacuum Evaporation   Molecular Beam Epitaxy
    Electron-beam Evaporation   Reactive Evaporation

Glow-discharge Process

Sputtering   Plasma Processing
    Diode Sputtering   Plasma Enhanced CVD
    Reactive Sputtering   Plasma Oxidation
    Bias Sputtering   Plasma Anodization
    Magnetron Sputtering   Plasma Polymerization
    Ion Beam Deposition   Plasma Nitridation
    Reactive Ion Plating   Microwave ECR Plasma CVD
    Cluster Beam Deposition (CBD)   Cathodic Arc Deposition

Gas-Phase Chemical Process

Chemical Vapor Deposition(CVD)   Thermal Forming Processes
    CVD Epitaxy   Thermal Oxidation
    Atmospheric Pressure CVD (APCVD) Thermal Nitridation
    Low-pressure CVD (LPCVD)   Thermal Polymerization
    Metalorganic CVD (MOCVD)
    Photo-enhanced CVD (PECVD)
    Laser-induced CVD (LICVD)
    Electron-enhanced CVD (EECVD)   Ion Implantation

Liquid-phase Chemical Techniques

Electro Processes   Mechanical Techniques
    Electroplating   Spray Pyrolysis
    Electroless Plating   Spray-on Techniques
    Electrolytic Anodization   Spin-on Techniques
    Chemical Reduction Plating
    Chemical Displaement Plating
    Electrophoretic Deposition   Liquid Phase Epitaxy

or strip. The container is heated by a current flow. A large number of containers, called evaporation sources have been devised to accommodate and evaporate a variety of materials of varying vapor pressures. If the evaporants react with the source material, a protective layer of alumina coated on the source or an alumina crucible, surrounded by a tungsten heater, can be used.

The sample charged to a boat or crucible is heated to a required temperature for deposition in indirect resistant heating methods. In flash evaporation, fine powder of the sample is continuously dropped in a predetermined rate from a vibration feeder on a sufficiently heated crucible or boat so that the droplets get evaporated instantly. This method, in certain cases, allows homogeneous film compositions even with material components of dissimilar vapor pressures.

In electron beam evaporation, the evaporant in a crucible or hearth of electrically conducting material is bombarded with a beam of electrons so that the evaporant is directly heated and vaporized.The portion of the evaporant that is heated, is in the centre of the exposed surface, and there is a long thermal conduction path through the material to the hearth. Therefore, the hearth can be maintained at lower temperatures compared to the melting point of the evaporant without prohibitive heat loss. That means, the reaction between the hearth and the evaporant is virtually inhibited. This technique is particularly suitable for materials that react with the source or require very high evaporation temperatures or both.

In all these evaporative methods, to ensure that the path of liberated vapors from source to substrate is not impeded by collisions with other gas atoms present, a good vacuum is necessary. Hence, the deposition is carried out in a sealed chamber that is evacuated to a low pressure of the order of 10^{-5} torr or better. Such a vacuum level can be attained easily with an oil diffusion pump backed by a mechanical rotary pump.

When the evaporated particles reach the substrate, they condense on the surface. The parameters like deposition rate, surface mobility, ordering processes and grain size are influenced by the substrate temperature. Hence, the later plays an important role on the quality and morphology of thin films deposited.

Sputtering Techniques: Sputtering process is the ejection of individual surface atoms from the solid surface by momentum transfer from high energy bombarding ions to surface atoms. Sputtering is one of the oldest vacuum process for producing thin films. However, it became an important technique only in 1950's when intensive studies of the sputtering phenomenon enabled better control of the process. An important advantage of sputtering over evaporative methods is that high quality films with good adhesion can be obtained. On the other hand, its low deposition rate of 20-100A/min is its main drawback, as long hours of sputtering is required for preparing films of significant thickness.

In cathode sputtering technique, the material to be deposited forms the cathode of the system in which a glow discharge is established using a working gas such as argon at a pressure of 10^{-1} to 10^{-3} torr and at a voltage of several kilovolts. The substrate on which the film is to be formed is placed on the anode of the system. Positive ions of the gas created by the discharge are accelerated towards the cathode. Under the bombardment of these ions, particles are removed from the cathode in the form of neutral atoms and ions. The liberated components then condense on the surrounding areas, including the substrate.

In the conventional dc cathode sputtering, as discussed above, a large negative potential is applied to the conductive target. This technique therefore is not suitable for insulating materials. The problem can be partly solved by placing the insulator on a conducting base plate. This arrangement can accelerate the gas ions towards the conducting plate and impinge on the insulator causing sputtering. However, an immediate build-up of the positive charges on the the surface of the insulator constitutes another problem as it will prevent further bombardment. To remedy the situation, a positive potential is applied to the conducting plate by reversing the power connections which will result in electrons accelerating towards the insulator. Negative charges will now buildup at the insulator surface which requires another power reversal and so on. This principle is essentially used in RF-sputtering. The rate of reversal would be sufficient to attain a practical rate of sputtering by employing a radio frequency generator operating at a frequency of several MHz.

Magnetron sputtering, frequently used for thin film deposition, is an RF-sputtering technique in which a magnetic field transverse through the electric field applied at the target surface. Target generated secondary electrons are hence trapped in cycloidal trajectories near the target. Since these electrons do not bombard the substrate, they do not contribute to increased substrate temperature and radiation damage. This allows the use of low temperature substrates (eg. plastic) and surface-sensitive substrates (eg. metal-oxide semiconductor devices) without adverse effects. In addition, this technique offers higher deposition rates compared to conventional RF-sputtering.

For electrical conductivity measurements, electrolyte film is deposited between gold or platinum electrodes as a planar or sandwich structure. The planar arrangement, shown in Fig. 1a, is useful when the conductivity of the electrolyte is high such as in the case of sulphide based glasses [2]. Sandwich configuration can be used if the conductivity of the electrolyte is not high as in the case of lithium oxide based glasses. A typical configuration used for ac conductivity measurements in our laboratory is shown in Fig. 1b. In the case of moisture sensitive films, deposition and electrical characterization is usually carried out in a controlled environment. Vacuum equipment can be arranged such that the deposition chamber is within a glove box.

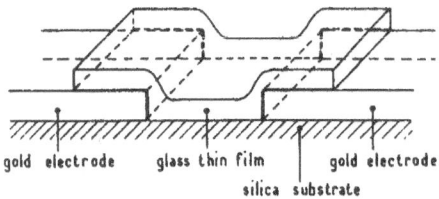 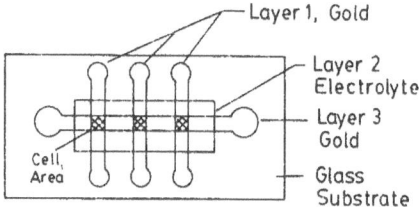

Fig. 1a Planar arrangement of the electrodes for electrical measurements.

Fig. 1b Sandwich configuration for electrical measurements.

3. THIN FILMS OF SOLID ELECTROLYTES

Successful deposition in thin layers of several Na^+, Ag^+ and Li^+ ion conducting solid electrolytes have been reported [2-19]. The first Na^+ ion conducting solid electrolyte (prepared by RF sputtering) has been reported in the system $Na-\beta GaO_2$, a chemically analogous material to $Na-\beta Al_2O_3$ [4]. These films exhibit conductivity values as high as $4 \times 10^{-2} \Omega^{-1} cm^{-1}$ at 300 °C. $Na-\beta Al_2O_3$ could not be prepared in thin film form because of considerable alkali loss during deposition as well as difficulty in thin film synthesis [3]. Lithium based thin films are attractive since they offer higher energy densities and open circuit voltages. Thin films of the well known lithium conductor, LISICON [$Li_{14}Zn(GeO_4)_4$], have been prepared by RF-sputtering [7]. The crystalline phase of the film is generally influenced by the sputtering conditions, especially the target composition and the ambient gas. High ionic conductivity values observed for the composition $Li_{3.4}V_{0.6}Si_{0.4}O_4$ in the solid solution $Li_3VO_4-Li_4SiO_4$ has led to the preparation of thin films by RF-sputtering in the system $Li_2O-V_2O_5-SiO_2$ [17]. Electrical conductivity values of the order of $1 \times 10^{-6} \Omega^{-1} cm^{-1}$ is obtained at 25°C by complex plan analysis.

The ease of fabrication in thin film geometry together with the possibility to prepare in a wide range of compositions, researchers have investigated thin film electrolytes based on glassy compounds. Many microelectronic techniques have been applied to realize them in thin film form. Although the starting compositions for most of the studies discussed below might be glass before thin film deposition, the films obtained need not be glass. However, they are, in general, X-ray amorphous solids.

Several Ag$^+$ion conducting amorphous thin films prepared by thermal evaporation, flash evaporation and RF-sputtering have been reported [5]. Films prepared by flash evaporation has been shown to give higher conductivities and a transport number value of unity, compared to other preparation techniques. Room temperature conductivities of the order of 2x10^{-2} Ω-^{1}cm^{-1} and 5x10^{-3} Ω-^{1}cm^{-1} have been measured for the glasses 60AgI.40Ag$_2$MoO$_4$ and 30AgI.23.3Ag$_2$O.46.7B$_2$O$_3$, respectively.

Levasseur et al have reported the preparation conditions using vacuum evaporation and RF-sputtering of films in the glasses Li$_2$O-B$_2$O$_3$, Li$_2$O-B$_2$O$_3$-SiO$_2$ and Li$_2$-B$_2$O$_3$-Li$_n$X (n = 1, X = I; n = 2, X = SO$_4$) and their electrical and chemical properties [8]. For the binary Li$_2$O-B$_2$O$_3$ glasses, it was observed that Li$_2$O content is higher in thin film material than those in the starting composition. Ito et al have studied the structure and the ionic conductivity of thin films in the system Li$_2$O-B$_2$O$_3$ [9]. In their study, structural change with variation in σ has been qualitatively demonstrated. Amorphous films of Li$_{3.6}$Si$_{0.6}$P$_{0.4}$O$_4$, prepared by RF-sputtering have been investigated by Kanehori et al for application to lithium secondary cell [10]. Realization of the fact that some of the films prepared earlier such as Li$_{14}$Zn(GeO$_4$)$_4$, Na-βAl$_2$O$_3$ and Li$_3$N are thermodynamically unstable or that they are incompatible with the cathode materials has resulted in the search for new materials for thin film studies. Amorphous films in the system Li$_2$O-SiO$_2$-ZrO$_2$ have been found to be highly conducting and thrmodynamically stable with Li [11]. Preparation of films in the system Li$_2$O-SiO$_2$ and Li$_2$O-B$_2$O$_3$-SiO$_2$ with high electrical conductivities ($\sigma_{rt} = 10^{-4} - 10^{-7}$ Ω-^{1}cm^{-1}) have been reported by Machida et al [12].

As mixed glasses containing more than one glass former gives higher conductivity compared to glasses containing single glass former, several investigation have been carried out. In this direction, FIC glasses based on Li$_2$O-B$_2$O$_3$-P$_2$O$_5$ and Li$_2$O-P$_2$O$_5$-Nb$_2$O$_5$ "mixed glass former" systems have been found to be good candidates for battery applications [20,21]. In a further study, amorphous thin films of Li$_2$O-P$_2$O$_5$-Nb$_2$O$_5$ glassy materials have been prepared by thermal evaporation and ionic conductivity values of the order of 10^{-5} Ω-^{1}cm^{-1} have been measured at 100°C [15]. Films were deposited by varying the source composition systematically according to the formula: xLiPO$_3$:(1-x)60Li$_2$O-30P$_2$O$_5$-10Nb$_2$O$_5$. This is necessary because of the fact that conventional thermal evaporation technique often does not produce thin film of the starting composition due to dissimilar vapor pressures of different components present in the starting composition.

Thin films prepared by RF-sputtering , however, have shown better electrical properties for the same system, Li$_2$O-P$_2$O$_5$-Nb$_2$O$_5$ [16]. Sputtering conditions for deposition are summarized in Table 2 and typical complex impedance plots at different temperatures are shown in Fig. 2 for the film having the composition 67.5Li$_2$O-28.3P$_2$O$_5$-4.2Nb$_2$O$_5$.

Temperature dependence of the conductivity has been compared for the RF-sputtered films, and the evaporated film showing the highest conductivity, in Fig. 3; the data for the bulk glass has also been included for comparison. Chemical composition and electrical properties of different thin film samples are listed in Table. 3. It can be seen from Fig. 3 that the conductivity at any

Table 2. Typical Sputtering Conditions	
Sputtering Gas:	Ar + O$_2$ (1:1)
Sputtering Pressure:	1.4x10^{-2} torr
Target-Substrate Distance:	35 mm
Substrate Material:	Corning Glass
Substrate Temperature:	180°C
Sputtering Rate:	0.9 μm/hr
Power:	3.5 W/cm2

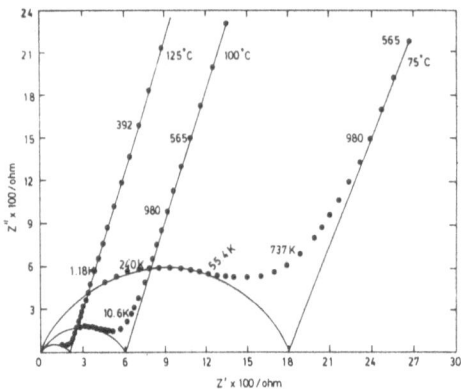

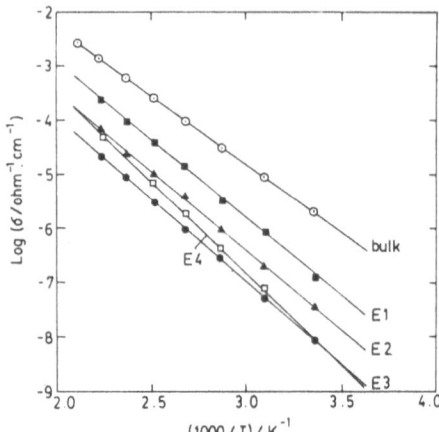

Fig. 2 Typical complex impedance plots at different temperatures for the film E1, 67.5Li₂O-28.3P₂O₅-4.2Nb₂O₅. Numbers indicate frequency in Hz.

Fig. 3 Temperature dependence of electrical conductivity for bulk glass and thin films (E1 — E3). Data from Ref. 15 for the evaporated film is also included for comparison (E4).

temperature shows an increasing trend with the increase in lithium content among the sputtered films. The highest conductivity is measured for the film E1 with a lithium content of 67.5 mol%, $\sigma = 1.55 \times 10^{-5}\ \Omega^{-1}\text{cm}^{-1}$ at 100°C. Electronic conductivity of these films measured by Wagner's polarization technique yielded values approximately four orders of magnitude less compared to ionic conductivity values at all the temperatures studied. Low electronic contribution is one of the main requirements for solid electrolytes in practical applications. Electrochemical decomposition potential, the value of the applied voltage for which there is a significant passage of current in a polarization cell has been determined to be ~2.7V for these electrolytes. Electrical properties of some of the silver and lithium ion conducting solid electrolytes studied in thin film form are summarized in Table 4.

Table 3. Chemical Composition and Electrical Properties of Different Thin Film Samples

Sample No	Li₂O mol%	P₂O₅ mol%	Nb₂O₅ mol%	E (eV)	$\sigma(o)$ $\Omega^{-1}\text{cm}^{-1}$	σ at100°C $\Omega^{-1}\text{cm}^{-1}$
E1	67.5	28.3	4.2	0.57	595	1.55×10^{-5}
E2	58.9	29.9	11.2	0.58	225	3.89×10^{-6}
E3	54.4	35.3	10.3	0.60	129	9.12×10^{-7}
E4	60.9	31.1	8.0	0.66	1625	1.82×10^{-6}
Bulk	60.0	30.0	10.0	0.51	680	9.30×10^{-5}

Studies on glasses have shown that the conductivity can be enhanced markedly by substituting oxygen with highly polarizable sulphur. This has led to the preparation of lithium chalcoborate thin films based on the system Li₂S-B₂S₃ and Li₂S-B₂S₃-LiI by Kabla et al [13]. Unlike the oxide systems, thin films of these glasses have shown similar composition with that of the starting proportions. By annealing the films containing LiI, it has been shown that the conductivity can be improved by a factor of 10. This behavior was

Table 4

Ionic Conductivity,σ and Activation Energy,E of Li^+ and Ag^+ Ion Conducting
Amorphous Thin Film Solid Electrolytes

ELECTROLYTE	DEPOSITION METHOD	σ $(\Omega.cm)^{-1}$	E (eV)
$AgI-Ag_2MoO_4$	Flash Evaporation	2×10^{-2}	0.2-0.3
$AgI-Ag_2O-B_2O_3$	RF-Sputtering	6×10^{-3}	-
$Ag_2O-P_2O_5$	RF-Sputtering	$10^{-7}-10^{-8}$	0.50
$Ag_2O-B_2O_3-P_2O_5$	RF-Sputtering	$10^{-7}-10^{-8}$	0.50
Ag_2S-GeS_2	Flash Evaporation	10^{-3}	0.30
$Li_2O-B_2O_3$	Thermal Evaporation	$10^{-8}-10^{-10}$	0.6-0.9
$Li_2O-B_2O_3-X$	RF-Sputtering	10^{-6}	0.60
$(x = SiO_2, Li_2SO_4, LiI)$			
$Li_2O-B_2O_3-P_2O_5$	Thermal Evaporation	10^{-10}	0.80
$Li_2O-SiO_2-ZrO_2$	Magnetron Sputtering	10^{-6}	0.57
$Li_{3.6}Si_{0.6}P_{0.4}O_4$	RF-Sputtering	5×10^{-6}	0.50
$Li_2S-B_2S_3$	Thermal Evaporation	10^{-3}	0.20
Li_2S-SiS_2	Thermal Evaporation	10^{-5}	0.40
$Li_2O-P_2O_5-Nb_2O_5$	Thermal Evaporation	10^{-8}	0.66
$Li_2O-P_2O_5-Nb_2O_5$	RF-Sputtering	2×10^{-7}	0.57

attributed by them to the quick ionic diffusion along the film-substrate interface, called the Phipps effect. For sulphide films in Ag^+ ion based systems, films of Ag_2GeS_3 have been prepared by evaporation and sputtering and their properties compared [14]. Both the sputtered and the evaporated films showed similar composition as the bulk glass (σ = $5\times10^{-4}\Omega^{-1}cm^{-1}$ at 20°C). However, the presence of a few percent of metallic silver was detected in the RF-sputtered films.

4. APPLICATIONS OF SOLID ELECTROLYTES

An earliest development in the fabrication of microcells based on Ag^+ ion conduction was due to Yamamoto and Takahashi in 1966 [22]. Microcells of the configuration $Ag/Ag_3Si/I_2$,C showed an open circuit voltage (OCV) of 0.67V at 29°C. Although a high current of $100\mu A/mm^2$ could be drawn from these cells, the main drawback was the problem due to iodine oxidizing the solid electrolyte. The replacement of Ag_3Si with $RbAg_4I_5$ was met with limited success. High energy density, long shelf-life, high reliability, low cost and greater choice of electrode materials have stirred interest in lithium batteries.

Kanehori et al have fabricated thin film secondary lithium cells using amorphous $Li_{3.6}Si_{0.6}P_{0.4}O_4$ as solid electrolyte and TiS_2 films deposited by low pressure CVD as cathode [10]. Fig. 4 illustrates the charge/discharge curves for two cells having the cathodes of different thickness prepared under different gas pressures. The cells were discharged at different current drain rates as indicated in the figure. Cell A showed discharge capacities of about 45 μAh/cm^2 (80% of the theoretical value for TiS$_2$ single crystal) while cell B showed discharge capacities of about 150 μAh/cm^2 (80% of the theoretical value for TiS$_2$ single crystal). Higher discharge capacities for cell B was attributed to higher diffusion co-efficient of lithium and thicker cathode films. It was also demonstrated by the authors that these cells could endure 2000 charge/discharge cycles.

Thin film cells based on Li$_2$O-P$_2$O$_5$-Nb$_2$O$_5$ solid electrolyte and TiS$_2$ pellet or amorphous V$_2$O$_5$ film as cathode have been constructed recently [15,16]. TiS$_2$ based cells showed OCVs in the range of 2.36-2.40V at room temperature. The discharge capacity computed was approximately 120μAh/cm^2 for a current drain of 4μA/cm^2 from OCV to a cut-off voltage of 1V. For the Li/V$_2$O$_5$ cells, a OCV value of 2.58V was observed, and a discharge capacity of

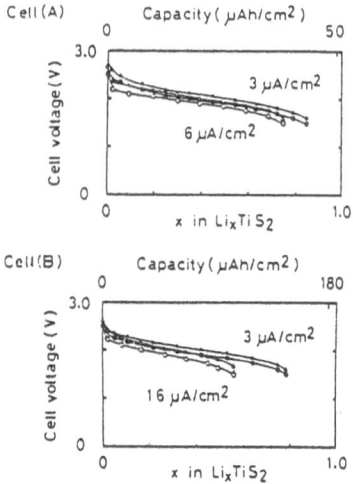

Fig. 4 Discharge-charge curves of thin film cells (●,o:discharge; ▲,△ :charge).

112μAh/cm^2 has been calculated for the similar current drain conditions. The lower OCV value (theoretical value for Li/V$_2$O$_5$ is 3.85V) was attributed to nonstoichiometric V$_2$O$_5$ film obtained by thermal evaporation process. In order to get a better electrode/electrolyte interface by having a common glass network former in the electrode and the electrolyte cells of the configuration Li/Li$_2$O-P$_2$O$_5$-Nb$_2$O$_5$/V$_2$O$_5$-P$_2$O$_5$ are being investigated [23].

Electrochemical cells comprising glassy thin films of both the solid electrolyte (Li$_2$O-B$_2$O$_3$-P$_2$O$_5$) and the mixed conducting oxides (V$_2$O$_5$-P$_2$O$_5$ or V$_2$O$_5$-TeO$_2$) as cathodes have been fabricated by evaporation and RF-sputtering [14]. These cells, called "microgenerators" contain stored energy of 7.5 Wh/mm^2 for a 10μm thick cell. Microbatteries consisting of a set of twenty cells have been deposited on a single alumina substrate as shown Fig. 5a. Fig. 5b shows the cross section of a single cell. Discharge characteristics of these cells at different constant current drain levels are given in Fig. 6. The stored energy of the cells, it is claimed, is sufficient to supply power for a dynamic memory for many years.

In a recent work, Ohtsuka et al have fabricated Li/MoO$_{3-x}$ and Li/MnO$_x$ microcells by RF-sputtering technique using Li$_2$O-V$_2$O$_5$-SiO$_2$ (film is not amorphous)as the electrolyte [17,18]. Li/MoO$_{3-x}$ cells showed good cathode utilization and rechargeability. The electrolyte film was found to be stable with the electrodes. However, the cells suffered from self-discharge due to high electronic conductivity of the electrolyte film. Li/MnO$_x$ cells have also suffered from self-discharge, but the problem was identified with the cathode layer rather than the electrolyte. Although the cells showed good rechargeability, as shown in Fig. 7, the discharge capacity was small, decreasing from 18μAh/cm^2 at the first cycle to 14μAh/cm^2 at the 98th cycle.

Another area of research where solid electrolytes are found to be of increasing interest is in the development of electrochromic windows and display devices. For such applications, an ion conducting layer is required to ensure that oxidation and

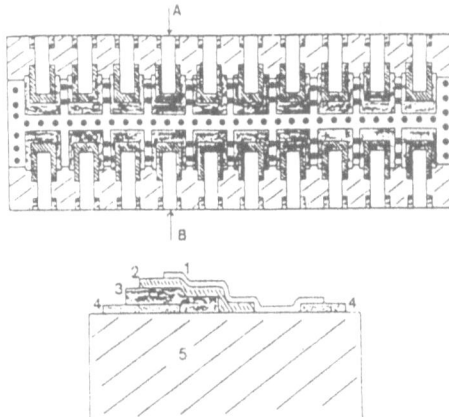

Fig. 5 (a) A set of 20 microgenerators on a single alumina substrate. (b) Cross section of a cell: (1) Li anode, (2) electrolyte, (3) cathode, (4) current collector and (5) alumina substrate.

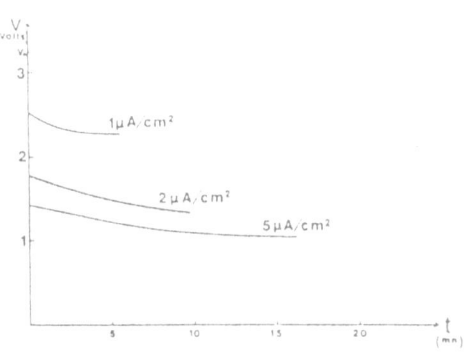

Fig. 6 Discharge characteristics of the $Li/Li_2O\text{-}B_2O_3\text{-}P_2O_5/V_2O_5\text{-}TeO_2$ micro-generator at room temperature.

reduction reactions are restricted to separate compartments. This layer should have good transparency throughout the visible and IR regions, a high electronic resistivity ($\rho > 10^{10}$) in order to promote good open circuit memory and a high ionic conductivity for good response. Number of proton conducting materials have been investigated. As the devices based on H^+ ion conduction depend on the water content present in the structure, they are found to be very sensitive to atmospheric conditions. This has created interest in develop-

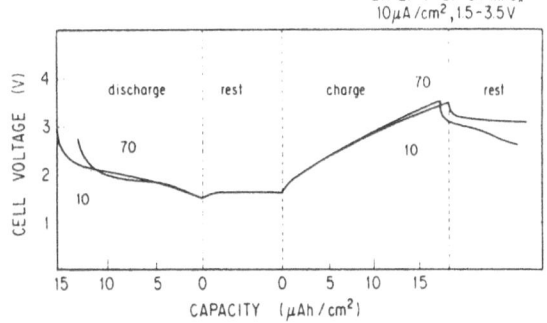

Fig. 7 Typical discharge-charge curves of the solid state cell.

ing alternative materials for solid electrolyte layer. Amorphous thin films of glasses based on $LiNbO_3$ and $LiTaO_3$ have been recently tested for the construction of electrochromic windows [24]. In this direction, Oi and his co workers have synthesized a series of fluoride compounds in amorphous thin film forms [25,26].

In general, preparation of a material in thin film form involves control over many process parameters. Geus surveyed several aspects of thin films such as structure of film surface, surface mobility of adsorbed atoms, growth of metal films on non metallic substrates, and charge transport [27]. The composition and structure of the prepared thin films play a significant role in determining the conductivity. Thus, it is important to optimize several process parameters during deposition. In vacuum deposition, for instance, the control of the vapor species is critical in precise control over the composition of thin films [28]. Perhaps, these are some of the reasons for the limited number of investigations in this technologically important area.

ACKNOWLEDGEMENTS

It is a pleasure to thank and gratefully acknowledge Prof B.V.R. Chowdari for his interest and for a number of helpful discussions which made this study possible. I would also like to thank Dr R. Gopalakrishnan and Ms. Ong Bee Hoo for their participation in Thin Film Research Program.

REFERENCES

1. W. Kern and K.K. Schuegraf, 'Hand Book of Thin-Film Deposition Processes And Techniques, ed. K.K. Schuegraf' Noyes Publications, USA, p4, 1988.
2. A. Levasseur, 'Materials for Solid State Batteries, eds. B.V.R. Chowdari and S.Radhakrishna' World Scientific Publ.Co., Singapore (1986).
3. J.H. Kennedy, Thin Solid Films 43 (1977) 41.
4. K. Miyauchi, T. Kudo and T. Suganuma, App. Phys. Lett 37 (1980) 799.
5. T. Minami, 'Materials for Solid State Batteries, eds. B.V.R. Chowdari and S.Radhakrishna' World Scientific Publ.Co., Singapore (1986).
6. M. Ribes, in 'Solid State Ionic Devices, eds. B.V.R. Chowdari and S. Radhakrishna' World Scientific Publ. Co., Singapore (1988).
7. H. Ohtsuka and A. Yamaji, Solid State Ionics 8 (1983) 43.
8. A. Levasseur, M. Kabla, P. Hagenmuller, G.Couturier and Y. Danto, Solid State Ionics 9 & 10 (1983) 1439.
9. Y. Ito, K. Miyauchi and T. Oi, J. Non-cryst. Solids 57 (1983) 389.
10. K. Kanehori, K. Matsumoto, K. Miyauchi and T. Kudo, Solid State Ionics 9 & 10 (1983) 1445.
11. K. Miyauchi, K. Matsumoto, K. Kanehori and T. Kudo, Solid State Ionics 9 & 10 (1983) 1469.
12. N. Machida, M. Tatsumisago and T. Minami, Yogyo-Kyokai-Shi 95(1) (1987) 146.
13. M. Kabla, M. Makyta, A. Levasseur and P. Hagenmuller, Solid State Ionics 15 (1985) 163.
14. L. Jourdaine, J.L. Souquet, V. Delord and M. Ribes, Solid State Ionics 28-30 (1988) 1490.
15. B.V.R. Chowdari and K. Radhakrishnan, Solid State Ionics 40/41 (1990) 680
16. B.V.R. Chowdari and K. Radhakrishnan, Solid State Ionics 44 (1991) 325
17. H. Ohtsuka and J. Yamaki, Solid State Ionics 35 (1989) 201.
18. H. Ohtsuka, S. Okada and J. Yamaki, Solid State Ionics 40/41 (1990) 964.
19. V. Delord, L. Jourdaine, M. Ribes and J.L. Souquet, 'Electrical Properties of Silver Vitreous Solid Electrolyte Thin Films' 6th Riso Int. Symp. (1985).
20. A. Magistris and G. Chiodelli, J. Power Sources 14(1985) 87.

21. B.V.R. Chowdari and K. Radhakrishnan, J. Non-Cryst. Solids 110 (1989) 101.

22. O. Yamamotto and T. Takahashi, Denki Kagaku, 34 (1966) 833.

23. K. Radhakrishnan and B.V.R. Chowdari, 'The Lithium Diffusion Co-efficient in V_2O_5-P_2O_5 and V_2O_5-TeO_2 Glasses' Submitted for 8th Int. Conf. on Solid State Ionics, Lake Louise, Canada, Oct. 20-26, 1991.

24. T.E. Hass, R.B. Goldner, G.Seward, K.K. Wong, G. Foley and R. Kabani, Proc. SPIE, 823-16, San Diego (1987).

25. T. Oi, Mat. Res. Bull., 19 (1984) 1343.

26. T. Oi, K. Miyauchi and K. Uehara, J. Appl. Phys. 53 (1982) 1823.

27. J.W.Geus, Proc.Fast Ion Transport in Solids-Solid State Batteries and Devices, Ed. W.Vangool. Nato Summer School, Belgirate, Italy (1972), P 331.

28. Li-Wei Zhang, M. Yahagi and K.S. Goto, Solid State Ionics 18 & 19 (1986) 1163.

Properties and Structure of Oxyfluoride Glasses

R. Gopalakrishnan
Department of Physics, National University of Singapore, Singapore

Fluoride ions are important in oxide glasses as it can modulate the electrical and structural properties owing to their size similar to that of the oxide ions. In the present paper the results from X-Ray photoelectron spectroscopy (XPS), Nuclear magnetic resonance (NMR), Infrared, Raman and Complex impedance (CI) studies of glassy materials encompassing oxylurophosphates, oxylfuroborates, oxylfurosilicates and oxylfurogermanates are presented and discussed. The use of XPS technique as a tool in studying these glasses is highlighted.

1. INTRODUCTION

A great deal of interest has been focused on the subject of oxide glasses containing fluoride ions because of their potential application in the manufacture of ultra low-loss optical fibers and domes, laser windows and high power laser host materials [1].Among these oxides, some exhibit relatively high anionic conductivities and they can be considered as possible candidates for electochemical applications [2]. Recent studies on the ionic transport of fluoride glasses have indicated that the bond length between glass former cations and fluoride ions is the decisive factor of prime importance in determining the magnitude of conduction [3,4]. However, because of the inadequate study of the influence of oxide ion on the mechanism of conduction of fluoride ions and the structure of those glasses, to the best of my knowledge, no conclusive agreement regarding the bonding state of fluorine in oxyfluoride glasses has been arrived.

Lead fluoride belongs to a group of superionic fluoride conductors (CaF_2, Baf_2 and SrF_2)in which ionic transport is related with a disordering in the anionic lattice whereas the noble glass formers are insulators owing to their covalent bondage. Among these fluorites, PbF_2 exhibits the lowest supreionic transition temperature (Tc=700K) coupled with high conductivity. Thus from the chemical standpoint the glass formation between the noble glass formers and PbF_2 will be interesting and may be expected to be characterized by a decrease in covalent linkages at ultra rich fluoride compositions.

XPS technique with its ability to distinguish different chemical states of atoms has recently proved to be a powerful tool in the structural analysis, an extension of the studies employing this technique has been attempted in order to have a better understanding of the mechanism of ionic transport in the oxyfluoride glassy system [5].

In the present paper results from the electrical and spectroscopic studies for the lead fluoride based silicates (FS),borates (FB),phosphates (FP)and germanates (FG)are only presented and discussed. However, some reference is made on other fluoride glass system.

2. EXPERIMENTAL PROCEDURE

2.1 General aspects of glass synthesis

Generally the stages in a oxyfluoride glass preparation may be categorized as follows: melting, cooling, casting and annealing. Glass samples of a given stochiometric composition are prepared by melting the calculated quantities of guaranteed reagent grade chemicals in an appropriate crucible. In practice, platinum, gold and silica can be used. However, these materials sometimes absorb the melt/get dissolved and subsequently they may have a dramatic influence on the physiochemical properties. The melt can be cooled in different ways with correspondingly various cooling rates. These rates range from several thousand kelvin per second when the molten glass is poured into a fine gap between two chromium-plated stainless steel cylinders,rotating rapidly in opposite directions, to 1 K per second if squeezed onto a preheated mould. The former is generally referred as " twin-roller rapid quenching " and the later as " conventional/classical quenching " techniques.

The quenching rate plays a significant role in the investigation of glass-forming systems and the magnitude of the vitreous domain. Unlike the fluoride glasses, the oxyfluoride glasses need not be prepared in a controlled/inert atmosphere. The next requirement is the chemical analysis of the glass samples especially for the fluorides as the ultra high fluoride compositions tend to vaporize more readily during the melting process. However, weight loss on melting measurements can be used to obtain qualitative values of volatilization losses and thus compositional variance within samples.

3. RESULTS AND DISCUSSION

3.1 Transport studies

In examining the influence of fluorides on oxide glasses in the ternary systems, the structure and conduction mechanism in binary lead systems need to be analyzed initially. Earlier spectroscopic studies have suggested that lead acts as a network modifier in high silica compositions and the Pb-O bond is ionic. At high lead oxide concentrations Pb-O bond is covalent as the lead takes network forming positions in $Pb_{4/2}$ units [6]. The possible charge carriers in binary lead silicate glasses has been proposed to be free O^{2-} as well as Pb^{2+}, H^+, electrons. In the ternary flurosilicate glasses the conductivity could thus be governed both by the binary species and the fluoride ions. However, the rapid increase in conductivity to three orders of magnitude with the increase in lead fluoride content in these glasses may be indicative of the change in conduction mechanism. Based on the well-known Tubantd test for the transport number, Schultz and Mizzoni have concluded that the conduction in halosilicate glasses is predominantly due to the halide ions [7].

Subsequent research works carried out in oxyfluoride glasses usually indicate the identical behavior similar to that of FS [7] glasses suggesting the fluorine as mobile species. Table I summarizes the physical data such as density, glass transition temperature (Tg), conductivity(σ) and activation energy(Eact) for various binary and ternary glasses. In both the FS [18] and FB [17] glasses, the conductivity appears to have an inverse relationship with activation energy when the PbF_2 content is systematically increased. Nevertheless, the σ values appear to saturate in the FS glasses at ultra fluoride rich compositions. This has been speculated due to the bonding of fluorine with silicon rather than to lead ions. The conductivity interestingly goes through a maximum in the $MnF_2:SiO_2$ system with the variation in the lead fluoride content leading to a "mixed anionic effect"[14]. Similar results have also been observed in the FG glasses, $PbF_2:MnF_2$ based glasses [10,11]. It appears that the enhancement of conductivity in PbF_2 based glasses may not directly be linked with the amount of lead fluoride content unlike the AgI-based glasses [21]. Another interesting feature in these glasses is the glass transition temperature, which appears to follow the behavior of σ values. The variation in Tg with the PbF_2 content supports that the structure is the prime factor in determining the magnitude of fluoride ion conduction. The variation of Tg and σ in FG glasses is shown in Fig. 1.

Table 1. Physical properties of some oxyfluoride glasses.

Compound (mol%)	Density (g/cm³)	Tg (K)	-log σ at T=473K	Eact (eV)	Ref.
$68.6PbO.31.4SiO_2$	7.15		10.3		7
$29.2PbF_2.48.5PbO.22.2SiO_2$	7.13		7.0 at T=410K		
$80PbO.20PbF_2$	7.88	545	7.7	0.91	8
$30PbO.70PbF_2$		460	5.3	0.67	
$Pb(PO_3)_2$		581			
$20PbF_2.80Pb(PO_3)_2$		611			9
$40PbF_2.60Pb(PO_3)_2$		633			
$85PbF_2.5MnF_2.10Al(PO_3)_3$		476	4.01[a]		10
$30PbF_2.50MnF_2.20Pb(PO_3)_2$		515	6.1[a]	0.55	
$45PbF_2.35MnF_2.20Pb(PO_3)_2$		540	4.8[a]	0.62	
$65PbF_2.15MnF_2.20Pb(PO_3)_2$		494	6.8[a]	0.65	11
$20PbF_2.50MnF_2.30Pb(PO_3)_2$		531	7.3[a]	0.84	
$35PbF_2.35MnF_2.30Pb(PO_3)_2$		583	6.0[a]	0.71	
$55PbF_2.15MnF_2.30Pb(PO_3)_2$		530	4.9[a]	0.56	
$55SnF_2.45P_2O_5$	4.49	432			
$60SnF_2.40P_2O_5$	4.50	418			12
$65SnF_2.35P_2O_5$	4.63	385			
$70SnF_2.30P_2O_5$	4.78	362			
$80PbF_2.20Al(PO_3)_3$		558	7.9[a]	0.96	13
$90MnF_2.10SiO_2$			13.0		
$80MnF_2.20SiO_2$			12.0		14
$70MnF_2.30SiO_2$			7.2		
$64MnF_2.36SiO_2$			9.0		
$2.1PbF_2.70.4PbO.27.5SiO_2$	7.31	588	8.7	0.96	15
$17.2PbF_2.54.3PbO.28.5SiO_2$	7.16	508	4.6 at T=400K	0.73	
$20PbF_2.10ZnF_2.70GeO_2$	5.23	739			
$30PbF_2.70GeO_2$	5.6	716			16
$40PbF_2.60GeO_2$	6.15	685			
$70PbO.30B_2O_3$			9.3	1.08	
$5PbF_2.65PbO.30B_2O_3$			7.3	0.96	17
$20PbF_2.50PbO.30B_2O_3$			5.6	0.55	
$40PbF_2.50PbO.10SiO_2$		451	3.7 at T=373K	0.41	18
$ZnF_2.PbF_2.SiO_2$					
$BiF_3.PbF_2.SiO_2$			>4.0		19
$MnF_2.PbF_2.SiO_2$					
$30SnF_2.40SnO.30P_2O_5$	3.85	410			20
$30SnF_2.20SnO.50P_2O_5$	3.12	393			

a: Conductivity measured by impedance technique.

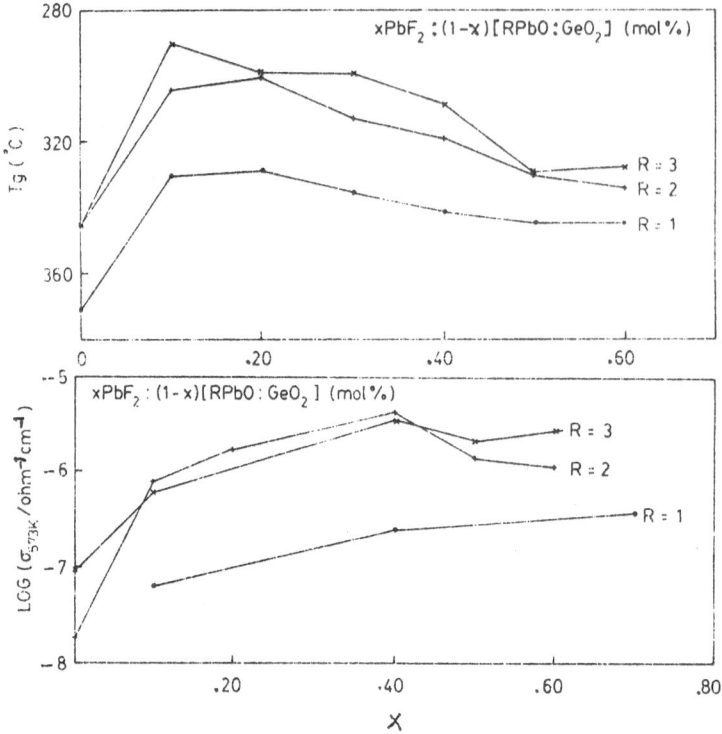

Fig. 1. Variation of Tg and σ with lead fluoride content.

3.2   XPS studies

Surface analysis using electron spectroscopes is presently a well-established technique in many different areas of materials science. XPS or electron spectroscopy for chemical analysis (ESCA) is one of the widely used surface analytical techniques and can provide reliable qualitative and semiquanditative characterization of the near-surface region (top 1-100A) of most solids. Besides, this technique also provides information about the chemical environment of the observed atoms [5,22].

3.2.1 Basic principles and theory

XPS is basically a photo emission process in which the primary excitation is carried out using an X-ray source. The specimen surface is irradiated under ultra-high vacuum conditions (to avoid multiple collisions) with low energy x-rays from either an Al Kα (1486.6 eV) or Mg Kα (1256.6 eV) source. Based on the well-known photoelectric effect, the X-ray photons with energy hγ,

126

interacting with an electron orbital with binding energy E_b, the detected electron kinetic energy is given by $E_k = h\gamma - E_b - \phi_{anal}$ where E_k = photo emitted electron kinetic energy (with reference to the Fermi level) and ϕ_{anal} = analyzer work function.

XPS typically concerns ionization of core and valence electrons with binding energies (BE) of less than 1000 eV. As the kinetic energy of the emitted electron from the solid is specific to the element of origin, XPS thus provides a valuable chemical information. Such chemical shifts (usually in the range 0.1 to 10 eV) may be used to identify the various oxidation states. Further details regarding the sample preparation, BE calibration and peak synthesis can be obtained from references 5 and 22.

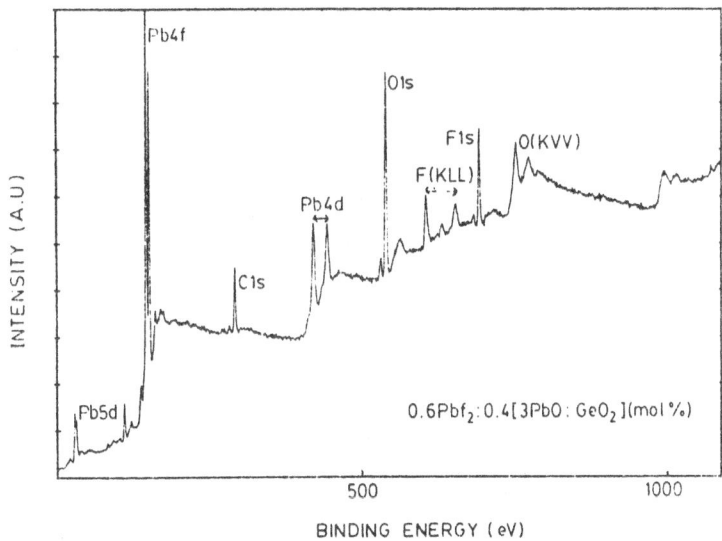

Fig. 2. Wide scan spectrum of lead flurogermanate glass.

A representative XPS survey spectrum for a FG glass is shown in Fig. 2. It indicates a series of sharp intense peaks arising from the direct excitation of electrons from core levels. The observed lead, oxygen, fluorine and carbon peaks are identified in the figure. In addition to surface core levels lines, X-ray induced Auger peaks [F(KLL) and O(KVV)] (XAES) are also observed. In recent years XAES coupled with the corresponding core level peaks have been utilized to derive some information on local coordination of the respective atom [24]. The binding energies of the electrons from the core levels of the constituent elements for various compositions have been measured and the data are summarized in Table II. Results obtained from the other research groups are also included for comparison.

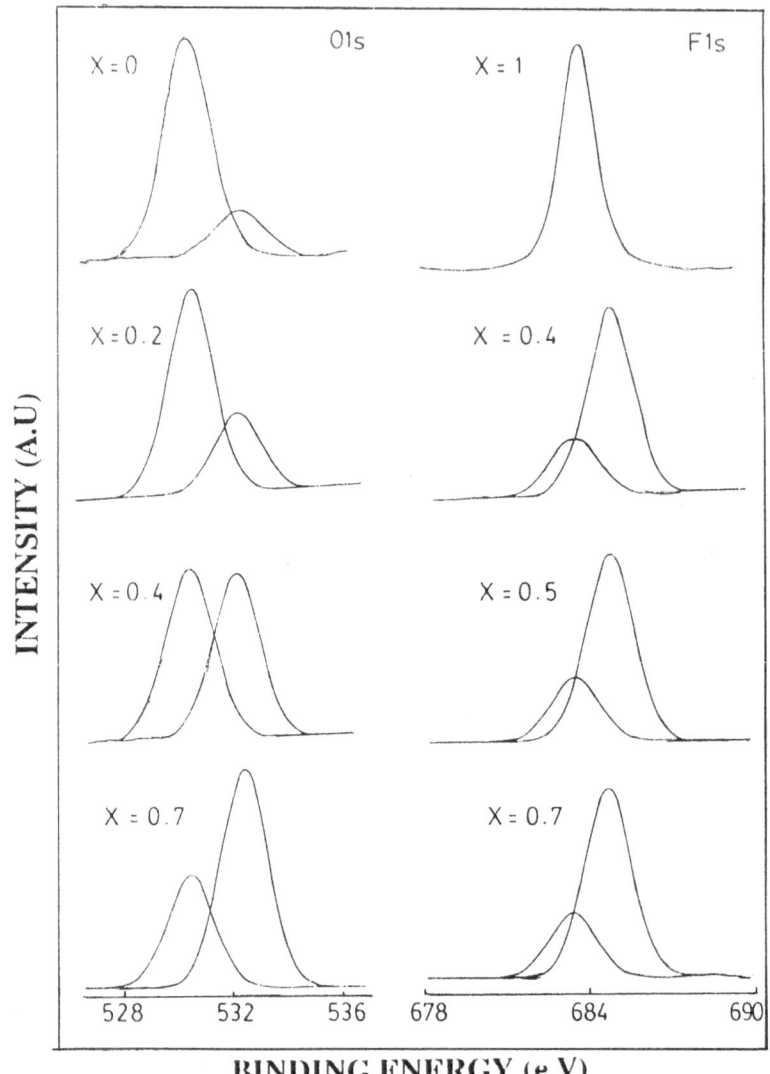

Fig. 3. Peak synthesized XPS O1s and F1s spectra in the XPbF2:(1-X)[PbO:GeO2] system.

Among the peaks in Fig.1, the O1s, F1s and Ge3d have been extensively studied because of their involvement in the conduction process. Fig. 3 represents the peak synthesized O1s and F1s spectra for some compositions in the FG series. With increasing PbF_2 cotent, the lower BE peak contribution clearly decreases. The decrease in area ratio between O1s(1) and O1s(2) shown in Table II indicates the possible substitution of fluoride in the modified oxide network. The shift in the UV edge with

Table II. Peak positions and the area ratios of some XPS transitions in the oxyfluoride system.

COMPOUND	x(mol %)	ELECTRON BINDING ENERGY (eV)						AREA RATIOS OF		
		F1s	B1s	Ge3d	O1s	Pb4f 7/2	Sn3d 5/2	O1s(1)/O1s(2)	F1s(1)/F1s(2)	Ge3d(1)/Ge3d(2)
xPbF$_2$.(70-x)PbO.30B$_2$O$_3$ [a]	5	687.8	191.4							
	25	683.6, 687.8[e]	191.4							
xPbF$_2$.(70-x)PbO.10B$_2$O$_3$.20AlF$_3$ [a]	20	684.3								
	60	684.3								
x/3(2SnF$_2$.PbF$_2$).1/2(100-x)Pb(PO$_3$)$_2$ [b]	45				531.5	138.7	486.9			
	55				531.5	138.7	486.8			
xMnF$_2$.(100-x)SiO$_2$ [c,d]	24	685.7, 687.9[e]							1	
	100	685.5								
xPbO.(100-x)GeO$_2$	0			32.7	531.8			4.68		
	50			31.6	530.3, 532.2[g]	138.1				
	75			31.6	530.2	138.0				
xPbF$_2$.(100-x)GeO$_2$	50	684.7		29.3, 32.4[f]	530.4, 532.2[g]	138.8		0.30		0.12
	80	684.7		29.4, 32.3[f]		138.8				0.53
	100	683:5				138.8				
xPbF$_2$.(100-x)[PbO.GeO$_2$]	10					138.2				
	20			29.3, 31.8[f]	530.4, 932.1[g]	138.1		2.69		
	40	683.4, 684.7[e]			530.4, 532.1[g]	138.3		1.09	0.31	
	50	683.4, 684.7[e]		29.3, 31.9[f]		138.8			0.35	
	70	683.3, 684.6[e]		29.4, 32[f]	530.4, 532.3[g]	138.8		0.52	0.34	
xPbF$_2$.(100-x)[2PbO.GeO$_2$]	30	683.3, 684.6[e]		29.3, 31.7[f]	530.3	138.4			0.28	
	50	684.4		29.2, 31.7[f]	532.2	138.8				
xPbF$_2$.(100-x)[3PBO.GeO$_2$]	30	683.3, 684.4[e]		29.3, 31.5[f]	530.1	138.3			0.4	
	40	684.4		29.5, 31.9[f]	530.5	138.7				
	50	684.3		29.5, 32.7[f]	532.4	138.8				

a from reference (17).

b from reference (13).

c from reference (14).

d C1s standard value is not specified.

e denoted as F1s(1) and F1s(2) respectively.

f denoted as Ge3d(1) and Ge3d(2) respectively.

g denoted as O1s(1) and O1s(2) respectively.

lead fluoride content in FS glasses suggested that the fluorine replaces an oxygen defect. In addition to, the conductivity saturation was attributed to the possible formation of Si-F bonds [18].

The variation in conductivity in the FB glasses coupled with the XPS results agree with the model proposed by Coon and Shelby[18]. However, the σ values did not show any saturation. F1s [area ratio between F1s(1) and F1s(2)] spectra did not follow the conductivity behavior shown in FiG. 1 in the case of FG glasses. Similar to the earlier analysis, attempts have been made to understand the binary PbF$_2$:GeO$_2$ initially rather than the ternary oxyfluoride glasses. Earlier XPS results from the binary MnF$_2$:SiO$_2$ systems indicated two types of fluorine atoms [14].

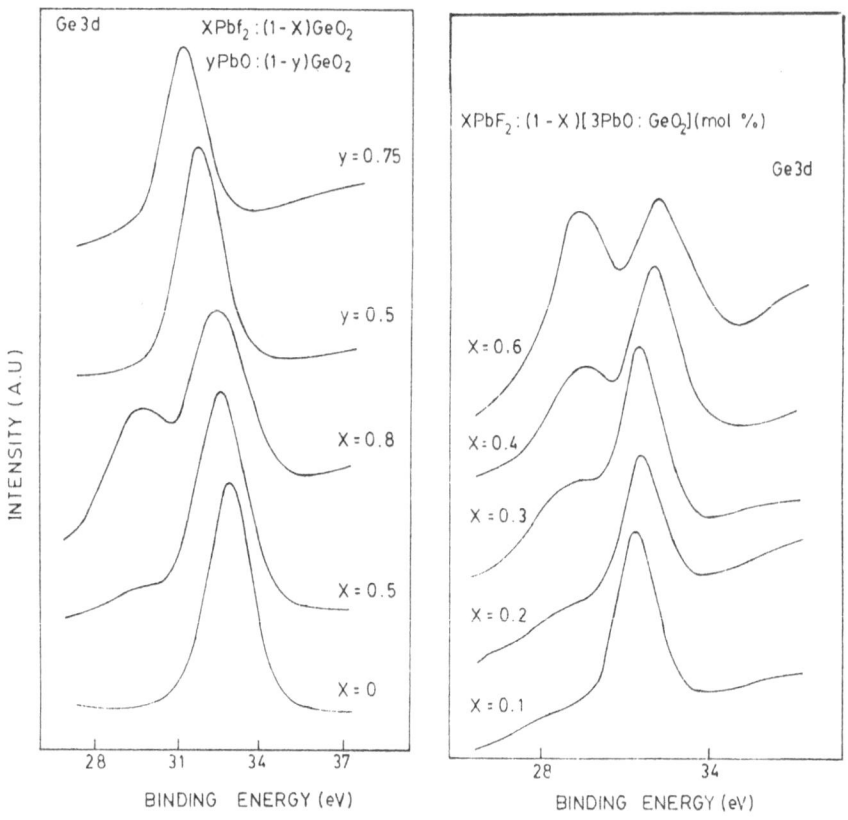

Fig. 4 Ge 3d core XPS spectra for FG glasses.

Nevertheless, a correlation between the F1s spectra and the Mixed anionic effect could not be made.

The glassy GeO2 network interestingly appears to indicate an useful result in XPS Ge3d spectra. Fig. 4 shows the variation of XPS Ge3d spectra with PbO and PbF2 content. As evident from Fig. 4, the Ge3d profile changes appreciably with PbF2 content and the appearance of two peaks suggests the presence of two kinds of Ge atoms. The binding energy values suggest the presence of Ge^{2+} and Ge^{4+} in the binary glasses [25]. The F1s XPS both in binary and ternary FG glasses has a doublet peak: one of the component peaks at 683.4 eV in BE has been attributed to the fluorine in Pb-F bonds, and the other at 684.4 eV to the fluorine in Ge-F bonds in accordance with the single bond strength data [26]. Similar observations have been made in other systems using infrared, Raman and NMR techniques [12,17 and 27].

The formation of Ge-F bonds coupled with the changes in germanium coordination may be responsible for the conductivity variation in the FG glasses shown in Fig. 1. Presently, it is rather difficult to propose a possible conduction mechanism in the oxyfluoride glasses for the following reasons.

1. Very few binary and ternary systems have been examined so far.

2. The conductivity has been measured in most cases by dc resistance measurements, rather than the impedance technique.

3. Although XPS appears to be compatible for structural analysis, an extensive investigation of O1s and cation core level spectra of glass former is not reported.

ACKNOWLEDGEMENTS

The author would like to thank Assoc. Prof. K.L.Tan and Assoc. Prof. B.V.R. Chowdari for their encouragement and support.

REFERENCES

[1] Inorganic Solid Fluorides, Chemistry and physics Ed. by P. Hagenmuller (Academic Press, New York, 1985).
[2] J.M. Reau M. Poulain, Mater. Chem. and Phy. 23 (1989) 189.
[3] Y. Kawamoto, I. Nohara, J. Fujiwara and Y. Umetani, Solid State Ionics 24 (1987) 327.
[4] P. Hagenmuller, J.M. Reau, C. Lucat, S. Matar and G. Villeneuve, Solid State Ionics 3 & 4 (1981) 341.
[5] Practical Surface Analysis by Auger and X-ray Photoelectron spectroscopy, Ed. by D. Briggs and M.P. Seah (Wiley, Newyork, 1983).
[6] B.M.J. Smets and T.P.A Lommen, J. Non-Cryst. Solids 48 (1982) 423.
[7] P.C. Schultz and M.S. Mizzoni, J. Am. Ceram. Soc. 56 (1973)65.
[8] B. Govinda Rao, H.G.K. Sundar and K.J. Rao, J. Chem. Soc., Faraday Trans. 80 (1984) 3491.
[9] H.G.K. Sundar, S.W. Martin and C.A. Angell, Solid State Ionics 18 & 19 (1986) 437.
[10] A.R. Kulkarni and C.A. Angell, Mat. Res. Bull. 21 (1986) 1115.
[11] A.R. Kulkarni, H.G.K. Sundar and C.A.Angell, Solid State Ionics 24 (1987) 253.
[12] C.M. Shaw and J.E. Shelby, Physics Chem. Glasses 29 (1988) 49.

[13] A.R. Kulkarni and C.A. Angell, J. Non-Cryst. Solids 99 (1988) 195.

[14] K. Hirao, A. Tsujimura, S. Tanabe and N. Soga, Mat. Sci. Forum 32 & 33 (1988) 415.

[15] Y. Wong, M. Kobayashi, A. Osaka, Y. Miura and K. Takahashi, J. Am. Ceram. Soc. 71 (1988) 864.

[16] H. Shuping, X. Chao and G. Fuxi, J. Non-Cryst. Solids 112 (1989) 151.

[17] Y. Wang, A. Osaka and Y. Miura, J. Non-Cryst. solids 112 (1989) 323.

[18] J. Coon and J.E. Shelby, J. Am. Ceram. Soc. 71 (1988) 354.

[19] K. Hirao, A. Tsujimura and N. Soga, J. Soc. Mater. Sci. Japan 39 (1990) 438.

[20] X.J. Xu and D.E. Day, Physics Chem. Glasses 31 (1990) 183.

[21] T. Minami, J. Non-Cryst. Solids 95 & 96 (1987) 107.

[22] K.W. Nebesny, B.L. Maschhoff and R. Armstrong, Anal.chem. 61 (1989) 469A.

[23] A. Osaka, Y. Miura and T. Tsugaru, J. Non-Cryst. Solids 125 (1990) 87.

[24] C.D. Wagner, Anal. Chem. 44 (1972) 967.

[25] R. Bouwman and P. Biloen, J. Catal. 48 (1977) 209.

[26] R. Gopalakrishnan, B.V.R. Chowdari and K.L. Tan, to be published.

[27] D. Kline and P.J. Bray, Phy. Chem. Glasses 7 (1966) 41.

Laser Materials

Quantum Microstructures and New Solid State Materials

Fabio Beltram*
AT & T Bell Laboratories, Murray Hill, NJ 07974, USA

INTRODUCTION

Modern epitaxial growth techniques such as Molecular Beam Epitaxy (MBE) have provided an unprecedented ability to artificially structure new materials on an atomic scale. These techniques, in particular applied to the growth of III-V materials have made possible the birth of *band-gap engineering*[1] through which the band diagram of a structure can be tailored with great freedom to *build in the material itself* the desired properties or effects.

In particular, the ability to grow almost atomically flat interfaces and to control the composition on the scale of the electron de Broglie wavelength has made possible the study of quantization phenomena in structures like quantum wells and superlattices and has opened the way to the study of low dimensional systems. Additionally, the ability to continuously tune the composition during the growth, particularly with MBE, yields the ability of independently tuning the transport properties of electrons and holes, using quasi-electric fields in graded-gap materials and the difference between conduction- and valence-band discontinuities in a given heterojunction.

These new materials have had a great impact on semiconductor physics and electronics. In the following I shall consider both these aspects through some case studies. In the next section the work done on bipolar transistors to improve and enhance their characteristics and functionality will be reviewed as an example of the impact of artificial heterojunctions on electronics. In the final section recent results in the study of transport in heterostructures will be reviewed to show how the exploitation of band-gap engineering can be an essential tool for the study of fundamental electronic properties.

* Present address: International Centre for Science and High Technology, Via Grignano 9, 34136 Trieste, Italy.

Several reviews have appeared in the literature both on the device and on the more fundamental aspects.[2,3]

1. HETEROJUNCTION BIPOLAR TRANSISTORS (HBTs)

1.1 Graded-gap base bipolar transistors

In a bipolar transistor, a current injected in the base layer controls the flow of electrons from the emitter to the collector layer. In conventional (homojunction) bipolar transistor, the base has a uniform band gap and no electric field. Therefore, electrons traversing the base travel relatively slowly by diffusion.

One way to speed up electrons is to use a graded-gap material for the base. This concept was pioneered by Kroemer,[4] who showed that, as a result of compositional grading, electrons and holes experience quasi-electric fields (F) of different intensities

$$F_e = -\frac{dE_c}{dz}$$

$$F_h = +\frac{dE_v}{dz},$$

(1)

where $E_c(z)$ and $E_v(z)$ are the conduction and band edges, and z is the growth direction. Notably the forces resulting from these fields accelerate electrons and holes in the same direction. In the case of a p-doped material (the case of interest for the p-doped base of a n-p-n transistor, see Fig. 1) the situation is different. The valence-band edge is now horizontal, and no field acts on the holes, while the effective field for the electrons is $F_e = -dE_g/dz$ (see Fig. 1). In other words, all the band gap grading is transferred to the conduction band. This can be

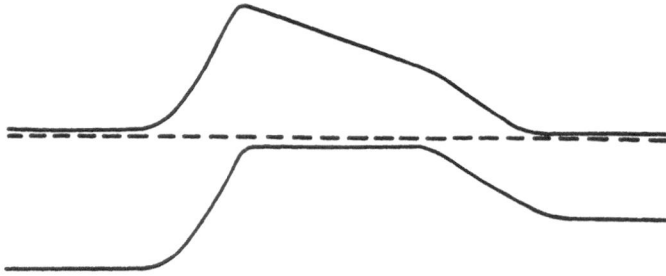

Fig. 1. Energy-band diagram of a n-p-n graded-gap base heterojunction bipolar transistor.

136

understood along the following lines. Consider the effect of p doping on an initially intrinsic material. The acceptor atoms will introduce holes that under the action of F_h given by Eq. 1, will be spatially separated from their negatively charged ionized parent acceptor atoms. This separation will produce an electrostatic (space-charge) field. Equilibrium is reached when the electrostatic field exactly cancels the quasi electric field F_v. The same field, however, will act on electrons as well, therefore the total field experienced by electrons will be $F_e = -dE_c/dz - dE_v/dz = -dE_g/dz$. Under the action of this field, which is partly electrostatic (space-charge contribution) and partly non-electrostatic (quasi-electric field), electrons drift through the base much faster than in a conventional one. The concept was demonstrated experimentally at AT&T Bell Laboratories.[5]

The interested reader is referred to the reviews in Refs. 1 and 3 for more material.

1.2 Abrupt junction HBTs

Another approach now actively pursued is shown in Fig. 2. The essential features of this structure are the use of part of the energy-gap difference between the wide-gap emitter and the base to suppress hole injection (as in the graded-gap base transistor) and part to launch electrons into the base with some excess kinetic energy (arising from the emitter-base conduction-band discontinuity). The former

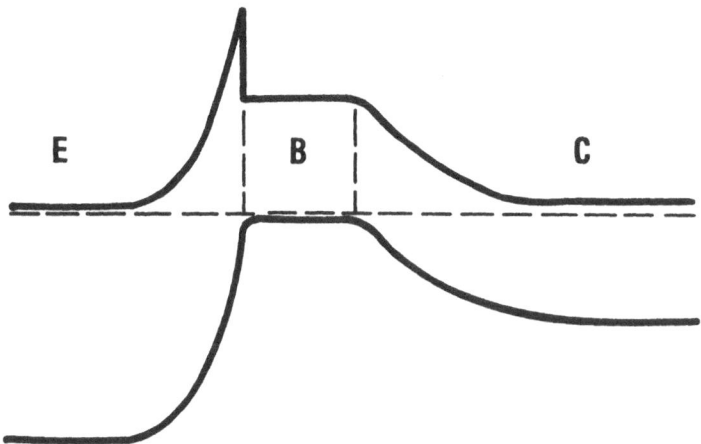

Fig. 2. Energy-band diagram of an abrupt emitter-base heterojunction n-p-n heterojunction bipolar transistor.

leads to increased current gain and allows to more heavily dope the base than the emitter, yielding the low base resistance and low emitter-base capacitance necessary for high-frequency operation. The latter can significantly increase device speed by shortening the transit time across the base and collector space-charge region.

HBTs have been studied mostly in the GaAs/AlGaAs and InGaAs/InAlAs/InP systems, much attention is now given to $Si_{1-x}Ge_x$ HBTs. Bipolar transistors are the devices of choice for high-speed applications, it seems probable that HBTs with measured cutoff frequencies already approaching 200 GHz will extend the limit of performance of bipolar devices. An up-to-date review can be found in Ref. 6.

1.3 Quantum functional transistors

In the previous sections, band-gap engineering was used to improve the performance of transistors by enhancing some figure of merit like cutoff frequency or current gain. A radically different approach can be taken by creating devices that are inherently *functional* in the sense of Morton.[7] J.A. Morton defined the concept of functional device as one that "reduces greatly the number of elements and process steps per function". Resonant tunneling (RT) devices are probably the best candidates for the implementation of this concept. One significant example is the multistate RT transistor demonstrated at Bell Laboratories. With one such device a four bit parity generator was demonstrated.[8] If we compare this with a standard circuit, we see that the latter requires 24 transistors to perform the same function ("reduced number of elements"). Moreover with the RT device all four bits are processed in parallel contrary to the traditional circuit where two bits at a time are compared sequentially ("reduced number of process step").

A number of RT transistors have been proposed and demonstrated, both unipolar and bipolar. The interested reader is referred to the very comprehensive reviews contained in Ref. 2. Here I shall briefly present the operation of the multi-state Resonant Tunneling Bipola Transistor (RTBT).

The concept of a RTBT originated with the general idea of associating to each state of a quantum system, for example the energy levels of a quantum well, a corresponding logic level.[9] However, since most circuit applications require characteristics in which the current peaks occur at the same current level and with similar peak-to-valley ratios,[10] subsequently it has become clear that the best approach is that of using the analogous resonance of different quantum wells.

The first step in building the multi-state RTBT is the vertical integration of RT diodes which is achieved by stacking a number of double barriers (DBs) in series, separated by heavily doped cladding layers to quantum mechanically decouple the adjacent DBs from each other.[11,12] The DBs are designed so that the ground state in each quantum well (QW) is substantially above the Fermi level in the adjacent cladding layers. The band diagram of the structure under bias is shown in Fig. 3. When bias is applied, the electric field is higher at the anode end of the device (Fig. 3a) because of charge accumulated in the QWs under bias. Quenching of

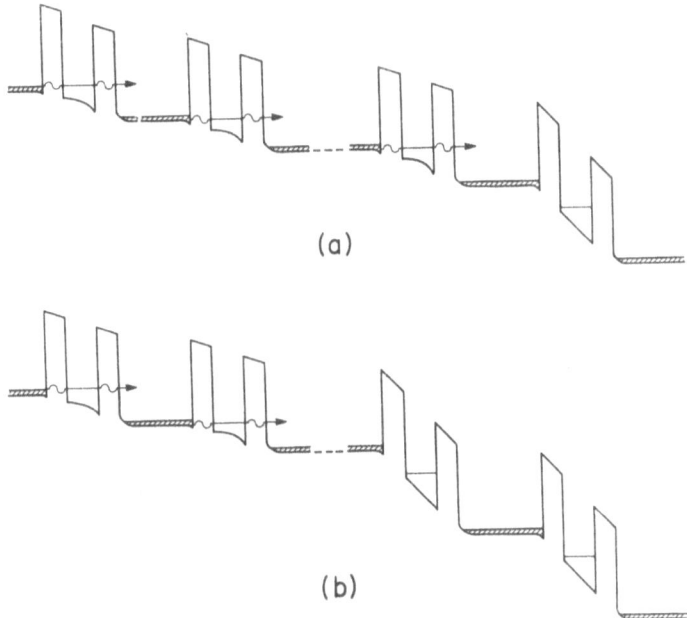

(a)

(b)

Fig. 3. Vertical Integration of Resonant tunneling diodes. Conduction-band diagram under applied bias (a) with RT quenched through the DB adjacent to the anode and (b) after expansion of the high-field region to the adjacent DB with increasing bias. The arrows indicate the RT component of the current.

RT is thus initiated across the DB adjacent to the anode and then sequentially propagates to the other end, as the high-field region widens with increasing applied voltage, as shown in Fig. 3a and b. Once RT has been suppressed across a DB, the voltage drop across it quickly increases with bias because of the increased resistance. The non-RT component through this DB provides continuity for the RT current through the other DBs on the cathode side. A negative differential resistance (NDR) region is obtained in the current-voltage characteristic (I-V), corresponding to the quenching of RT through each DB. Thus with n diodes, n peaks are present in the I-V.

Generating multiple peaks by combining tunnel diodes in series is well known.[13] However the mechanism in that arrangement is different. The tunnel diodes used

139

in such a combination must have different characteristics with successively increasing peak currents, so that each of them can go into the NDR region only when the corresponding current level is reached. [13]

Devices consisting of two, three and five RT $Al_{0.48}In_{0.52}As$ (50Å)/ $Ga_{0.47}In_{0.53}As$ (50Å) DBs in series, separated by a 1000 Å thick n^+ $Ga_{0.47}In_{0.53}As$ regions were tested. The I-V characteristics taken in both polarities of the applied voltage for the case of three DBs are shown in Fig. 4 at room temperature and at liquid nitrogen. Positive polarity here refers to the top of the mesa being biased positively with respect to the bottom.

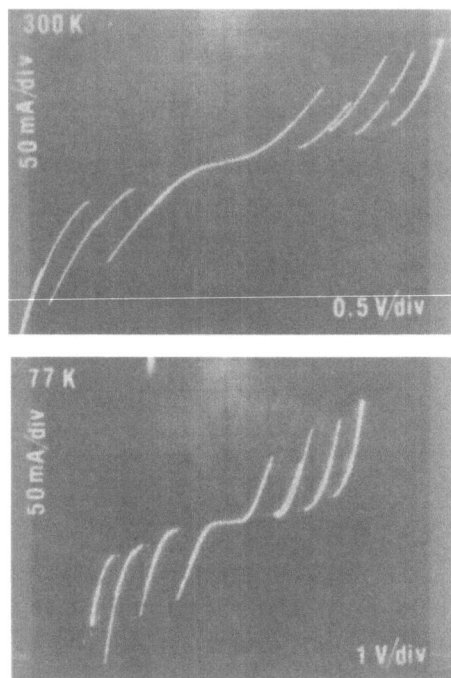

Fig. 4. Current-voltage characteristics of the device
with three vertically integrated RT double barriers,
taken for both bias polarities at 300 K and 77K.

This stacked RT structure was used to design a RTBT exhibiting multiple NDR and negative transconductance characteristics. [14] The device essentially consists of a $Ga_{0.47}In_{0.53}As/Al_{0.48}In_{0.52}As$ n-p-n transistor with a stack of two $Ga_{0.47}In_{0.53}As$ (50 Å)/$Al_{0.48}In_{0.52}As$ (50 Å) RT DBs, as discussed before, embedded in the emitter. Details of the structure and its fabrication are described in Ref. 14. The operation of the transistor can be understood from the band-diagrams in the common-emitter configuration shown in Fig. 5. The collector-emitter bias (V_{CE}) is kept fixed and the base-emitter voltage (V_{BE}) is increased.

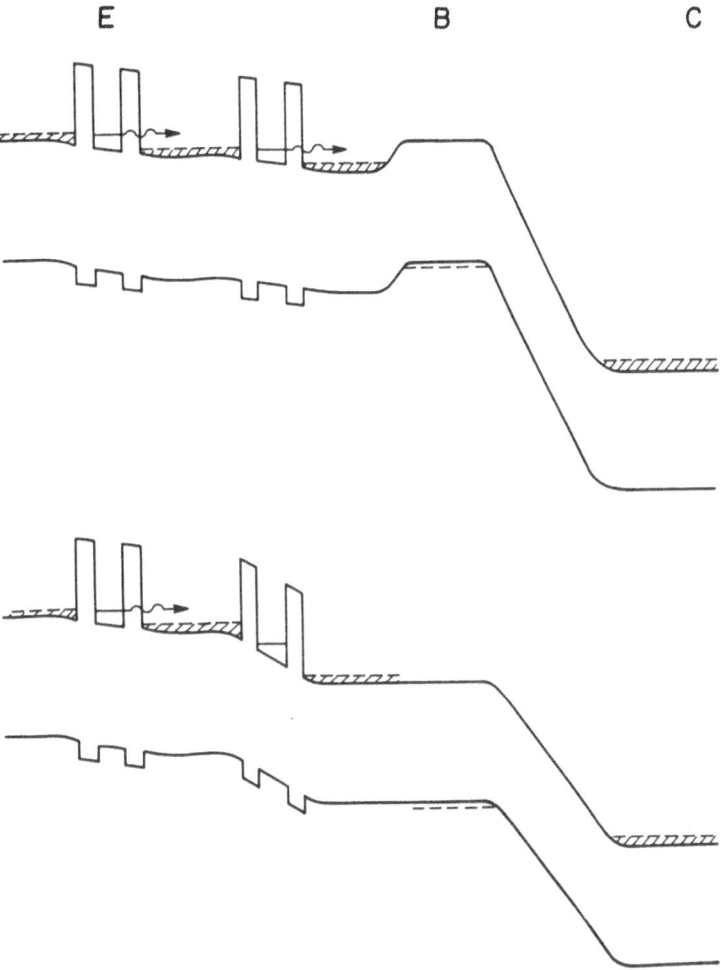

Fig. 5. Energy-band diagram of the multiple-state RTBT in the common emitter configuration for different base-emitter bias conditions. (a) Electrons resonantly tunnel through both DBs; in this regime the transistor operates as a conventional bipolar. (b) Quenching of RT through the DB adjacent to the pn junction, gives rise to a negative differential resistance region in the collector current. Quenching of RT through the other DB, produces a second peak in the I-V.

For V_{BE} smaller than the built-in voltage ($V_{Bi} \simeq 0.7$ eV at 300 K) of the $Ga_{0.47}In_{0.53}As$ p-n junction, most of the bias voltage falls across this junction (Fig. 5a), since its impedance is much greater than that of the two DBs in series, both of which are conducting via RT. The device in this region behaves as a conventional bipolar transistor with the emitter and hence the collector current increasing with V_{BE} (Fig. 6) until the base emitter junction reaches the flat-band condition. Beyond flat-band, most of the additional increase in V_{BE} will fall across the DBs (Fig. 5b), and as RT through these quenches sequentially by the mechanism of Fig. 3, abrupt drops in the emitter and hence the collector current are observed (Fig. 6). The highest peak-to-valley ratio in the transfer

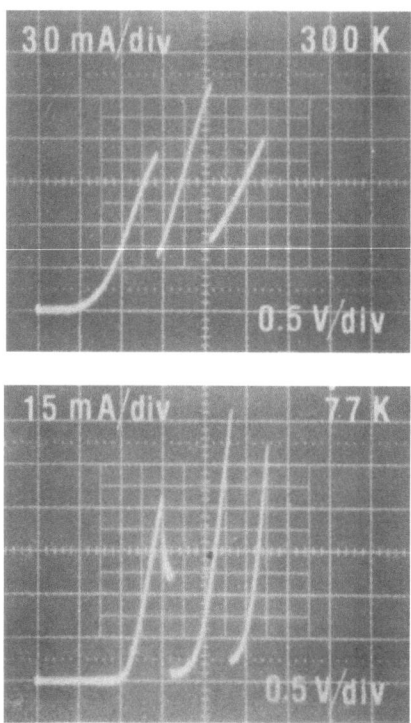

Fig. 6. Collector current vs base-emitter voltage, in the common emitter configuration for $V_{CB} = -0.1$ V at 300 K (top) and 77 K (bottom).

characteristics at room temperature was 4:1 while it increased to about 20:1 at 77 K.[14]

142

Fig. 7 shows the common emitter output characteristics of the transistor (I_C vs. V_{CE} at different I_B) at room temperature and 77 K. At low base currents I_B (and hence low base-emitter voltages, V_{BE}), the device behaves as a conventional bipolar transistor as discussed before, with large current gain (200 at 77 K and 70 at 300 K). With increasing I_B (V_{BE}) beyond the flat-band condition, the excess

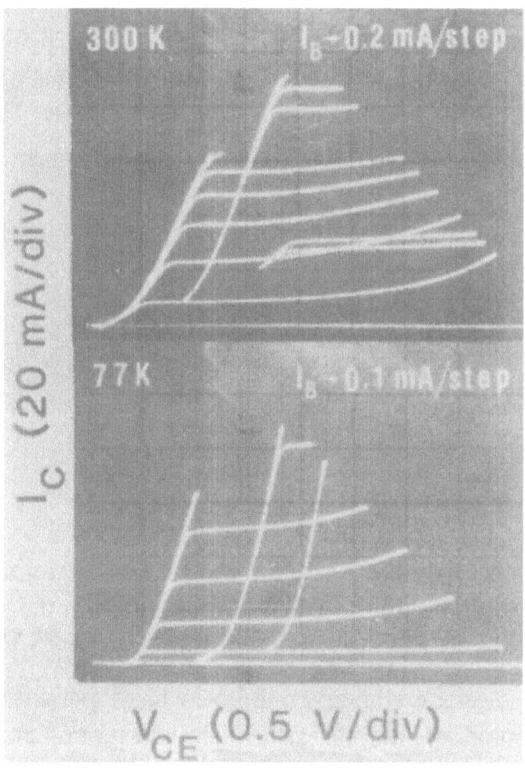

Fig. 7. Common-emitter output characteristics of the multiple-state RTBT. Collector current vs. collector emitter voltage for different base currents, at 300 K (top) and 77 K (bottom).

applied voltage V_{BE} starts appearing across the series of DBs in the emitter and as RT through them quenches sequentially, at some threshold base currents, the electron current across the base-emitter junction drops abruptly while the hole current, flowing by thermionic emission, continues to increase. This results in sudden quenching of the current gain at these threshold base currents and hence the collector current I_C also quenches, giving rise to two NDR regions (Fig. 7). The highest peak-to-valley ratio observed are 6:1 at room temperature and 22:1 at 77 K. It should be noted that the small signal current gain of the transistor at room temperature in its second (1.2 mA $< I_B <$ 1.6 mA) and third

($I_B > 1.6$ mA) operation regions are reduced to 40 and 20 respectively. This is expected since the hole current flowing from the base towards the emitter increases with increasing V_{BE}, thus reducing the injection efficiency. This reduction of the current gain is less pronounced at 77 K, since the thermionic flow of holes is much lower at that temperature.

Fig. 8 shows the common-emitter transfer characteristics of a similar transistor with three DBs in the emitter at 77 K. The third peak is shifted out to a significantly higher voltage compared to the other two. Such behavior is not

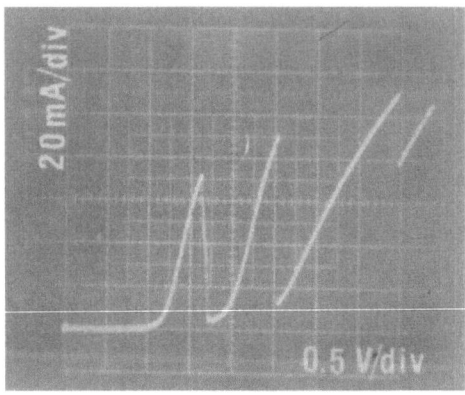

Fig. 8. Common-emitter transfer characteristics of a multiple-state RTBT with three DBs in the emitter at 77K. I_C vs. V_{BE} is shown for $V_{CE} = 4.75$ V.

uncommon in RT devices whenever there is large parasitic resistance.[15] When three DBs are put in series, the parasitics also add up and enhance the effect. The structure with three peaks has to be optimized to minimize these effects.

The microwave performance of a multi-state RTBT has been studied and a 24GHz cutoff frequency at room temperature has been reported. Experimental results on frequency multipliers, parity-bit generators and multi-state memories are available. The interested reader is referred to Ref. 2.

2. HETEROJUNCTION SUPERLATTICES

2.1 Field-induced localization in superlattices

In this section I shall review some recent results[16] that show how band-gap engineering can be the key to the experimental study of more fundamental problems. I shall consider the case of electron transport in a superlattice (SL) in the presence of a uniform electric field, the object of intense investigations since the original proposal of SLs by Esaki and Tsu.[17]

When an electric field F is applied to a SL of period a, some different transport regimes are commonly identified. At low fields, the current increases linearly with field (mobility regime). The current is expected to decrease with increasing field when the electron distribution probes the negative-mass region of the miniband, i.e., according to Esaki and Tsu[17], for

$$F > \frac{\hbar}{ea\tau},$$ (2)

where τ is the scattering time and e the electron charge. This behavior is caused by the fact that an increasing fraction of the carriers approaches the minizone boundary therefore undergoing Bragg diffraction.

Another regime was studied by Tsu and Döhler[18] who considered the case of strong localization in a tight-binding SL (i.e. a SL with weak coupling between wells). They showed that due to the decreasing overlap between the wavefunctions of adjacent wells, the transition rate due to hopping (and therefore the current) decreases with increasing field for

$$F > \frac{\Delta E}{ea},$$ (3)

where ΔE is the miniband width.

One must not be led to the incorrect notion that Eq. 3 is a *necessary* requirement for the observation of localization in transport.[19-21] Even for fields much lower than those causing the complete localization mentioned above, a progressive localization of the electronic states is to be expected. As originally discussed by Wannier,[22] and subsequently by Kazarinov and Suris[23] in the context of transport in SLs, in an electric field the electronic wavefunctions extend over a number of periods of the order of $\Delta E/eaF$ and are separated in energy by eaF (the so called Wannier-Stark ladder, Fig. 9a). Thus, as the field is increased the wavefunctions become increasingly localized in space up to the extreme point where they are shrunk to one well. This is the limit of Eq. 3 in which the SL consists of a "ladder" of identical isolated quantum wells (Fig. 9b). A decrease in the current is expected throughout this regime since the spatial overlap between the Stark-ladder states decreases with increasing field and with it the matrix element for transitions. The question then arises, what is the threshold for localization?

Localization will occur when the energy levels of the Stark ladder can be resolved. In the presence of collisions this happens when their separation is

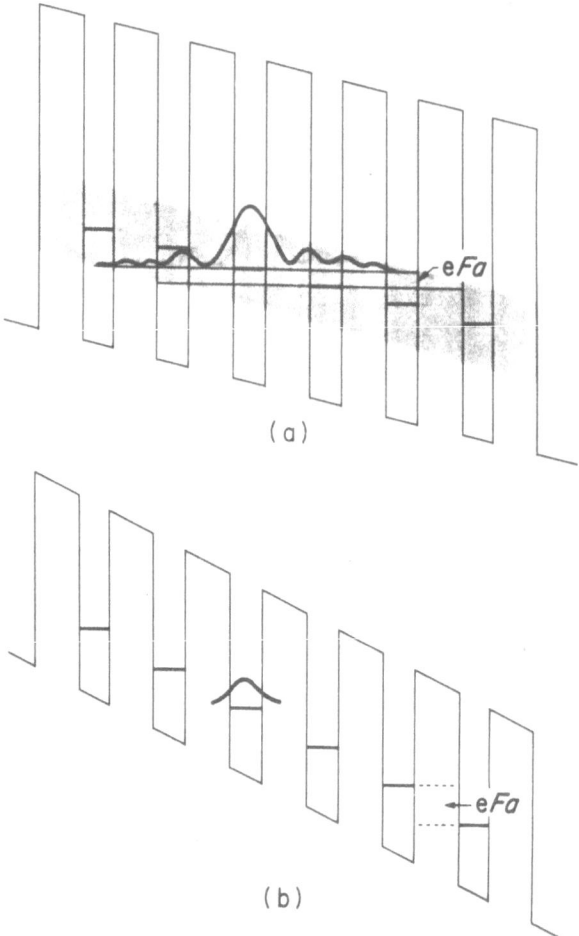

Fig. 9. Schematic conduction-band diagram of a heterojunction superlattice with an applied field, (a) electronic states extend over several periods and can be broadened by scattering into a band (shaded region) if Eq. 2 is not satisfied, (b) at very high biases (defined by Eq. 3), electronic states are confined to single wells.

greater than the collision broadening, i.e. $eaF > \hbar/\tau$. Therefore, in this physical picture, the threshold for the observation of negative differential conductance (NDC) is $F > \hbar/ea\tau$. This is the same field calculated by Esaki and Tsu for the onset of NDC in a SL (Eq. 2). In fact these two pictures for NDC are equivalent since the Stark-ladder states (Fig. 9a) arise from the interference between the forward-propagating and the Bragg-reflected wave. Yakovlev[24] investigated the conductivity of electrons in semiconductors with narrow (~ 0.1 eV) bands in

146

strong electric fields. Although this paper was written before the introduction of the SL concept, its results are applicable to describe transport in these structures. Yakovlev indeed found that for fields satisfying Eq. 2, the conductivity decreases with increasing field.

These theoretical predictions are experimentally verifiable by simply measuring the current-voltage characteristics of SLs: starting at a bias defined by Eq. 2, a wide region where the current decreases with increasing voltage would be the signature of the localization. Some care, however, must be taken.

One of the main problems hindering the experimental study of electronic transport in SLs is the interdependence of the intensity of the current injected and the field present in the SL which is unavoidable in 2-terminal structures. At higher fields the large current densities injected make the field in the SL nonuniform and cause the formation of high field domains.[25,26] Therefore a a 3-terminal structure was designed that could simulate as closely as possible the "ideal" situation: a monoenergetic beam of electrons of constant flux current impinging on the SL, independent of the value of the field in the SL.[16] The equilibrium band-diagram of the structure is schematically drawn in Fig. 10a. The forward-biased p-n "emitter" heterojunction provides a controllable source of current independent of the reverse bias applied to the "collector" heterojunction which in turn controls the field in the SL. By measuring the "collector" current at constant "emitter" current, the "intrinsic" SL transport properties as a function of a uniform SL field can be tested, avoiding the complications mentioned above.

The structures were grown by molecular beam epitaxy lattice matched on an undoped InP substrate. The growth started with a 4955-Å $Ga_{0.47}In_{0.53}As$ buffer layer doped $n^+ = 1 \times 10^{18}$ cm^{-3} followed by a 5945 Å thick $Ga_{0.47}In_{0.53}As$ undoped collector region. The SL was then grown followed by a 5000-Å $Al_{0.163}Ga_{0.312}In_{0.525}As$ layer doped $p^+ = 5 \times 10^{18}$ cm^{-3}. Finally 850-Å $Al_{0.48}In_{0.52}As$ $n^+ = 5 \times 10^{17}$ cm^{-3} was grown separated by a 300-Å quaternary graded region from the $Ga_{0.47}In_{0.53}As$ $n^+ = 1 \times 10^{19}$ cm^{-3} 2000-Å-thick cap contact layer. Two different undoped SL were studied. Structure A consists of 14 periods of 17–Å $Al_{0.48}In_{0.52}As$/37–Å $Ga_{0.47}In_{0.53}As$. Structure B of 9 periods 23–Å $Al_{0.48}In_{0.52}As$/36–Å $Ga_{0.47}In_{0.53}As$.

In structure A the calculated ground state miniband dispersion is $\Delta E_A \approx 115$ meV and its bottom lies at $E_0^A \approx 130$ meV while in structure B, $\Delta E_B \approx 80$ meV and $E_0^B \approx 154$ meV.[16] Therefore, in order to inject electrons into the miniband, the composition of the p-doped quaternary layer was chosen to be $Al_{0.163}Ga_{0.312}In_{0.525}As$. Assuming a linear x-dependence for the $Al_xGa_{0.475-x}In_{0.525}As$/$Ga_{0.47}In_{0.53}As$ conduction band discontinuity, this corresponds to a conduction band offset of ≈ 180 meV, roughly the center of the minibands (see Fig. 10). The thickness of this layer is such that all the emitter electrons thermalize before impinging on the SL. Note that for both samples, the

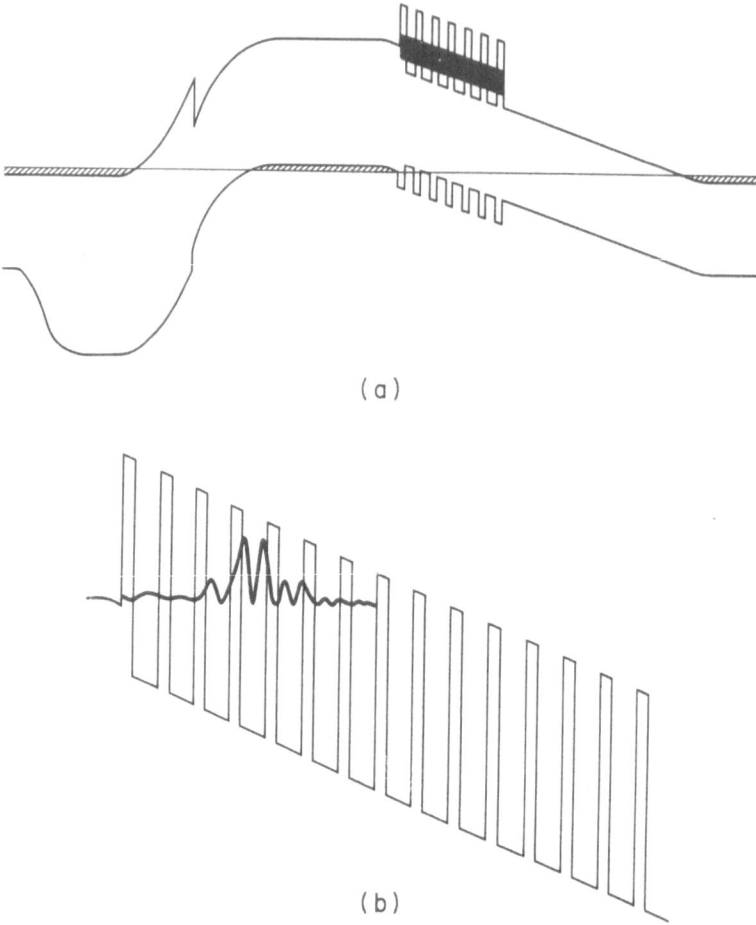

(a)

(b)

Fig. 10. (a) Energy-band diagram (not to scale) of the samples studied, (b) Conduction-band diagram of sample A at a bias such that a quasi-state supported by a subset of the superlattice of thickness equal to the electron coherence length enhances electronic transport. The solid curve represents the calculated wavefunction corresponding to the peak at 10.9 V.

ground-state miniband is wide enough to ensure the absence of localization due to fluctuations.[27] Moreover, the large energy separation ($\Delta E_s \approx 200$ meV in both samples) between the ground and first excited minibands insures that Zener tunneling will not be possible in these structures for $F < \Delta E_s/eL \approx 30$ kV/cm (for both samples), where L is the total length of the SL.

The collector-current density as a function of the collector-junction bias at constant emitter current is shown in Fig. 11 for both samples. (The amplified scale has been chosen to clearly show the structure presented by the curves. The

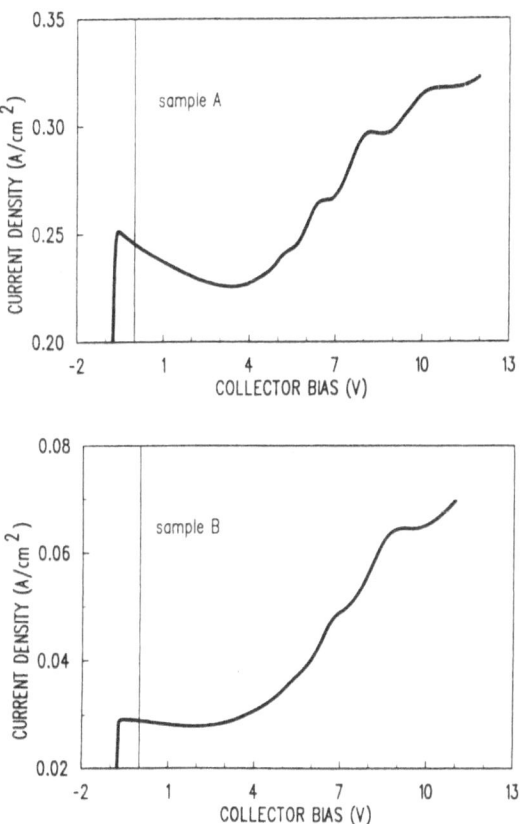

Fig. 11. Collector current density as a function of the collector bias at constant emitter current ($I_E = 0.39$ A/cm^2) for samples A (top) and B (bottom). Both measurements were performed at $T = 15$ K. The features presented by the characteristics were visible up to temperatures as high as 200 K. The arrows indicate the calculated bias positions of the resonances.

initial steeply rising region shows no features.) No space charge effects are present in the structure. In fact, by taking into account the electron velocity in the miniband regime, one obtains an upper limit for the carrier density of 10^{12} cm^{-3}.

149

Moreover, this is proved by the experimental fact that by varying the emitter current and therefore the electron flux incident on the SL, the collector current is simply scaled while the voltage positions of the relevant features are not altered. The current monotonically decreases in a wide bias range. As discussed above, this is the experimental manifestation of localization. The threshold for the onset of NDC can be determined by subtracting the forward-biased collector-junction dark current to the measured characteristics; it is $F \approx 3$ kV/cm, for both samples. This value is consistent with Eq. 2 taking $\tau \approx 4 \times 10^{-13}$ s, an adequate value for intrasubband scattering time at these low fields.[28] This threshold value is much lower (a factor of 10) than the minimum field required for interminiband tunneling, as required for the observability of NDC.[17] It is also important to note that in this bipolar-transistor structure one can not observe NDC by intervalley transfer in conditions of constant current injection. In these conditions, in fact, the decrease in velocity caused by the higher effective mass of the satellite valleys is compensated by an increase in the carrier density; the collector current is therefore not altered. On the other hand, the present NDC mechanism can be observed since the Bragg-reflected electrons in the negative mass region of the minizone give rise to an opposite flux so that the collector current decreases while the base current increases to maintain a constant emitter bias. No NDC is observed in similar AlInAs/GaInAs structures without a SL in the collector.

The slope continues to be negative for fields corresponding to wavefunction confinement over few (~3-4 for both samples) wells where the characteristics exhibit a minimum.

The presence of the minimum indicates that, at sufficiently high bias, other paths can enhance electronic transport and become dominant. Firstly, an exponentially increasing Fowler-Nordheim current through the SL seen as an effective medium is to be expected. More importantly, on this monotonically increasing background, several peaks are observed. To understand their physical origin, let us recall that in this field range, new states can arise from the mixing between Stark-ladder states originally belonging to different minibands or between the latter and resonances in the classical continuum above the SL barriers. In the case of transport in a SL without collision, these transmission resonances are determined by the whole SL. In the presence of collisions, however, they are determined by subsets of the SL of thickness equal to the electron coherence length. This is illustrated in Fig. 10b where one such state is shown for sample A. The electron coherence length must therefore be explicitly considered and can be included through the electron mean free path λ. In fact, the fraction of electrons that will tunnel coherently for exactly n periods of the SL is equal to $e^{-\frac{na}{\lambda}} (1 - e^{-\frac{a}{\lambda}})$, and the corresponding transmission will be indicated by $T_n(F)$. The overall transmission coefficient is therefore given by:

$$T(F) = \sum_{n=1}^{n_T-1} e^{-\frac{na}{\lambda}} \left(1 - e^{-\frac{a}{\lambda}}\right) T_n(F) + e^{-\frac{n_T a}{\lambda}} T_{n_T}(F), \qquad (4)$$

where n_T is the total number of wells. A fraction $1\text{-}T(F)$ of the impinging electrons are therefore reflected back into the p-doped layer and recombine. The remaining electrons are injected into the SL and reach the collector contact. Excellent agreement was found between calculated and experimental position of the peaks with $\lambda \approx 300$ Å. On the contrary, allowing for an infinite coherence length (i.e., electrons traversing ballistically the entire SL) leads to a much larger number of calculated resonances.

2.2 Interaction phenomena between deep levels and minibands

In this final section I shall present some calculations showing that deep levels introduced in the barriers of tight-binding SL can have intriguing effects on the SL states.[29] These results suggest the possibility of performing some type of *defect engineering* that could provide even more degrees of freedom to the solid state scientist.

Let us consider the periodic structure whose conduction band is sketched in Fig.12a. I shall examine the introduction of a sheet of defects in the center of the barriers corresponding to a deep center lying in energy between the two original bands. Such deep levels could be obtained also by the deposition of sheets of shallow impurities at high surface concentration as shown by Hjalmarson.[30] Fig. 12b shows the enhanced width of the bands in the resulting structure and the creation of one extra band due to the states introduced by the defects. This enhancement is maximized when (Fig. 12c) the energy level of the defect is matched to the ground state of the corresponding isolated quantum well.[31]

In an actually grown structure with weak coupling between wells (tight-binding SL) the states of the SL easily become localized. This occurs when the combined energy level broadening due to intralayer thickness fluctuations and collisions is greater than the miniband widths. Therefore we can have the interesting fact that the periodic introduction of a highly localized state in a periodic heterostructure causes a delocalization in the resulting states with consequent dramatic change in the electronic properties.

One should also consider another variable: the position of the layer in the barrier. Fig. 12d shows the case of energy matching of 12c, but with the defect layer displaced significantly from the central position in the barrier. The ground-state miniband of Fig. 12c splits into two bands considerably narrower and shifted. This shift can be such that, as shown, a new SL band is formed lying in energy below the bottom of the quantum wells. This means that the resulting SL band-

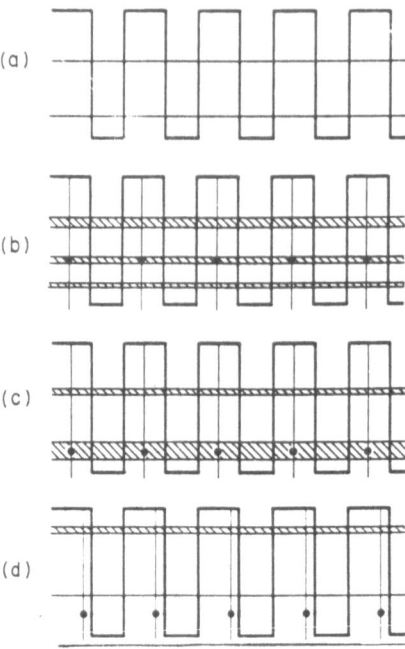

Fig. 12. Energy band diagrams of multiquantum well structure without defects (a) and after their introduction (b) (c) (d). In (b) the defect sheets, shown as vertical lines, are in the middle of the barriers and their energy level, shown as a dot, is intermediate between the two original bands. In (c) the defects level is lined up with the ground level of the wells. In (d) with energies still matched, the defects are displaced from the central position.

gap can be made smaller than that of the bulk of both materials.

The band structure calculation of heterojunction SLs is often performed via a Kane-type formalism based on the solution of an effective Hamiltonian as studied in detail by Bastard.[32] The results obtained with this technique have been shown to be in excellent agreement with linear combination of atomic orbitals (LCAO) calculations.[32] The structure was studied within the same formalism by using a model potential to describe the defect.[29] Since the qualitative results of the calculations are not influenced by the details of the potential chosen (only its symmetry and weight, i.e. its integral, are important) a delta function was used. This model is adequate for several type of deep centers as suggested by photoionization cross section measurements[33] and allows one to express the results with a compact analytical formula.

As an example, a $GaAs/Ga_{0.7}Al_{0.3}As$ periodic structure with $w=80\overset{\circ}{A}$ and $b=100\overset{\circ}{A}$ was considered.[29] The first two bands of the unperturbed periodic structure are shown in the reciprocal space in the extended zone scheme by the dashed curves in Fig. 13. The first band is centered at $E_0=42.5$ meV (which is the first energy level of the corresponding isolated quantum well[31]) and its width is $\Delta E_0 = 4.4 \times 10^{-2}$ meV. The second band is characterized by $E_1 = 165.6$ meV and $\Delta E_1 = 1.1$ meV.

Referring to this periodic structure, let us consider now the case of defects centered in the barriers with energy levels matched to those of the isolated wells (sketched in Fig.12c). The model gives for the first two bands of the new structure the results shown by the thick solid curves in Fig. 13 ($\Delta E_0 = 9.1$ meV, $\Delta E_1 = 2.4$ meV). The first band of the resulting structure is more than 200 times wider than without deep levels. The broadening depends on the SL parameters: the effect is larger for narrower starting bands (keeping $w=80\overset{\circ}{A}$, for $b=140\overset{\circ}{A}$, for instance, brings ΔE_0 down to 2.1×10^{-3} meV, and the miniband is broadened about 10^3 times). This broadening is maximum with the defect layer centered in the barrier. In this case in fact the overlapping between the defect and SL eigenstates is maximum. It is worth mentioning here that using the well width as a parameter, the energy levels of the well and of an available deep center can be very effectively matched.

Another noticeable aspect is that there is no energy gap at π/a in the dispersion relation. The periodic introduction of the defects (matched in energy and centered in the barriers) compensates the effect of the periodic SL potential and closes the gap. It is easy to show that *any* symmetric potential centered in the barriers can give the same result provided it has the proper weight.

For what concerns the second band, it is approximately doubled in width and is shifted toward higher energies.

Let us consider now the effect of a displacement s of the defect from the central position in the barrier. In Fig. 13, the thin solid curve refers to the case of energy matching but with $s=10\overset{\circ}{A}$. The gap re-opens at π/a (again this is a property of any symmetric potential). The first band narrows considerably ($\Delta E_0 = 1.9$ meV) and shifts toward lower energies.

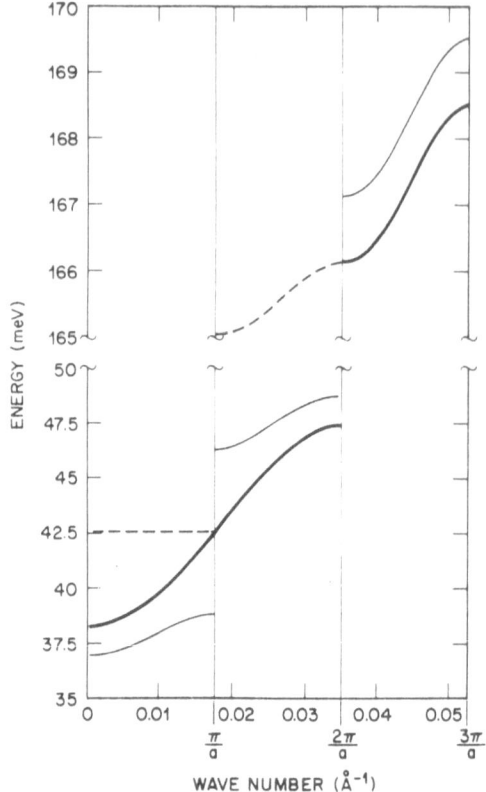

Fig. 13. Energy dispersion relation in the extended zone scheme for a 80Å well, 100Å barrier GaAs/$Ga_{0.7}Al_{0.3}$As superlattice: (dashed line) without defects, (thick solid line) with defects centered in the barriers and energy matched with the ground state in the wells, (thin solid line) with defects displaced 10Å from the center and energies matched.

This suggests that we are able to control the band gap of the SL by the introduction of defects. It has been shown[29] that a miniband below the bottom of the quantum well (Fig. 12d) can be introduced.

Acknowledgments: It is a pleasure to acknowledge my colleagues and collaborators at Bell Laboratories, in particular Federico Capasso, Susanta Sen, Deborah L. Sivco, Albert L. Hutchinson, Alfred Y. Cho, and Roger J. Malik. I would like also to thank Janice A. Preckwinkle for her skillful preparation of this camera-ready manuscript.

REFERENCES

1. *Semiconductors and Semimetals*, Vol. 24, R. K. Willardson and A. C. Beers eds. (Academic Press, New York, 1987).

2. *Physics of Quantum Electron Devices*, F. Capasso ed. (Springer Verlag, Berlin, 1990).

3. F. Beltram, and F. Capasso, S. Sen, in *Electronic Materials: A New Era in Materials Science*, J.R.Chelikovsky and A.Franciosi eds. (Springer Verlag, *in press*) Ch. 10.

4. H. Kroemer, RCA Rev. **18**, 332 (1957).

5. R.J. Malik, F. Capasso, R.A. Stall, R.A. Kiel, R. Wunder, and C.G. Bethea, Appl. Phys. Lett. **46**, 600 (1985).

6. P.M. Asbeck, in *High-speed semiconductor devices*, S. Sze ed. (John Wiley & Sons, New York, 1990), Ch. 6.

7. J.A. Morton, IEEE Spect., p. 62, Sept. 1965.

8. S. Sen, F. Capasso, A.Y. Cho, and D.L. Sivco, Electron. Lett. **24**, 1506, (1988).

9. F. Capasso and R. A. Kiehl, J. Appl. Phys. **58**, 1366 (1985).

10. See, e.g., the circuit applications presented in Refs 2 or 3.

11. R. C. Potter, A. A. Lakhani, D. Beyea, E. Hempling, and A. Fathimulla, Appl. Phys. Lett. **52**, 2163 (1988).

12. S. Sen, F. Capasso, D. Sivco, and A. Y. Cho, IEEE Electron Dev. Lett., 402, **9** (1988).

13. *General Electric Tunnel Diode Manual, First Ed.*, 66 (1961).

14. F. Capasso, S. Sen, A. Y. Cho, and D. L. Sivco, Appl. Phys. Lett., 1056, **53** (1988).

15. M. Tsuchiya, H. Sakaki, and J. Yoshino, Japan. J. Appl. Phys. **24**, L466 (1985).

16. F. Beltram, F. Capasso, D.L. Sivco, A.L. Hutchinson, and S.-N. G. Chu, A.Y. Cho, Phys. Rev. Lett. **64**, 3167 (1990).

17. L. Esaki and R. Tsu, IBM J. Res. Develop. **14**, 61 (1970).

18. R. Tsu and G. Döhler, Phys. Rev. **B12**, 680 (1975).

19. G. Döhler, R. Tsu, and L. Esaki, Solid State Commun. **17**, 317 (1975).

20. F. Capasso, K. Mohammed, and A. Y. Cho, IEEE J. Quantum Electron **QE-22**, 1853 (1986).

21. A. Sibille, J. F. Palmier, H. Wang, and F. Mollot, Phys. Rev. Lett. **64**, 52 (1990).

22. G. Wannier, Phys. Rev. **117**, 432 (1960).

23. R. F. Kazarinov and R. A. Suris, Fiz. Tekh. Poluprov. **5**, 797 (1971) [Sov. Phys. Semicond. **5**, 707 (1971)]; R. F. Kazarinov, and R. A. Suris, Fiz. Tekh. Poluprov. **6**, 148 (1972) [Sov. Phys. Semicond. **6**, 120 (1972)].

24. V. A. Yakovlev, Fiz. Tverd. Tela **3**, 1983 (1962) [Sov. Phys. Solid State **3**, 1442 (1962)].

25. L. Esaki and L. L. Chang, Phys. Rev. Lett. **33**, 495 (1974).

26. K. K. Choi, B. F. Levine, R. J. Malik, J. Walker, and C. G. Bethea, Phys. Rev. **B 35**, 4172 (1987).

27. F. Capasso, K. Mohammed, A. Y. Cho, R. Hull, and A. L. Hutchinson, Phys. Rev. Lett. **55**, 1152 (1985).

28. D. C. Herbert, Semicond. Sci. Technol. **3**, 101 (1988).

29. F. Beltram and F. Capasso, Phys. Rev. **B38**, 3580 (1988).

30. H. P. Hjalmarson, J. Vac. Sci. Technol. **21**, 524 (1982); H. P. Hjalmarson, Superlattices and Microstructures **1**, 379 (1985).

31. By isolated quantum well, I mean one and a half period of the superlattice.

32. G. Bastard, Phys. Rev. **B 24**, 5693 (1981).

33. G. Lucovsky, Solid State Commun. **3**, 299 (1965).

$(C_{10}H_{21}NH_3)_2PbI_4$: A natural quantum-well material

Ryoichi Ito, Chang-qing Xu and Takashi Kondo

Research Center for Advanced Science and Technology
The University of Tokyo, Meguro-ku, Tokyo 153, Japan

and

Department of Applied Physics, Faculty of Engineering
The University of Tokyo, Bunkyo-ku, Tokyo 113, Japan

C_{10}-PbI_4, a layered perovskite-type material, has been studied with emphasis on its linear and nonlinear optical properties. It has been found that this material is a natural quantum-well system, in which excitons are confined in a thin inorganic layer sandwiched between organic barrier layers. Very clear two-dimensional characters of the lowest exciton, its large binding energy and large third-order optical nonlinearity have been found.

1. INTRODUCTION

Nonlinear optics (NLO) is expected to play a key role in such future systems as all-optical communications and computing. These systems will be based on those NLO devices in which laser light is controlled by another laser light, typical examples being optically gated optical switches and optical bistable devices.[1] In order to realize these devices, one must find materials with sufficiently large (third-order) optical nonlinearity, fast response, transparency and easy processability.

Toward this goal, a variety of materials have been studied, including semiconductors, doped and undoped glasses, organic crystals, polymers and liquid crystals. Among these, considerable attention has recently centered on low-dimensional systems, such as semiconductor quantum-well structures (2-D systems), polydiacetylenes(1-D systems) and semiconductor-doped glasses(0-D systems). Their fundamental optical absorptions are excitonic and their large optical nonlinearities are believed to stem from the enhanced dipolar transition moments due to the confinement of the excitons.

In this paper, we report on our experimental studies of $(C_{10}H_{21}NH_3)PbI_4$, a new NLO material, which we will hereafter abbreviate to C_{10}-PbI_4. This material belongs to a large group of materials of the general formula, $(C_nH_{2n+1}NH_3)_2MX_4$, where M= Cu, Mn, Cd, Fe, Pd, Co, Zn, Pb, $\cdots$; X= Cl, Br, I. Most of them have the same layer-type perovskite structure as the high-T_c superconductor $La_{2-x}Sr_xCuO_4$, and have been known in the field of dielectrics and magnetism. Interest thus far has centered, however, around their numerous structural phase transitions and two-dimensional magnetic properties. Only recently, their optical properties have begun to be investigated. In particular, Ishihara et al.[2,3] found that C_n-PbI_4 (n= 4, 6, 8, 9, 10, 12) has excitonic fundamental absorption bands and their oscillator strength and binding energy are much larger than those of PbI_2, a layered semiconductor well known for its excitonic property.

Intrigued by the very interesting results obtained by Ishihara et al., we have started a systematic study of C_{10}-PbI_4, including structural phase transitions, electro-absorption, magneto-optics and third-harmonic generation measurements. We have found that the layered perovskite material has indeed many interesting properties, not the least of which is its large optical nonlinearity. This paper describes some of the results we have obtained so far on this novel compound.[4-6]

2. SAMPLE PREPARATION

Both single crystals and oriented polycrystalline films were used. C_{10}-PbI_4 was synthesized from $C_{10}H_{21}NH_3I_4$ and PbI_2 by a solvent method; $C_{10}H_{21}NH_3I_4$ and PbI_2 were made to react in an aqueous solution of KI containing acetone. The obtained polycrystalline powder of C_{10}-PbI_4 was dissolved in a mixed solution of acetone and nitromethane. A single crystal with a size as large as $2 \times 1 \times 0.1$ mm^3 resulted in about a week.

For optical measurements requiring uniform thin films, we used films, $50 \sim 200$ nm thick, spin-coated on quartz substrates. X-ray diffraction study showed that these films were highly oriented with the c-axis perpendicular to the substrate surface. We have confirmed that their optical properties are very similar to those of single crystals.

3. STRUCTURAL PHASE TRANSITIONS

The crystal structures were determined using a four-axis X-ray diffractometer and a Weissenberg camera. Figure 1 shows the crystal structure of C_{10}-PbI$_4$ at 265 K. This structure consists of two-dimensional layers built up from corner-sharing iodine octahedra, with the divalent Pb^{2+} ions in the centers; the cavities between octahedra are occupied by NH$_3$ polar heads of alkylammonium groups, which form NH...I hydrogen bonds with the iodine atoms. At room temperature, the space group of the crystal is D_{2h}^{15} with Z=4 formula units in a unit cell, and a=8.968 Å, b=8.667 Å, c=42.51 Å. The distance between the two axial I ions is 6.42Å and the diameter of the I ion is 4.32Å, and hence the sum of them, 10.84Å, may be tken to be the thickness of an inorganic layer.

C_{10}-PbI$_4$ exhibits three structural phase transitions below 540 K, its decomposition temperature. Figure 2 shows the temperature dependence of the lattice parameter along the c-axis. All the phase transitions are of the first order and are associated with the change of order in the alkylammonium chains. Upon transition from the III to IV phase, the crystal changes from the orthorhombic to monoclinic system and a single crystal becomes twins. This makes it extremely difficult to study the IV phase(low temperature phase) of single crystals.

The color of C_{10}-PbI$_4$ crystal changes from orange to yellow upon cooling around 275 K. This is due to a discontinuous blue shift of about 0.1 eV in the fundamental absorption at $\sim$2.5 eV, as can be seen in Fig.3.[3] This spectral change is, of course, a result of the structural phase transition (II$\rightarrow$III/ IV). Upon transition from the II to III phase, the [PbI$_6$] octahedra remain essentially unchanged, while the relative alignment of the neighboring

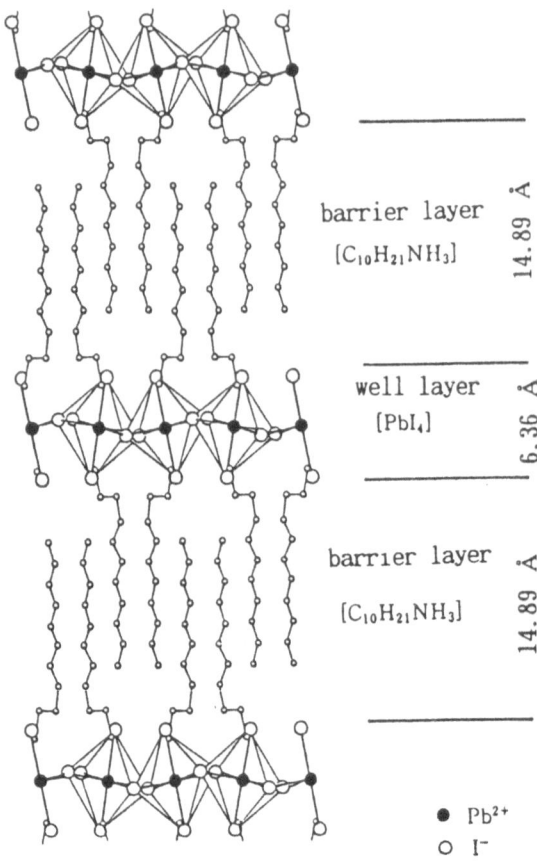

barrier layer
[$C_{10}H_{21}NH_3$]

14.89 Å

well layer
[PbI_4]

6.36 Å

barrier layer
[$C_{10}H_{21}NH_3$]

14.89 Å

● Pb^{2+}
○ I^-

Fig. 1. Unit cell structure of C_{10}-PbI_4 at 265 K.

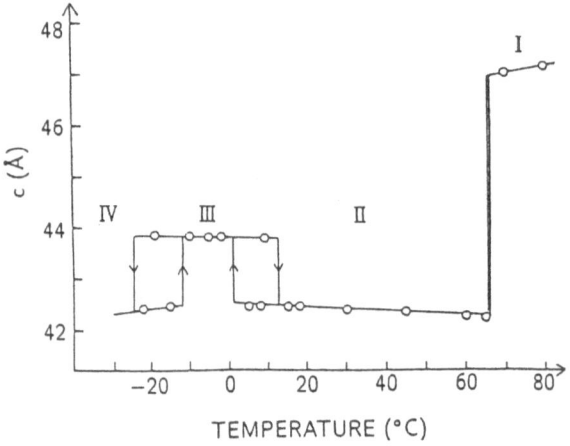

Fig. 2. Temperature dependence of the lattice
parameter along c-axis of C_{10}-PbI_4.

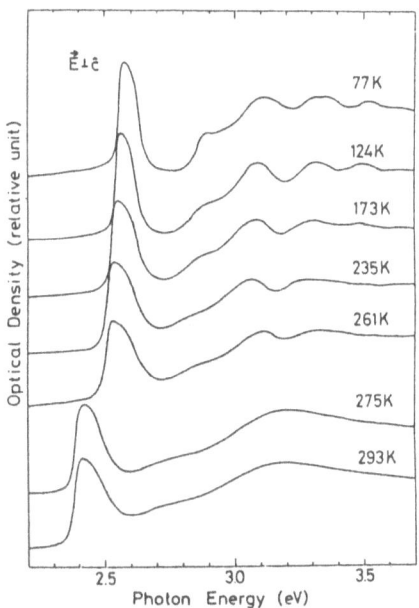

Fig. 3.
Absorption spectra of C_{10}-PbI$_4$.

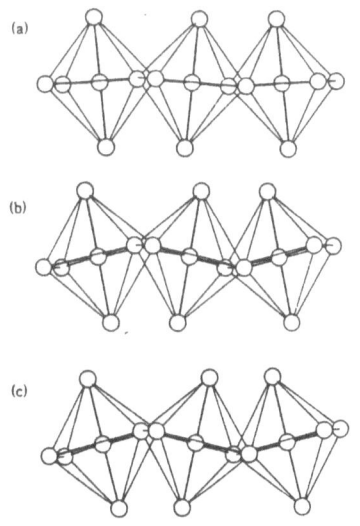

Fig. 4.
Structures of [PbI$_4$]$^{2-}$ layer of
C_{10}-PbI$_4$; (a) phase II; (b) phase III;
(c) phase IV.

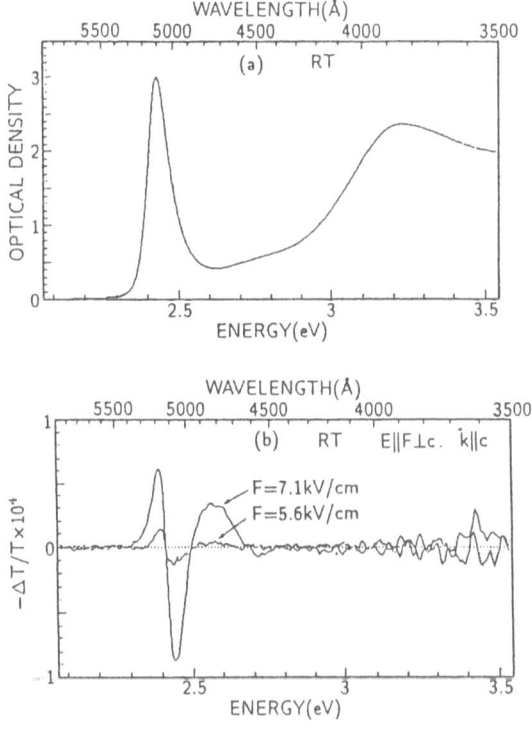

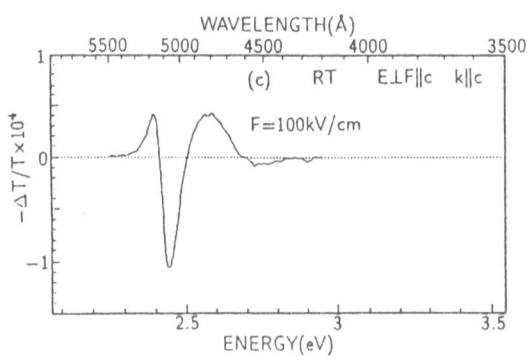

Fig. 5. RT-phase C_{10}-PbI$_4$ absorption spectrum
(a) and EA spectrum in the transverse geometry
(b) and in the longitudinal geometry (c).

octahedra is considerably changed, as is shown in Fig.4. This change is presumably responsible for the discontinuous blue shift accompanying the phase transition Ⅱ→Ⅲ.

4. TWO-DIMENSIONAL EXCITONS

The strong optical absorption around 2.5 eV was assigned by Ishihara et al.[2,3] to the lowest exciton state(which we will call A exciton). They deduced, from the temperature dependence of photoluminescence due to this exciton state, a binding energy as large as 370 meV, an extremely large value which may be compared to $30 \sim 60$ meV estimated for PbI_2,[7] a generic material of C_n-PbI_4. In order to study the physics underlying this enhancement of the exciton binding energy, we have measured the electro-absorption (EA)[4] and magneto-absorption (MA)[5] of C_{10}-PbI_4. The samples were spin-coated thin films with the c-axis aligned perpendicularly to the surface.

Figure 5 show the EA spectra near the A exciton absorption for the room-temperature (RT) phase (Ⅱ-phase in Fig.3); (a) is the absorption spectrum, (b) the EA spectrum in the transverse-field geometry and (c) the EA spectrum in the longitudinal geometry. From the shape and field- and configuration-dependence of the EA signal, we find that (1) the A exciton absorption is indeed due to an exciton with a Wannier character, and (2) the exciton has a strong 2-D character.

The EA measurement at 77 K for the low temperature (LT) phase (Ⅳ-phase in Fig.3) revealed another interesting phenomenon; the EA signal followed the applied field with a time constant of $\sim 30 \mu$s. This is an extraordinarily slow response which presumably involves a macroscopic lattice motion of unknown origin.

The magneto-absorption spectra were taken at 4.2 K under several magnetic fields (up to 40 T) applied perpendicular (Faraday configuration) and parallel (Voigt configuration) to the sample surface or the [PbI_4] planes. The A exciton peak splits into two components at 4.2 K probably owing to the reduced symmetry in the LT phase. Figure 6 shows the measured MA spectra. It should be noted that the A exciton peaks exhibit Zeeman splittings and diamagnetic shifts in the Faraday configuration, while, in the Voigt configuration, they exhibit neither any shift nor any splitting. This we interpret as a good evidence for the strong 2-D character of the A exciton.

Thus, we have seen that the A exciton of C_{10}-PbI_4 has a strong 2-D character. To our knowledge, no artificially fabricated semiconductor quantum-well structures have shown such clear 2-D characters. What is it that makes this material so special?

To answer this question, let us turn back to the crystal structure of C_{10}-PbI_4 shown in Fig.1. First note that [PbI_4] layers, in which electronic excitations take place, are very thin, as thin as 11Å, and the interface between the inorganic and organic layers is atomically flat. This should make C_{10}-PbI_4 an ideal quantum-well structure. As a result, the exciton in PbI_2 with a Bohr radius of 19Å[7] becomes considerably compressed two-dimensionally in C_{10}-PbI_4 on account of the potential step which is as large as 3 eV. Second, there exists a large dielectric-constant step between the inorganic and organic layers; it is estimated that $E_{well} \sim 6$, while $E_{barrier} \sim 3$. Such a dielectric step is expected to enhance the Coulomb interaction between an electron and a hole in the well layer, leading to an increase in excitonic binding energy and ocsillator strength.[8, 9]

The Bohr radius and the binding energy of the A exciton were estimated on the basis of the EA and MA measurements. From the quadratic Stark shift of the EA signal, a_B (Bohr radius)=7.5 Å and E_b (binding energy)=590 meV have been estimated. On the other hand, the diamagnetic shifts observed in the MA measurement give a_B=12Å and E_b=360 meV for the low-energy component of the A exciton, and a_B=18Å and E_b=240 meV for the high-energy component. Thus, although the exact values of the Bohr radius and the binding energy of the A exciton state are still to be settled, it has become unequivocally clear that C_{10}-PbI_4 has a binding energy much larger than that of PbI_2.

The dramatic enhancement of the excitonic binding energy (and oscillator strength) is probably understood as the result of the dielectric confinement of the exciton which was first proposed by Keldysh[8] and recently elaborated by Hanamura et al.[9] In this model, the Coulomb interaction between an electron and a hole confined in a thin layer becomes stronger when the surrounding (barrier) material has a smaller dielectric constant. As a result, the exciton becomes more tightly bound. The present material seems to offer the first example supporting the dielectric enhancement model.

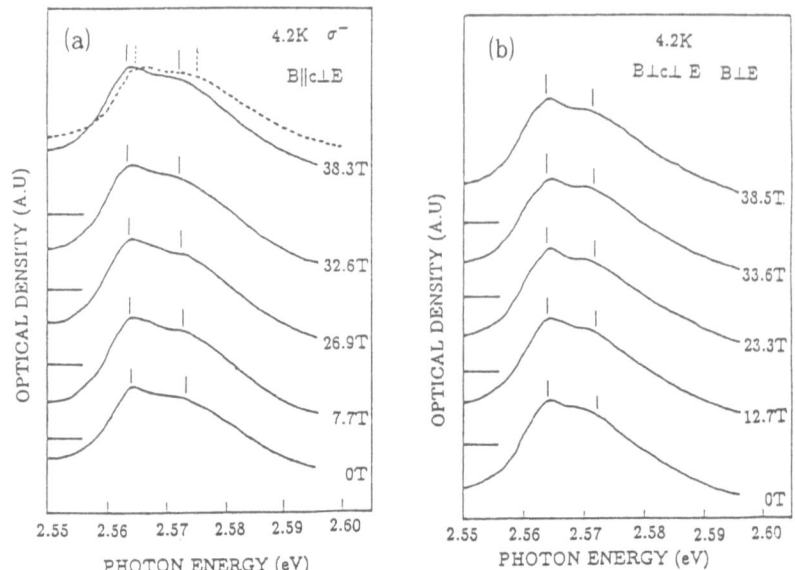

Fig. 6. MA spectra of C_{10}-PbI_4 at 4.2 K.   (a) Left-hand circular (σ^-) polarization in Faraday configuration. The broken line is for σ^+ polarization. (b) σ polarization in Voigt configuration.

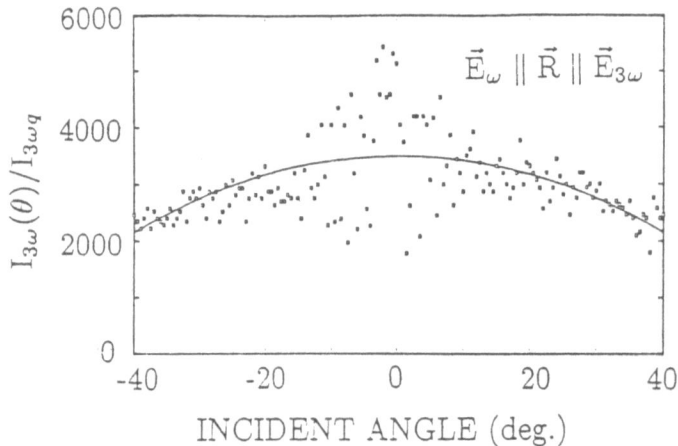

Fig. 7. Relative THG intensity from C_{10}-PbI_4 film at RT and the fundamental wavelength of 1.53 μm.

5. THIRD-ORDER OPTICAL SUSCEPTIBILITY

Excitonic systems have been known to exhibit interesting optical nonlinearities of the third order. In particular, excitons in confined systems have shown large nonlinearities largely because of concentrated densities of states and enhanced excitonic binding energies. Since C_{10}-PbI_4 has been shown to have a very stable lowest exciton state, we may expect to see large optical nonlinearity in this material.

We have studied third-harmonic generation (THG) to investigate $\chi^{(3)}$, the third-order optical susceptibility, of C_{10}-PbI_4.[6] With THG, we can study the resonance effect with the fundamental frequency far apart from the exciton absorption peak. In fact, our attempts to observe degenerate four-wave mixing have not been successful thus far in the vicinity of resonance on account of severe optical damage caused by the fundamental beam.

The spectrum of $\chi^{(3)}(-3\omega; \omega, \omega, \omega)$ was measured from 1.77μm to 1.50 μm. The sample was a spin-coated film, 150 nm thick, on a quartz substrate. The fundamental beam was obtained from a $LiNbO_3$ difference frequency generator using a Q-switched Nd:YAG laser and a tunable dye laser. The fundamental beam, with a peak power density of 50 MW/cm^2, first passed through the sample film and then the substrate.

Figure 7 shows the THG intensity from the C_{10}-PbI_4 film normalized to that from a fused silica reference at 1.53μm as a function of the incident angle. I_{3w_q} is the maximum of the envelope of a Maker fringe pattern for the fused silica reference. In order to obtain the absolute values of $\chi^{(3)}$ for C_{10}-PbI_4, $\chi_q^{(3)} = 2.82 \times 10^{-14}$ esu and the known data on the refractive index of fused quartz were used. The refractive index and the extinction coefficient of C_{10}-PbI_4 were determined from ellipsometry, optical reflectivity and absorption measurements.

$\chi^{(3)}(-3\omega; \omega, \omega, \omega)$ at 1.53 μm was determined to be 0.7×10^{-9}esu. This value is very large for an organic material;[10] in fact, it is almost comparable to that of polyacetylene at resonace,[11] the largest reported $\chi^{(3)}$ value in organic polymers.

1.53μm is nearly three-photon resonant to the A exciton absorption at 512 nm. In order to study the resonance enhancement effect, THG measurements were also performed at other off-resonant wavelengths over the wavelengths

1.77~1.50 μm. Figure 8 shows the obtained spectrum (open circles and solid line). It can be seen that $\chi^{(3)}$ exhibits pronounced enhancement at λ =1.53 μm which is three-photon resonant to the A exciton absorption.

In general, $|\chi^{(3)}(-\omega; \omega, -\omega, \omega)|$, which is important in optical processing such as optical bistability, has a dispersion similar to that of $|\chi^{(3)}(-3\omega; \omega, \omega, \omega)|$. Therefore, we can expect a large resonance enhancement of $|\chi^{(3)}(-\omega; \omega, -\omega, \omega)|$ at the position of the A exciton. It is worth noting that the off-resonance values of $\chi^{(3)}$ are also very large in C_{10}-PbI$_4$. Assuming $\chi^{(3)}$ to be real, the nonlinear index n_2 relating the refractive index variation to the light intensity ($\Delta n = n_2 I$) can be estimated to be about 10^{-9} cm^2/kW in the approximation $\chi^{(3)}(-\omega; \omega, -\omega, \omega)| \fallingdotseq \chi^{(3)}(-3\omega; \omega, \omega, \omega)$, which is true if the fundamental photon energy is considerably smaller than the band gap energy and the dispersion of $\chi^{(3)}$ is neglected. This impressive large off-resonance value of the nonlinear index may open an attractive prospect for the application of this new class of materials in the field of nonlinear optics.

6. CONCLUSIONS

C_{10}-PbI$_4$, which belongs to a large group of layered perovskite-type materials, has been studied with emphasis on its linear and nonlinear optical properties. It has been found that the lowest exciton state has a binding energy as large as 300~500 meV, and has a very clear 2-D character. Large third-order optical nonlinearity has been observed, showing a distinct resonance enhancement due to the lowest exciton. These unique preperties indicate that this material is a natural quantum-well system in which excitons are confined by both potential and dielectric steps within a very thin inorganic well layer.

Acknowledgments: This work was done with the cooperation of many people. Most important contributions came from S.Fukuta, H.Sakakura of the Department of Applied Physics, S.Takeyama and N.Miura of the Institute for Solid State Physics, both of the University of Tokyo, and K.Kumata and Y.Takahashi of Tokyo Research Laboratory, IBM Japan, Ltd. Stimulating discussions with T.Koda,

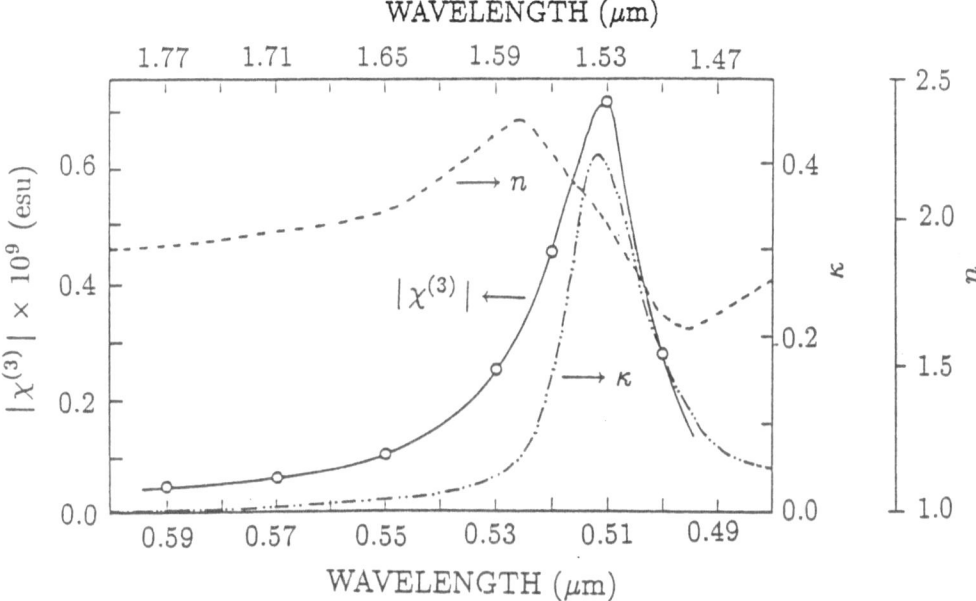

Fig. 8. | χ$^{(3)}$ |(open circles and solid line), n(broken line) and κ (dash-dotted line) of C_{10}-PbI_4 at RT. The upper scale gives the wavelength of the fundamental wave.

E. Hanamura and M. Gonokami, all our colleagues at the Department of Applied Physics, are deeply appreciated.

REFERENCES

1) Gibbs, H.M. (1985). Optical Bistability: Controlling Light with Light, Academic Press.

2) Ishihara, T., Takahashi, J. and Goto, T. (1989). Solid State Commun. Vol. 69, p. 933.

3) Ishihara, T., Takahashi, J. and Goto, T. (1990). Phys. Rev. B, Vol. 42, p. 11099.

4) Xu, C., Fukuta, S., Sakakura, H., Kondo, T., Ito, R., Takahashi, Y. and Kumata, K. (1991). Solid State Commun. Vol. 77, p. 923.

5) Xu, C., Sakakura, H., Kondo, T., Takeyama, S., Miura, N., Takahashi, Y., Kumata, K. and Ito, R. (1991). Solid State Commun. To be published.

6) Xu, C., Kondo, T., Sakakura, H., Kumata, K., Takahashi, Y., and Ito, R. (1991). Solid State Commun. To be published.

7) Nagamune, Y., Takeyama, S. and Miura, N. (1989). Phys. Rev. B, Vol. 40, p. 8099.

8) Keldysh, L.V. (1979). Pis'ma Zh. Eksp. Teor. Fiz. Vol. 29, p. 716 [JETP Lett. Vol. 29, p. 658].

9) Hanamura, E., Nagaosa, N., Kumagai, M. and Takagahara, T. (1988). J. Mater. Sci. Engng. Vol. 1, p. 255.

10) Kobayashi, T. ed. (1989). Nonlinear Optics of Organics and Semiconductors, Springer Proc. in Phys. Vol. 36, Springer.

11) Fann, W.-S., Benson, S., Madey, J.M.J., Etemad, S., Baker, G.L. and Kajzar, F. (1989). Phys. Rev. Lett. Vol. 62, p. 1492.

Physics and Applications of Electroluminescent Materials

Hiroshi Kobayashi

Department of Electrical and Electronic Engineering, Tottori University, Koyama, Tottori 680, Japan

The features of thin-film electroluminescence (EL) are reviewed; such as differences between high field- and injection-EL, classification of EL, structure of device and panel, electric operation and characteristics, host materials and luminescent centers. In addition, as applications, the possibility of multi- and full-color EL panels are discussed.

1. ELECTROLUMINESCENCE — EL AND LED

The term "electroluminescence" implies the phenomena of light emission from materials, mainly semiconductors, when electric voltage or electric current are applied. The terms of the electroluminescence, today, are grouped as follows.

Electroluminescence (EL) — (a) High Electric Field EL
                               ELD Electroluminescent Display
                  (b) Injection EL
                               LED Light Emitting Diode
                               LD   Laser Diode

(a) High Electric Field EL — At present, the high electric field electroluminescence (EL) is made use of information- and illumination-electronics. As information electronics, a flat type of display called "Electroluminescent Display Panel (ELD)" is developed. As an illumination use, the EL is used for back light of "Liquid Crystal Display (LCD)".

(b) Injection EL — A Light Emitting Diode (LED) is a well-known example of injection EL. The LED's are commonly used in daily electronics; such as audio video system, surroundings of our everyday life. The LED's in the visible light region are able to emit red, yellow and green, but not blue with enough luminance. The LED's in the infra-red (IR) region are now classified into two types; one is usual type of diodes and the others are semiconductor laser diodes. The latter are used in compact discs and laser discs and are going to be used in optical communications.

In the following, we will consider the physics of high electric field EL and of injection EL. In Fig.1 (a)(b), the devices and energy band models of the high electric field EL and injection EL are shown.

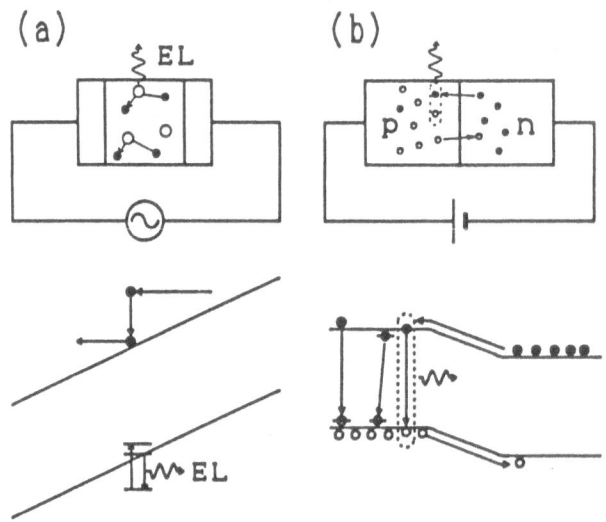

Fig.1 Schematic device models and energy band models of
(a) high electric field EL and (b) injection EL.

In Fig.1 (a), the high electric field EL of thin-film type EL is described. When a high voltage is applied, a high electric field is introduced into emitting layer. The voltage is about 200 - 300 V and the corresponding electric field is $2 - 3 \times 10^6$ V/cm, assuming that the emitting layer is ~ 1 μm thick. The electrons in the emitting layer are accelerated by the electric field. They make hot electrons and collide with the light emitting centers, resulting electroluminescence.

In Fig.1 (b), the injection EL is described. such as LED. When a voltage is applied, electrons are injected from n- into p-region or holes from p- into n-region, that is, minority carriers are injected. The voltage is 1 - 3 V. Subsequently, the minority carriers recombine at various recombination centers, which lead to light emission.

The further details of the two types of EL are summarized in the following.

(a) High Electric Field Electroluminescence

· extremely high electric field
$$d = 1 \ \mu m$$
$$V = 200 - 300 \ V \quad > 10^6 \ V/cm$$

· majority carrier device

· poly-crystalline thin-film

· light emission due to luminescent centers
($3d^5$(Mn) or $4f^n$(Re) electron configuration)

· IIb-VIb ZnS (3.8 eV)   bandgap energy
   IIa-VIb CaS (4.4 eV), SrS (4.3 eV)   (3.8 – 4.4 eV)

(b) Injection Electroluminescence

· p-n junction, (conduction-type control)   1 – 3 V

· minority carrier injection

· single crystal

· light emission due to
      trapped exciton
      free to bound
      donor-acceptor pair

· IV-IV SiC   bandgap energy
   III-V $Ga_xAl_{x-1}As$, $GaAs_xP_{1-x}$, GaP, GaN   (2 – 3 eV)
   II-VI ZnSe

2. CLASSIFICATION OF HIGH ELECTRIC FIELD ELECTROLUMINESCENCE

As is already described in section 1, the electroluminescence (EL) is classified into two groups. (a) one is high electric field EL and (b) the other is injection EL. And then, the high electric field EL is classified further into several types, as is shown below.

```
EL — (a) inorganic EL ┌— thin-film EL — ac-thin-film EL
                      │                  dc-thin-film EL
                      │
                      └— powder EL —— ac-powder EL
                                        dc-powder EL

      (b) organic EL  ————————          dc-thin-film EL
```

As shown above, considering from material aspects, EL's are sorted out (a) inorganic and (b) organic, and then thin-film or powder types. They are further classified into ac- and dc-types. Here, we take up only the problem of ac-thin-film EL, because this field of the ac-thin-film EL is the one that has made a great progress in this decade, and that includes various physics of interest and application use.

3. THIN-FILM EL STRUCTURE AND ELECTRIC CHARACTERISTICS

3.1 Thin-Film EL Structures

In Fig.2, the structures of a thin-film EL panel is shown. The structure consists of glass substrate / electrodes / insulating layer / phosphor layer / insulating layer / electrodes. This structure is called "a doubly insulating structure", because the phosphor layer is sandwiched in two insulating layers.

Phosphor Layer — The thickness is typically 0.5 - 1.0 μm. The II-VI compound semiconductors of ZnS or CaS, SrS are being used, which are activated with luminescent centers such as Mn^{2+} ions (transition metal) or Tb^{3+}, Ce^{3+}, Eu^{2+} ions (rare-earth metal).

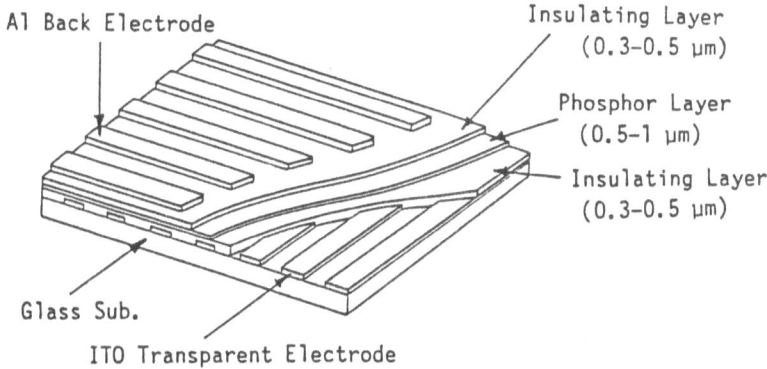

A1 Back Electrode
Insulating Layer (0.3-0.5 μm)
Phosphor Layer (0.5-1 μm)
Insulating Layer (0.3-0.5 μm)
Glass Sub.
ITO Transparent Electrode

Fig.2 Schematic structure of thin-film EL display panels.

Insulating Layer — The thickness is typically 0.3 - 0.5 μ m. The required conditions for the insulating layer are the following; a high breakdown electric field E_b of over 10^6 V/cm, high dielectric constant ε of over 10. The breakdown mode is also a problem —— self-healing or propagating. By taking these requirements into accounts, such materials as Y_2O_3, Al_2O_3, Ta_2O_5 (oxides), SiO_2, Si_3N_4 (silicon system) and $BaTiO_3$, $PbTiO_3$ (perovskite system) are used.

Electrodes — The electrode consists of horizontal (X) and vertical (Y) electrodes. The horizontal are scanning electrodes and the vertical are data electrodes. Usually the data electrodes are on the glass substrate and should be transparent. Because EL light comes out through the glass substrate, the transparent In_2O_3:Sn (ITO) is used exclusively. The scanning electrode is usually metal thin-film.

Substrate — The substrate is a soda glass or a silicate glass, that is a quiet existence, but has a most important role. The following properties are problems; surface roughness, physical properties (melting point, softening point, thermal expansion coefficient, optical transparency, etc.); chemical properties (acid resistant, no alkali contaminant, etc.); mechanical properties (tension, compression, fragility, etc.)

As can be seen, EL emission occurs at any crossing point of X- and Y-electrodes. We, therefore, are able to obtain character- and picture-information on the EL panel by scanning all points sequentially .

3.2 Electric Characteristics

In Fig.3 (a), temporal behaviors of driving pulse voltage V_p, conduction current i_c, and luminance L are shown. The value of V_{0-p} of V_p is about 200 - 300 V. The applied voltage should be ac, because the device has a structure of doubly insulating layers. The actual currents are mostly displacement (capacitive), and small conduction-current (dispersive) are on it. In the figure, only conduction current, which is effective to give rise to light emission, is shown.

In Fig.3 (b), applied voltage dependences of luminance L, efficiency η, electric field E are shown. Above a threshold voltage, electroluminescence starts to occur where the electric field is as large as 10^6 V/cm, and that is very large. The efficiency η tends to saturate as L increases.

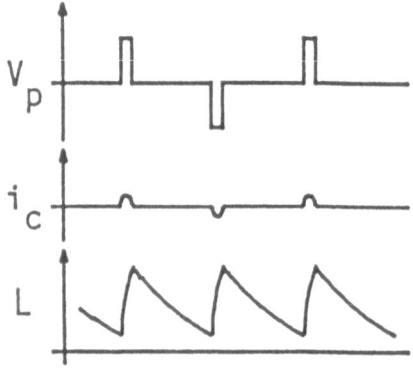

Fig.3 (a) Typical temporal variations of the applied pulse voltage V_p, conduction current i_c and EL intensity L.

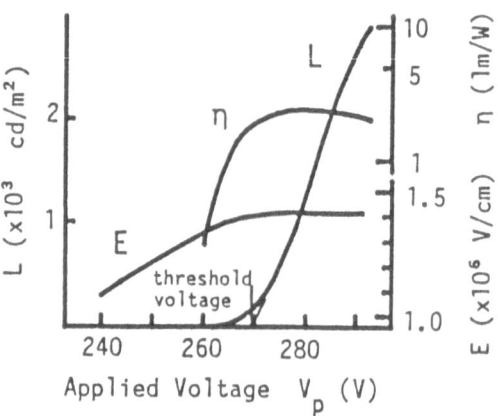

Fig.3 (b) Applied voltage dependences of luminance L, electric field E and efficiency η.

In Fig.4, some representative examples of experimental data are shown. Driving frequency is 5 kHz. The ZnS:Mn provides yellow EL, whose L and η are satisfactory for practical use for EL panels. The ZnS:TbF$_3$ is green EL, whose L and η satisfies the minimum requirements for practical use. The ZnS:SmF$_3$ provides red EL, whose L and η are still lower. The ZnS:TmF$_3$ provides blue EL, whose L and η are very low even for laboratory level. All these results are summarized in Table 1.

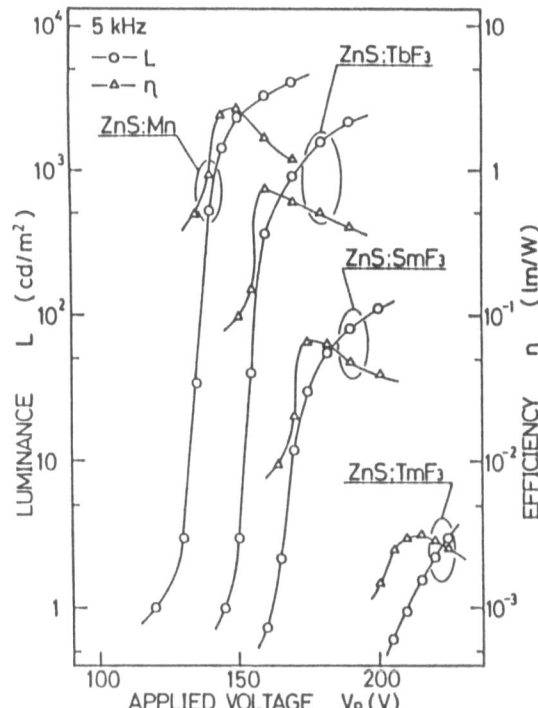

Fig.4 Voltage dependences of EL luminance L and efficiency η for ZnS:Mn and ZnS:ReF$_3$ (Re means rare-earth elements). Driving frequency is 5 kHz.

4. HOST MATERIALS AND LUMINESCENT CENTERS

The EL spectra and EL materials are summarized in Fig.5 (a)-(f).

4.1 ZnS (IIb-VIb) Host Materials and Luminescent Centers

The ZnS is most commonly utilized as EL host materials. The ZnS itself does not provide EL, and is, therefore, doped and activated with several elements that are called luminescent centers, as follows:
ZnS:Cu,X (X:Al,Cl,Br,I) — Fig.5 (a)
This type of luminescent center is called donor-acceptor pair. When ZnS is doped and activated with Cu together with Al, Cl, Br, I. EL emissions from blue ($\sim$400nm) to red ($\sim$700nm) is obtainable. Cu acts as acceptor and Al, Cl, Br, I as donors. These donor-acceptor (d-a) pair type luminescent centers are well known in the fields of luminescent materials. This EL can not be observed for thin-film EL, though can be obtained for powder EL.

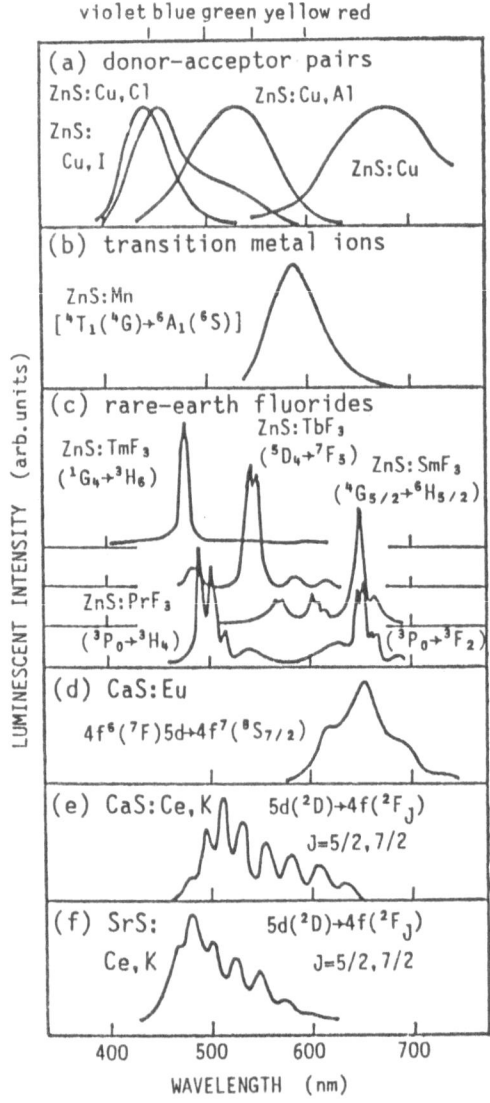

violet blue green yellow red

(a) donor-acceptor pairs

ZnS:Cu,Cl   ZnS:Cu,Al

ZnS:
Cu,I   ZnS:Cu

(b) transition metal ions

ZnS:Mn
$[^4T_1(^4G) \rightarrow ^6A_1(^6S)]$

(c) rare-earth fluorides

ZnS:TbF$_3$

ZnS:TmF$_3$ $(^5D_4 \rightarrow ^7F_5)$
$(^1G_4 \rightarrow ^3H_6)$ ZnS:SmF$_3$
                                    $(^4G_{5/2} \rightarrow ^6H_{5/2})$

ZnS:PrF$_3$
$(^3P_0 \rightarrow ^3H_4)$ $(^3P_0 \rightarrow ^3F_2)$

(d) CaS:Eu

$4f^6(^7F)5d \rightarrow 4f^7(^8S_{7/2})$

(e) CaS:Ce,K $5d(^2D) \rightarrow 4f(^2F_J)$
                      J=5/2,7/2

(f) SrS: $5d(^2D) \rightarrow 4f(^2F_J)$
Ce,K   J=5/2,7/2

LUMINESCENT INTENSITY (arb.units)

400   500   600   700

WAVELENGTH (nm)

Fig.5 Electroluminescent spectra for various materials.

ZnS:Mn — Fig.5 (b)

The Mn^{2+} has an electron configuration of $(3d)^5$. This gives yellow-orange ($\sim$ 580nm) EL in ZnS, coming from inner shell transition in Mn^{2+}. This EL is observable both for thin-film and powder EL. The Mn^{2+} substitute for Zn^{2+} in ZnS. This ZnS:Mn EL is very bright. This also has a high efficiency and long life. This EL, therefore, is already utilized as a commercially available EL panel.

176

ZnS:Re — Fig.5 (c)

Most of the rare-earth (Re) elements show EL in ZnS host. Re's in ZnS make usually trivalent (Re^{3+}) ions. Light emission arises from inner shell transition of electron configuration $(4f)^n$, whose spectrum is, therefore, sharp. Each rare-earth ion has its own luminescent color. Tb^{3+} emits green (~545nm), Sm^{3+} red (~650nm), Tm^{3+} blue (~480nm). Pr^{3+} emits white because of the two peaks at blue-green (~500nm) and red (~650nm).

However, at present, Tb^{3+} is only one candidate that has enough luminance which will be used, if we consider the luminance efficiency, color purity and life.

4.2 CaS, SrS (IIa-VIb) Host Materials and Luminescent Centers

The CaS and SrS have stronger ionicity as compared with ZnS. Re's are, therefore, likely to be doped easily into CaS, SrS host. In addition, radii of Sr^{2+} and Ca^{2+} are close to those of Re's. Re's in CaS or SrS are divalent or trivalent ions.

CaS:Eu — Fig.5 (d)

The Eu^{2+} ion shows a red EL, peaking at 650nm, whose spectrum is broad. The Eu^{2+} ion has $(4f)^7$ electron configuration. The emission arises from the transition between the ground state $(4f)^7$ $(^8S_{7/2})$ and the excited state $(4f)^6(^7F)5d$. As the 5d electron is involved in the excited state, the EL spectrum is broad.

CaS:Ce — Fig.5 (e)

The EL spectrum shows a green EL that has a peak at 505nm and a shoulder at 520 - 550nm. The luminescent center of Ce^{3+} has an electron configuration of $(4f)^1$. The emission arises from the transition between the ground state $4f(^2F_J)(J=5/2,7/2)$ and the excited state $5d(^2D)$. As 5d electron is involved, the spectrum is broad. The spectrum has a wavy form that results from an interference effect of thin-film.

SrS:Ce — Fig.5 (f)

The EL spectrum shows a blue-green color that peaks at 475nm and has a shoulder at 520 - 550nm. The Ce^{3+} is the same as in CaS:Ce. The difference in EL color between CaS:Ce and SrS:Ce, though they have the same luminescent centers of Ce^{3+}, results from the difference of a crystalline field between SrS and CaS.

5. THIN-FILM EL DISPLAY PANELS

In Table 1, the present status of EL phosphors and EL panels are summarized. Luminescent colors, luminance levels and efficiencies are listed. As to yellow ZnS:Mn EL, the EL panels are now already commercially available. Their luminance of about 200 cd/m^2 and efficiency of 2 - 5 lm/W are large enough for practical use. Driving frequency 60 Hz, that is the same as TV field frequency, was used. As to ZnS:Tb.F. a prototype of EL panels has already developed. The luminance of 100 cd/m^2 and efficiency of 0.5 - 1 lm/W satisfy the minimum requirements. A few more steps seem to be needed for practical use. In order to realize the multi- or full-color EL. many researches are now being made. For red EL, the luminance level of 10 cd/m^2 with ZnS:Sm,Cl and CaS:Eu appear

Table 1. Electroluminescent characteristics of thin-film EL devices

	Phosphors	Color	Luminance cd/m^2, 60 Hz	Efficiency lm/W
EL panels are now commercial-ly available	ZnS:Mn	Yellow-orange	200	2-5
Prototype of EL panels have been demonstrated	ZnS:Tb,F	Green	100	0.5-1
Research on materials				
for color EL	ZnS:Sm,Cl	Reddish-orange	10	0.05
	CaS:Eu	Red	10	0.05
	ZnS:Tm,F	Blue	0.01	<0.01
	SrS:Ce,K	Blue-green	30	0.1-0.2
for white EL	ZnS:Pr,F	White	5	0.02
	SrS:Pr,K	White	30	0.1-0.2
	SrS:Ce,K,Eu	White	30	0.1-0.2

to be lower by a factor of 2 - 3. The efficiency of 0.05 lm/W is in
the same situation. As to blue EL, the luminance level of 0.01 cd/m²
and the efficiency of 0.01 lm/W with ZnS:Tm,F are too low for
practical use. As to the blue with SrS:Ce,K, the luminance level is
satisfactory, but the color purity is a problem. As to white EL.
this problem is most interesting and receiving much attention in
view of application. A flat panel of white and black are called
"paper white", that is considered to be very comfortable for human
eyes. At present a considerable amount of researches are being
made. However, most of researches are in laboratory levels.

In Fig.6, color purity called CIE chromaticity diagram is shown.
When considering EL color. this color purity is as important as
luminance. The three primitive colors of red (R), green (G) and
blue (B) of cathode ray tube (CRT) are shown, which locate at the
vertexes of triangle. The phosphors for CRT are R-Y₂O₂S:Eu (red).
G-ZnS:Cu,Al (green) and B-ZnS:Ag,Cl (blue). Until now many efforts
have been made to develop new EL materials, which have possibility
of being used in practical use by referring this CIE diagram.

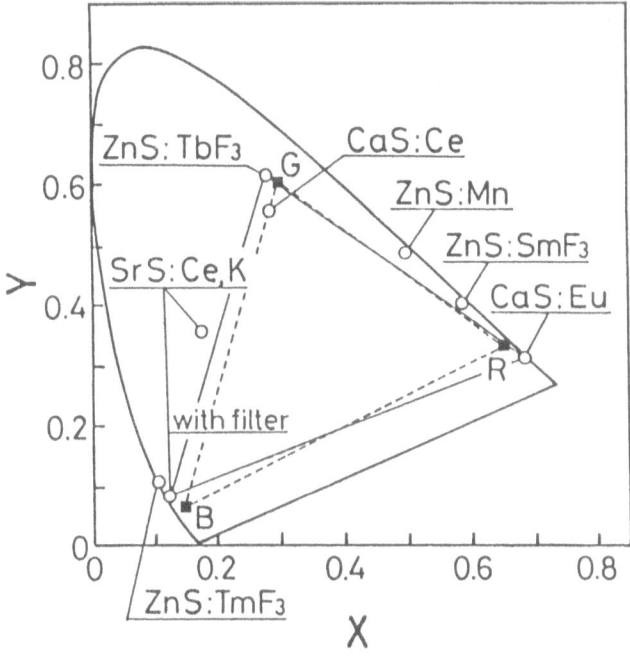

Fig.6 CIE chromaticity diagram. R,G,B are the points of CRT.
Color coordinates of EL phosphors are plotted.

6. SUMMARY

The features of thin-film electroluminescence are surveyed. As to the basic physics of thin-film EL, there are still many problems to be solved. As to application, concerning the yellow EL of ZnS:Mn, the research level has been matured up to the point that the yellow EL panel is already commercially available. However, as to other colors, the green EL of ZnS:Tb is at the point that luminance, efficiency and life have been almost solved and that it is close to panel development. The red EL of CaS:Eu, and ZnS:Sm has the problems that luminance and efficiency should be improved by a factor of 2 - 3. As to the blue EL of ZnS:Tm and SrS:Ce, new development is expected, since the luminance level and efficiency are quite low.

REFERENCES

[1] Kobayashi, H. and Tanaka. S. High Field Electroluminescence in Mn and Rare-earth Doped II-VI Compound Semiconductors, Proceedings of 1990 Seoul International Symposium on the Physics of Semiconductors and Applications. ed. Lee, C. (Korean Physical Society, 1990) pp.173 - 181.

[2] Kobayashi, H. Multicolor Thin Film Electroluminescence Devices, Proceedings of the Fourth Asian Pacific Physics Conference Vol. 2, ed. Ahn S.H., Choh S.H., Cheon Il-T, Lee C, (World Scientific Publishing Co., 1991) pp.1012 - 1017.

[3] Electroluminescence. ed. Shionoya, S. and Kobayashi, H. Springer Proceedings in Physics 38, (Springer-Verlag Berlin, Heidelberg, 1989).

[4] Handbook of Phosphors (Ohm Co., Tokyo, 1987) (in Japanese).

[5] Sze, D.M. Physics of Semiconductor Devices, 2nd edition (John Wiley & Sons, New York, 1981).

An Overview of Nonlinear Optics of Liquid Crystals

Iam-Choon Khoo

Department of Electrical and Computer Engineering, The Pennsylvania State University
University Park, PA 16802, USA

Owing to their complex physical structures, liquid crystals
exist in several mesophases and possess large optical
nonlinearities. An overview of the theories for the three
principle mechanisms for optical nonlinearities in the nematic
phase, namely, laser induced director axis reorientation, density
and temperature fluctuations, is presented. These nonlinear
effects are further discussed in the context of optical wave
mixings such as phase conjugation and real time holography, and
nonlinear optical switching processses.

Introduction to Nonlinear Optics

In the electromagnetic formulation appropriate for discussing
the nonresonant "classical" nonlinear optical effects in liquid
crystals, the distinction between linear and nonlinear optics may
be expressed in terms of the polarization $\vec{P}$ induced by an applied
optical electric field $\vec{E}$ in a material.

In linear optics, the polarization, in the simplest case of
isotropic material, is proportional to the applied optical

181

electric field E:

$$\vec{P}_L = \varepsilon_0 \chi^{(1)} \vec{E} \tag{1}$$

where $\chi^{(1)}$ is the electric susceptibility (at optical frequency).

From the constitutive relationship

$$\vec{D} = \varepsilon_0 \vec{E} + \vec{P} = \varepsilon \vec{E} \tag{2}$$

this defines the optical dielectric constant ε of the material by

$$\varepsilon = \varepsilon_0 (1 + \chi^{(1)}); \tag{3}$$

the refractive index of the material (assumed non-magnetic)
$n = \sqrt{\varepsilon/\varepsilon_0}$ is thus given by

$$n = [1 + \chi^{(1)}]^{\frac{1}{2}} = 1 + \frac{1}{2} \chi^{(1)} \tag{4}$$

In more general terms, particularly for anisotropic materials, it is necessary to express $\chi^{(1)}$ as a tensor, and equation (1) becomes, in spatial component form

$$\begin{bmatrix} P_x \\ P_y \\ P_z \end{bmatrix} = \varepsilon_0 \begin{bmatrix} \chi_{xx} & \chi_{xy} & \chi_{xz} \\ \chi_{yx} & \chi_{yy} & \chi_{yz} \\ \chi_{zx} & \chi_{zy} & \chi_{zz} \end{bmatrix} \begin{bmatrix} E_x \\ E_y \\ E_z \end{bmatrix} \tag{5}$$

From (2), one can see that, indeed, all the P's are still linearly related to the E's.

In almost all states of matter: atoms, molecules, gases, liquids, solids, plasmas, etc., the polarizations induced by the applied optical electric field, in general, are nonlinear; they assume a variety of functional forms depending on the exact

182

physical mechanisms underlying the light-matter interactions. A
example of nonlinear polarization is one where terms containing
higher powers of E appear, i.e.,

$$\vec{P}_{NL} = \varepsilon_0 \chi^{(1)} \vec{E} + \chi^{(2)} : \vec{E}\,\vec{E} + \chi^{(3)} : \vec{E}\,\vec{E}\,\vec{E} \qquad (6)$$

where $\chi^{(2)}$, $\chi^{(3)}$ are tensors of increasing orders.

Since the electric field $\vec{E}$ is a three-dimensional vector,
$\chi^{(2)}$ and $\chi^{(3)}$ are understandably very complex tensors with
numerous components.[2,3] A clearer understanding of nonlinear
optics, and considerable insights into the particular case of
nonresonant interactions involved in liquid crystals may be
gained, if we study the problem from the point of view of light
induced refractive index change. This would be the case, for
example, if the third order nonlinear susceptibility is of the
form

$$\vec{P}_{NL} = \chi_{NL} EE^* \vec{E} \qquad (7)$$

From the constitutive relationship (2), we have

$$\vec{D} = \varepsilon_0 \vec{E} + \vec{P}_{total}$$

$$= \varepsilon_0 \vec{E} + \vec{P}_L + \vec{P}_{NL} \qquad (8)$$

and using (4) in (5), one can easily show that the presence of $\vec{P}_{NL}$
is equivalent to modifying the optical dielectric constant of the
material from $\varepsilon_i = \varepsilon_0 (1 + \chi^{(1)})$ to ε_i' given by

$$\varepsilon' = \varepsilon + \chi_{NL} |E|^2 \qquad (9)$$

If $\chi_{NL}|E|^2 \ll \varepsilon$ (which is usually the case even for the most
nonlinear optical material), equation (9) yield for the refractive

index $n = \sqrt{\varepsilon'/\varepsilon_0}$ of the material

$$n = n_0 + n_2|E|^2$$
$$= n + n_2(I)I \tag{10}$$

where $n_2 = \chi_{NL}/2\varepsilon_0 n_0$, and $n_2(I)$ is related to n_2 through the definition for the optical intensity $I = \frac{\varepsilon_0 nc}{2}|E|^2$. In other words, one may say that a material is a nonlinear optical material if its refractive index (or equivalently, its optical dielectric constant) can be modified by the optical field.

The implications, effects and phenomena, and practical devices that may result from the fact that the refractive index of a material is modifiable by the impinging optical field are innumerable. Real time holography, opto-optical switching, optical bistability, phase conjugations, self-focusing and self-phase modulations, etc. are but a few examples [3].

If we refer back now to equation (4), we note that the generated polarization $\vec{P}$ has the same (temporal) frequency component as the incident optical field $\vec{E}$, i.e., the reradiated optical field therefore, is also of the same frequency. These processs are usually referred to as degenerate optical wave mixing effects.

On the other hand, and in general, $\vec{P}$ may be related to $\vec{E}$ in many other ways. For example, if we have $P = \chi^{(2)}$ EE or $P = \chi^{(3)}$ EEE, the generated fields will contain second (twice the frequency) or third harmonic (three times the frequency) components, respectively. The generation of such harmonics requires very fast, almost instantaneously electronic responses from the medium under study. In this paper, our focus is on the less-so-rapid responses of liquid crystals associated with molecular reorientations, density and temperature fluctuations, all of which create refractive index changes. It is important to note here that while such degenerate wave mixing effects do not allow (temporal) frequency change, they nevertheless allow for spatial

184

frequencies (propagation directions $\vec{k}$ of the optical field)
changes. For example, consider figure 1, where two coherent

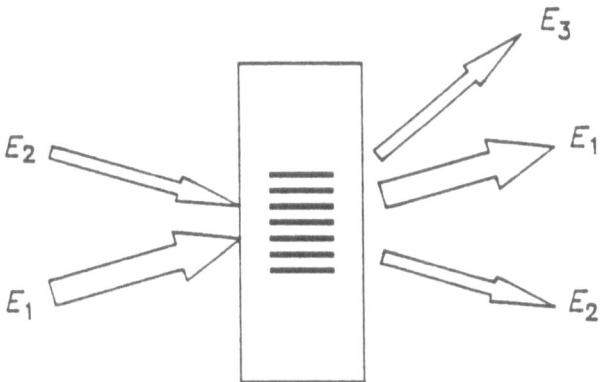

Figure 1

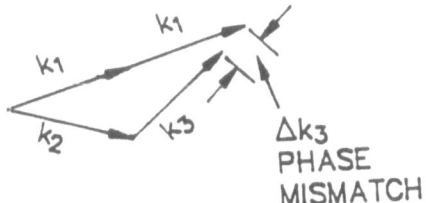

optical field, E_1, E_2 are incident on a nonlinear material. Owing
to a nonlinear response $\vec{P}$ of the form given in (4), a polarization
component P_3 may be generated in the form:

$$P_3 = \chi^{(3)} \, E_1 E_2^* E_1 \tag{11}$$

where for simplicity of discussion, we have assumed that all
fields are polarized in the same direction. This polarization
will radiate an optical field E_3 in the (nearly) phase matched

i.e., momentum matched direction $\vec{k}_4$ given by

$$\vec{k}_4 = 2\vec{k}_1 - \vec{k}_2 \tag{12}$$

185

Figure (2) shows another possibility, where the nonlinear medium

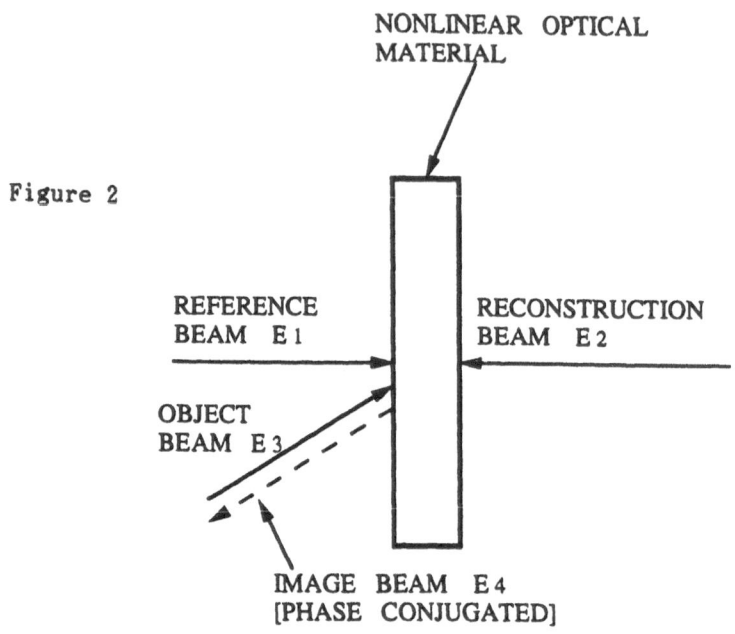

NONLINEAR OPTICAL
MATERIAL

Figure 2

REFERENCE
BEAM E_1

RECONSTRUCTION
BEAM E_2

OBJECT
BEAM E_3

IMAGE BEAM E_4
[PHASE CONJUGATED]

is irradiated with two counterpropagating beams E_1 and E_2 and E_3

($P = \chi^{(3)} E_1 E_2 E_3^*$. This produces a nonlinear polarization (besides

many others) that scatters a wave $\vec{E_4}$ in the (phase-matched)

direction $\vec{k_1} - \vec{k_3} + \vec{k_2} = \vec{k_4} = -\vec{k_3}$ (since $\vec{k_1} = -\vec{k_2}$).

It is important to note here that: the generated field E_4 is

counterpropagating to E_3 and that E_4 is a conjugate of E_3. If we

treat E_1 as the <u>reference beam</u>, E_3 the input object (i.e., E_1 and

E_3 interfere to form a "real time hologram"), E_2 the <u>reconstruction</u>

<u>beam</u>, then the <u>generated wave</u> E_4 is what one may call the

<u>reconstructed image beam</u>. Since $E_4 \sim E_3^*$, the image beam E_4 has

186

been termed the phase-conjugate of the object beam E_3, and possesses several interesting useful characteristics such as aberration correction and wavefront reversal properties, etc. [3].

There are many other examples, but the preceeding two examples typify these so-called degenerate multiwave mixing effects that could be mediated by an optically induced refractive index change. In equation (8), the refractive index change is associated with the interference term $E_1 E_2^*$ of the two incident fields E_1 and E_2, i.e., a <u>spatially modulated refractive index change</u>, sometimes termed refractive index grating.

Laser Induced Nonlinear Optical Effects in Liquid Crystals

Liquid crystals have proven themselves as one of the best low-cost, low-power-consumption [linear] electro-optic display and image processing materials. This is due to their large optical refractive index anisotropy and other unique physical properties.[5] They are composed of uniaxially birefriengent molecules (with typical chemical structure as depicted in figure 3).

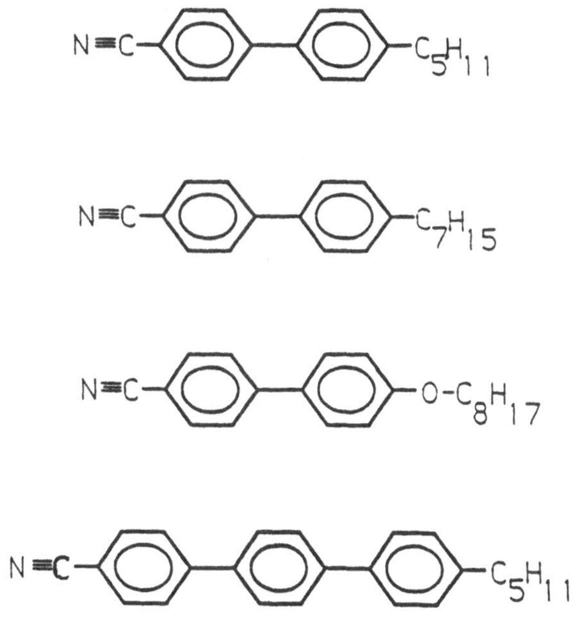

Figure 3

187

Typically, their refractive indices $n_\parallel \sim 1.7$ and $n_\perp \sim 1.5$ (where $n_\parallel$ and $n_\perp$ are the index corresponding to optical field polarized parallel and perpendicular to the crystal axis, respectively).

As a function of temperature below the nematic isotropic transition temperature T_c, liquid crystals can exist in several distinct phases, which are characterized by the degree of ordering among the liquid crystal molecules [5]. Broadly speaking, these phases are smectic, cholesteric, and nematic, with smectic being the most correlated phase before the crystalline state. In this paper, we shall concentrate on the nematic phase [4], where most nonlinear optical studies are conducted.

In the nematic phase, the molecules are directionally correlated (although they are positionally uncorrelated), i.e., when placed between cell plates with proper surface alignment forces, the molecules will align themselves in some preferential direction, which is denoted by the so-called director axis of the nematic crystal. Because of their susceptibility to molecular realignment by applied field (dc or low frequency ac) and therefore large refractive index change (In optics, a refractive index change of 0.2 is ENORMOUS!), various optical switching, imaging and modulation devices have been developed. Research and development in this area continue to be very active.

Since 1979, several groups have demonstrated that it is possible to induce liquid crystal director axis reorientation with moderate power laser [4,6]. Typically, laser intensities on the order of a few tens of Watts/cm^2 could induce a molecular reorientation of about 10^{-2} radian, and a consequent "change" in the optical refractive index of about 10^{-3}. This gives a so-called nonlinear coefficient $n_2(I)$ of around 10^{-5} cm^2/Watt, which puts (nematic) liquid crystals in the class "most nonlinear optical materials." However, under moderate laser power, the response times of the director axis reorientation is quite long, approaching seconds in some instances, and is therefore not useful for practical applications. Nevertheless, because of the slow response, and the extraordinary large nonlinearity, the system enables one to "visually" follow the dynamics of several new

188

aspects of some nonlinear optical phenomena such as optical
bistability and switching, self-focusing, etc. and gain further
insights into these processes [7].

By increasing the incident laser power (or intensity), the
process of molecular axis reorientation may be sped up (at the
cost of obtaining lower nonlinear coefficient n_2); with megawatt/
cm^2 laser pulses, several groups [8,9], were able to demonstrate
that the process could occur in the nanosecond time scale where n_2
now is on the order of 10^{-9} cm^2/Watt. In this time scales and,
with the use of intense laser pulses, however, several other
physical mechanisms are also involved besides molecular
reorientation. Principally, laser induced heating, density
changes, and flows begin to manifest; all these processes will
drastically alter the refractive index of the liquid crystal, and
therefore, also contribute to its optical nonlinearity.

Detailed theoretical and experimental studies of these laser
induced reorientation, temperature and density fluctuations, and
flow effects have been performed recently. The readers are
referred to reference 10 for more details. Figure 4 summarizes

Figure 4

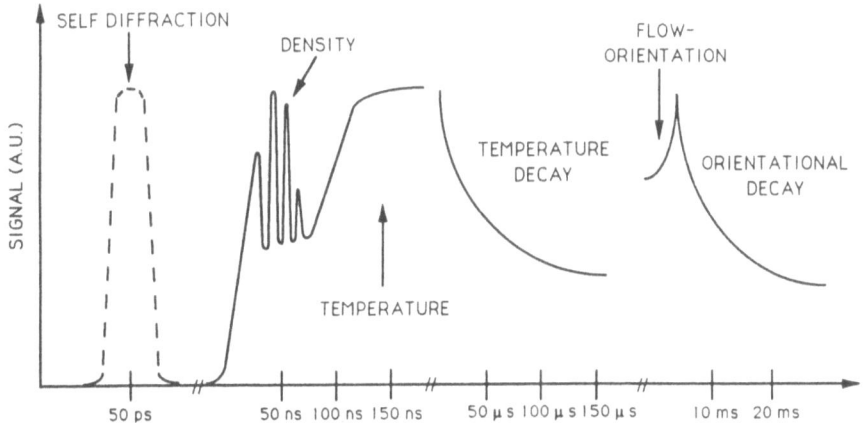

189

most of the principal nonlinear effects when a nematic (and to a large extent, isotropic and smectic phases as well) liquid crystal interacts with a laser; the dynamics of these nonlinear processes, i.e., their rise and decay, are also depicted. These results are obtained by the so-called dynamical grating technique, where two coherent short laser pulses (∼ picosecond in duration) creates an intensity grating in a nematic liquid crystal sample (c.f., figure 5). Via the intensity dependent refractive index, an index grating

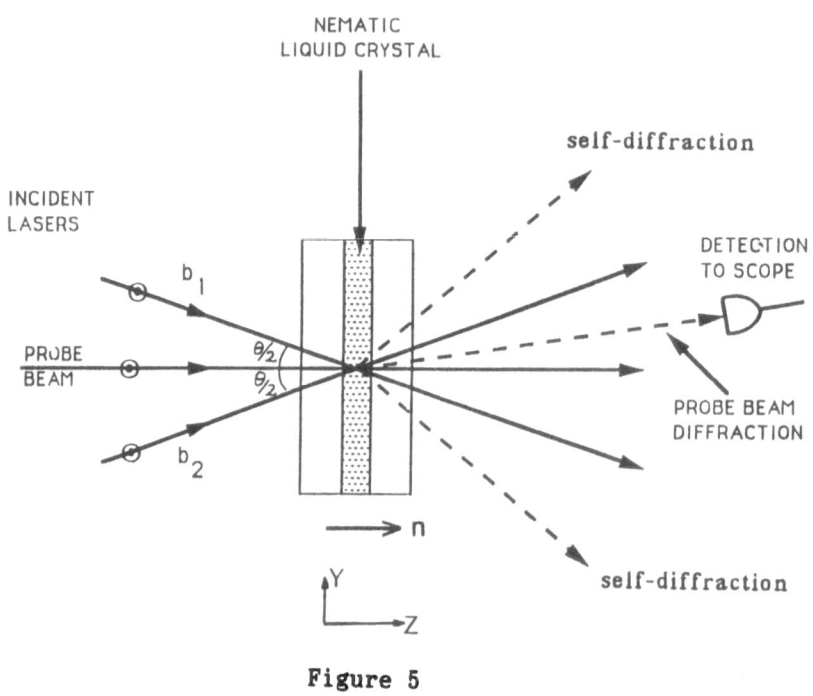

Figure 5

is induced, which will diffract the incident excitation pulses as well as a probe laser. If the underlying nonlinear mechanism is fast enough, an index grating component can be created within the picosecond time scale, giving rise to self- diffraction of the picosecond laser pulses. On the other hand, effects that are characterized by slower responses (nanosecond or longer) will be manifested in the diffraction of the cw probe beam.

In the picosecond time scales, the nonlinear optical responses are dominated by mechanisms of largely electronic in origins (e.g., two-photon absorption) or <u>individual</u> (as opposed to collective) molecular axis reorientation (or reordering). These optical nonlinearities are generally small, and comparable to typical organic liquids. They are therefore not of practical usefulness, although they provide valuable information on liquid crystal physics [11].

In the nanoseconds-and-longer time scales, large nonlinear optical effects are observed. These include molecular reorientation (θ), temperature rise (T), density change (ρ) and flows (ν). Using the notation of reference [10], the magnitude and time dependence of the refractive index Δn associated with these dynamical physical quantities are summarized here.

Reorientation nonlinearity

The reorientational process is characterized by a time constant Γ_θ given by

$$\Gamma_\theta = \frac{K_q{}^2 - (\Delta\varepsilon \ E \ op^2 \ \cos^2\beta)/4\pi}{\gamma} \tag{13}$$

The induced refractive index change Δn is given by:

(i) <u>Steady state</u> ($t \gg \Gamma_\theta{}^{-1}$; $\Gamma_\theta \sim 1$ sec)

$$\Delta n = \frac{\Delta\varepsilon \ E_{op}{}^2 \ \sin 2\beta}{8\pi\Gamma_\theta} \left[\frac{\partial n}{\partial \theta}\right]$$

$$= n_2^{ss}(\theta) I_{op} \tag{14}$$

Typically [4] observed steady state nonlinear coefficient $n_2^{ss}(\vartheta)$ associated with reorientation effect is on the order of 10^{-4} cm^2/Watt,

(ii) transient ($t \ll \Gamma_\theta{}^{-1}$; $\Gamma_\theta{}^{-1} \approx 1$ sec = τ_θ,

191

$$\Delta n(\theta) = \frac{\Delta \varepsilon \ E_{op}^2}{8\pi \Gamma_\theta} \ \sin 2\beta (\frac{\partial n}{\partial \theta}) t \ \Gamma_\theta$$

$$= n_2^{ss}(\theta) \ (t\Gamma_\theta) \ E_{op}^2$$

$$= n_2^t \ I_{op} \tag{15}$$

The "transient" nonlinear coefficient $n_2^t(\theta)$ is related to the steady state value by

$$n_2^t(\theta) = n_2^{ss}(\theta)(\frac{t}{\tau_\theta}) \tag{16}$$

For a microsecond laser pulse, therefore, we have $n_2^t(\theta) = 10^{-6}$ $n_s^{ss}(\theta) \sim 10^{-10} \ cm^2/Watt$

Thermal Nonlinearity

(i) Steady state ($t \ll \Gamma^{-1}$; $\Gamma^{-1} \approx ms$)

Typically observed thermal effect is on the order of the orientational effect or larger, depending on the absorption constant (α) and the thermal index gradient ($\frac{dn}{dT}$) and is on the order of $10^{-3} \ cm^2/Watt$ or larger, in the steady state. In this case,

$$\Delta n(T) = \frac{\alpha cn \ E_{op}^2}{4\pi \ \rho_o c_\rho \Gamma_R} \ (\frac{\partial n}{\partial T}) = n_2^{ss}(T) \ I_{op} \tag{17}$$

(ii) Transient ($t \ll \Gamma_R^{-1}$; $\Gamma_R^{-1} \approx ms = \tau_R$)

$$\Delta n(\theta) = \frac{\alpha cn \ E_{op}^2}{4\pi \ \rho_o c_\rho \Gamma_R} \ (\frac{\partial n}{\partial T}) = t\Gamma_R \tag{18}$$

$$= n_2^t(T) \ I_{op}$$

192

where the "transient" nonlinear coefficient $n_2{}^t(T)$ is related to the steady state value by

$$n_2{}^t(T) \simeq n_2^{ss}(T)(\frac{t}{\tau_R}) \qquad (19)$$

For a microsecond laser, therefore, $n_2(t)$ $(T) \approx 10^{-6}$ cm^2/Watt, which ranks amongst the largest of known nonlinear materials currently being investigated for optical switching applications. It is important to note here that the thermal index coefficient $(\frac{\partial n}{\partial T})$ is also highly sensitively dependent on the temperature. At temperatures near T_c, $(\frac{\partial n}{\partial T})$ can be <u>two orders of magnitude</u> larger than its value at temperatures far from T_c.

<u>Density Effect</u>

There are <u>two</u> time-scales involved in the relaxation of the density component, the Brillouin lifetime $\tau_B = \Gamma_B{}^{-1}$ and the thermal relaxation time $\tau_R{}^{-1}$. Studies[6,8] have shown that in the <u>transient case</u>, the density effect is comparable in magnitude to the purely thermal effect (mentioned above) at temperatures far from T_c. However, at temperatures nearer to T_c, the density effect is much smaller in magnitude. Hence they may not be very useful for practical nonlinear switching in the microsecond time scale. However, its relative shorter response ($\approx$100ns) and the fact that there are special chiral nematic liquid crystals in the isotropic phase (where large interaction length is possible) with much larger density coefficients indicate that this effect is a potentially ideal candidate for nanosecond laser pulse switching application.

<u>Conclusion</u>

We have presented an overview of the mechanisms for optical nonlinearities, their magnitudes and dynamical characteristics of nematic liquid crystals. We have also alluded briefly to some nonlinear optical processes such as degenerate optical wave mixings and optical phase conjunctions. Space does not allow us

193

to elaborate further on their details applications and some of the detailed novel physics. However, our discussion of the fundamental role played by the laser induced refractive index change in liquid crystals in these nonlinear optical processes should serve as a useful introduction to the myriad of nonlinear optical studies performed on liquid crystals. Some of these have appeared in the two reviews [4,6]. More recent work may be found in, for example, references [13-14].

References

1.   Yariv, A. [1991], "Quantum Electronics," (John Wiley & Sons, NY).
2.   Bloembergen, N. [1965], "Nonlinear Optics," (Benjamin, NY).
3.   Reintjes, John F. [1984], "Nonlinear Optical Parametric Processes in Liquid and Gases," (Academic Press, NY).
4.   Khoo, I. C. [1988], "Nonlinear Optics of Liquid Crystals," in Progress of Optics, ed. E. Wolf [North Holland, Amsterdam].
5.   de Gennes, P. G. [1974], "Physics of Liquid Crystals," (Oxford University Press, London).
6.   Tabiryan, N. V., Sukhov, A. V., and Zeldovich, B. Ya., [1986] Mol. Cryst. Liq. Cryst. 136, p. 1.
7.   Khoo, I. C. and Shen, Y. R., [1985], Optical Engineering, Vol. 24, p. 579.
8.   Hsiung, H., Shi, L. P., and Shen, Y. R. [1984], Phys. Rev. A30, p. 1453.
9.   Khoo, I. C.,
10. Khoo, I. C.,
11. Prost, J., and Lalanne, J. R. [1973], Phys. Rev. A8, 2090.

Material Characterisation for Infrared Hybrid Arrays*

K. Shivanandan

Radio, Infrared and Optical Branch, Center for Advanced Space Sensing
Naval Research Laboratory, Washington D.C., USA

Hybrid focal plane arrays using PV HgCdTe have been developed in 128x128 and 256x256 formats mated through indium interconnects to Si CMOS readouts. The band of response for these arrays range from 0.8 - 2.5 micróns with pixel sizes of 40microns x 40 microns for the 256x256 and 60x60 for the 128x128 arrays. Mean dark currents lower than 5e⁻/s at 77K and 0.2⁻/s at 60K have been measured. Detector noise is typically of the order of 2 E −16 A/Hz at 1Hz under bias of −500V. The arrays exhibit quantum efficiencies of 50%.

At longer wavelengths of between 8 - 13.5 microns MBE growth and fabrication of insitu doped p - on - n HgCdTe/GaAs focal plane arrays have been developed with a 64x64 format structure. Electronically scanned buffered direct injection readouts (ESBDI) have been interfaced with the arrays for low input impedance and low noise performance.

Utilising proven InSb detector technology and foundry silicon C-Mos processes high performance hybrid focal plane arrays (128x128, 256x256) with integrated signal processing and support electronics have been developed. The detector elements achieve a high quantum efficiency response from less than 1 micron to greater than 5 microns and can operate at a maximum frame rate of 1000 Hz.

Extrinsically doped silicon (Si:Ga, Si:As , Si:Sb) hybrid arrays have been fabricated with buried electrical contacts and multiplexed with readouts to operate at liquid Helium temperatures. Several types of digital shift registers have been developed for scanning the arrays. Shielded switches to prevent reset feed through and innovative fabrication methods to eliminate anomalies exhibited by MOSFETS at low cryogenic temperatures have been used.

---

* prepared from lecture notes supplied by the author.

The individual merits of these arrays will be presented. The problems of material characterisation for use at cryogenic temperatures and at different backgrounds will be emphasised. Some applications of the arrays will be highlighted.

Fig. 1 shows the range of materials used as infra red detectors in the region 1 - 14 microns. Over the last two decades new semiconductor materials and new techniques of preparation have provided a large number of innovative devices for IR detection. There are four major materials which are used in the preparation of Infrared detectors

* PtSi
* InSb
* HgCdTe
* Si:X (X = Ga , As, Sb)

each of these materials have characteristic features.

PLATINUM SILICIDE

* Fabricated by evaporation of Platinum onto Silicon by an annealing heat treatment

* Photons are absorbed in the metal, resulting in excitation of electrons above the Fermi level and creation of vacancies below the Fermi level

* Semiconductor Valence Electrons tunnel over the potential barrier between the metal and semiconductor.

* Array is back illuminated through the silicon

* Wavelength response 1 - 5 microns

* Quantum efficiency : 1%

INDIUM ANTIMONIDE

* Three - five semiconductor
* Low melting point (523 C) Makes it easy to handle
* Small energy gap
* High electron mobility because of reduced mass

* p-type InSb : Indium (In) and Antimony (Sb) mixed in stoichio-metric proportions and crystals grown from melt by pulling.
* Conduction mechanism: holes
* N-type InSb: Conduction mechanism- free electrons

There are three types of InSb detectors:

1. Photovoltaic : Diffusing a p type dopant into an n-type Single crystal. Contacts are made on both sides and p-layer is illuminated.

2. Photoelectromagnetic: Photon generated carriers generated in a fixed E-M field.

3. Photoconductive: Carrier generation in the intrinsic bands using high purity materials.

HYBRID ARRAYS

Advantages

*Independent optimisation of the detector material and CCD mutilplexer

* Single CCD/output circuit for many different detector materials.

* Increased signal processing area on focal plane

Disadvantages

* Many mechanical and electrical interfaces

* Thermal expansion mismatch limits size of array

* Input circuit into CCD multiplexer must be efficient and noiseless.

The infrared detectors and the hybrid array combination of infra-red detectors find applications in Remote sensing, Astronomy, Medical Diagnosis, and in military.

Focal Plane array Readouts have been used in several applications.

* Line address: One pixel from each column is shifted to the output multiplexer (MUX) and the entire Row readout before a second pixel is shifted to the output MUX.

* Frame Field transfer: Two sections with separate clocks. One to detect the image during integration and the other to store the image during the integration of next image field.

* Interline Transfer: Vertical readout image in columns and the charge transferred to an output multiplexer.

* photosensing is done with one material and the CCD readout is done with another chip, normally silicon. Indium used to bond detector array to CCD multiplexer.

Material	Operating temperature (K)	Spectral response (microns)
InAsSb	100	2 – 8
InSb	77	1 – 5.5
HgCdTe	40 – 77	2 – 14
Si : In	42	2 – 8
Si : Ga	14	2 – 16
Pyroelectric	300	1 – 30

FIG. 1. INFRARED DETECTOR MATERIALS

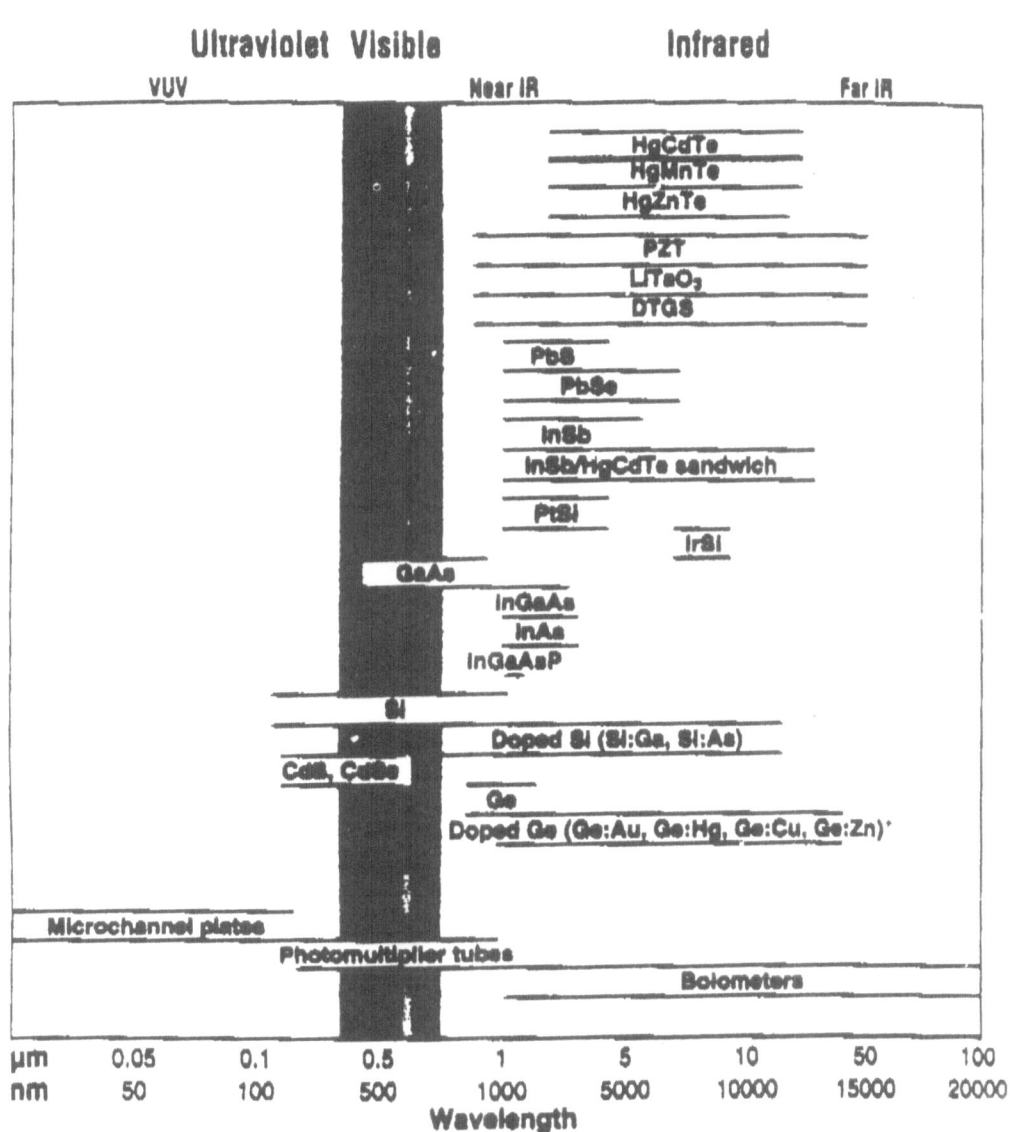

QUANTUM EFFECIENCY OF PtSI DETECTOR

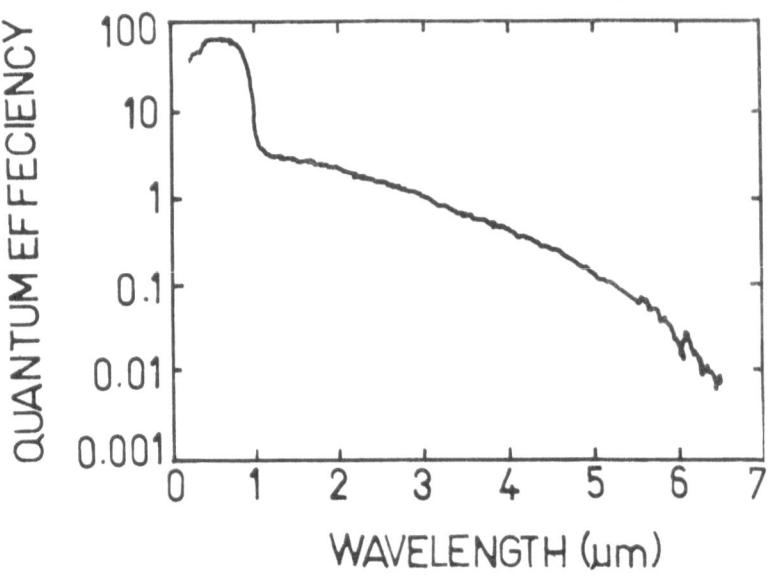

SCHOTTKY BARRIER CELL STRUCTURE

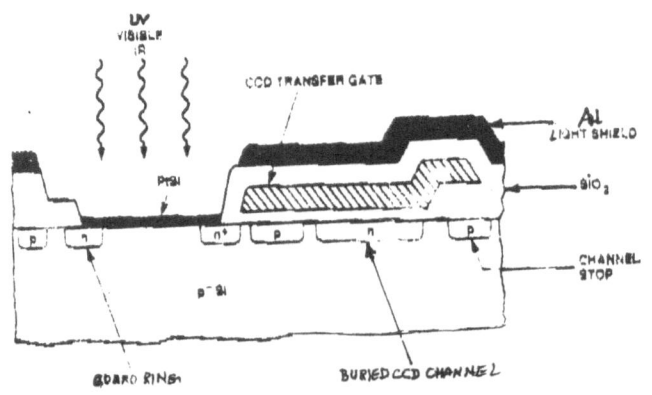

Manufacturer	Array Size
MIT Lincoln Laboratory	244x160
Aeronutronic Ford	256x256
Thompson CSF	256x256
Valvo/Philips	128x512
David Sarnoff Research Center	244x320
Fairchild	488x512
Mitsubishi	512x512
Reticon	512x512
Hughes	488x640
Kodak	488x640

SPECTRAL RESPONSE OF INSB PV/PC DETECTORS

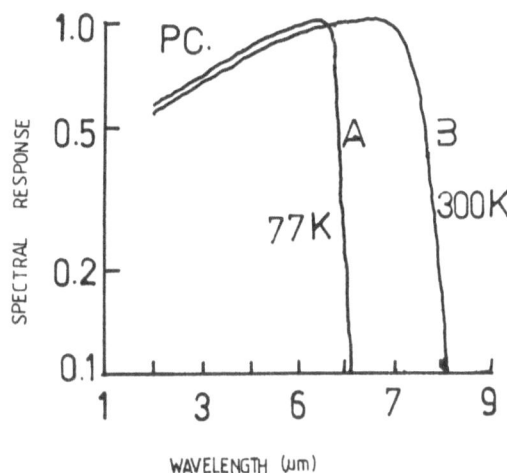

SPECTRAL RESPONSE OF TYPICAL InSb
PHOTOCONDUCTIVE DETECTORS A : 77 K,B : 300 K.

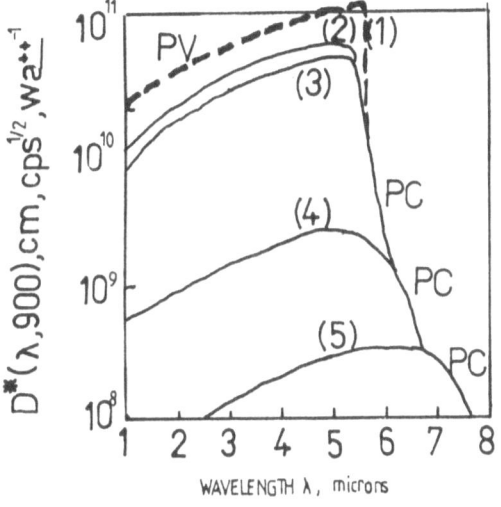

(1). IDEAL PHOTOVOLTAIC
DETECTOR (= 5.6u)
(180 FIELD OF VIEW)

(2). PHOTOVOLTAIC (77 K, 180 FIELD OF VIEW,
BEST AVAILABLE)

(3). PHOTOCONDUCTIVE (77 K,180 FIELD OF VIEW,
BEST AVAILABLE)
(4). PHOTOCONDUCTIVE (195 K,180 FIELD OF VIEW)

(5). PHOTOELECTROMAGNETIC (PEM)
(295 K, 360 FIELD OF VIEW)

SPECIAL DETECTIVITY OF InSb DETECTORS

201

Materials for Optical Fibre

Y.T. Chen, W. Chen, Harith Ahmad, K.H. Kwek and N.H. Tioh

Institute of Advanced Studies, University of Malaya, Kuala Lumpur, Malaysia

I. Introduction

Optical fibre is a circular dielectric wave guide. It has a central core which has a slightly higher refractive index than the surrounding material which is known as the cladding. With this kind of structure, light can be transmitted through it by the principle of total internal reflection.

Optical fibres have found wide applications not only in data transmission, e.g., communications, computer link, etc., but also in various sensor systems. As a matter fact, although the concept and the principles of optical fibre as a wave guide of light have been very well developed in the 1960's, the real applicable fibre only appeared in the middle of the 1970's. The main obstacle in making optical fibre is in its loss. In 1970, Kapron et al succeeded in using silica glass in making low-loss fibre: that started a new era. The major considerations in choosing the material for optical fibres are:

(1) Transparency
Transparency is usually a function of the wavelength of light. The transparent regions for SiO_2 is in the range of $0.3 - 3$ μm; for fluoride glass it is in the range of $0.5 - 5$ μm; and for KCl crystal, it is in the range of $0.4 - 18$ μm. Fig.1 gives the typical loss measured for a low-loss optical fibre (see J.M. Senior et al, 1989).

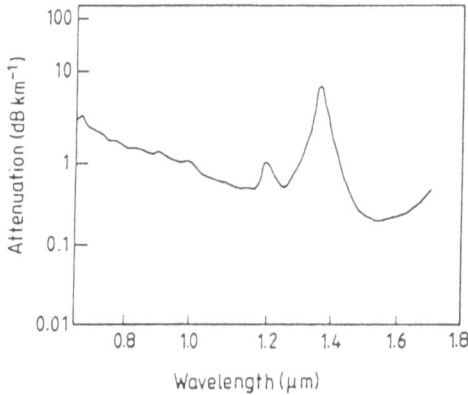

Fig.1 A typical attenuation curve of a low-loss optical fibre

The transparency is determined by many factors: molecular absorption, material scattering (Rayleigh scattering and non-linear scattering), macro bending, etc. The first two factors are determined by the material; therefore, the preparation and purification of the optical fibre material are very important.

(2) Material contribution to dispersion
The bandwidth of a signal is one of the most important factors to be considered for the quality of the optical fibre. The dispersion of light within a fibre will cause distortion of the wave form at the end of fibre, thus limiting the bandwidth. This dispersion is caused by the difference in the velocities of the different modes of light in a medium that is not a vacuum. Therefore, in choosing the material, dispersion should be as small as possible. The material dispersion is given by the second derivative of the index $n(\lambda)$ to the wavelength λ, as

$$D = \frac{\lambda}{c} \frac{d^2 \, n(\lambda)}{d \, \lambda^2}$$

where c is the speed of light.
However, to further reduce the dispersion, a single–mode fibre instead of multi–mode fibre can be used; moreover a grade–index fibre can provide a wider bandwidth than that of a step–index fibre.

(3) The ease in the control of refractive index
The precise control of the refractive indices of the core and the cladding is very important in the manufacture of high quality fibres. As we mentioned before, there are two kinds of index profile for the core: grade–index and step–index as shown in Fig.2. This is usually controlled by the impurity used for doping. The choice of material should, first of all, be based on the ease in obtaining the right profile in the radial direction; the index fluctuation along the axial direction has to be controlled also.

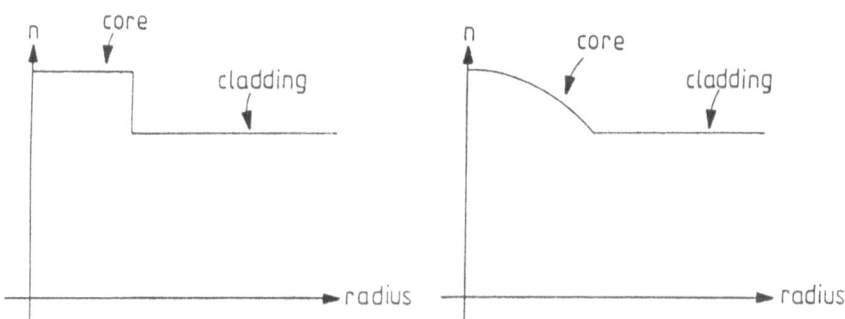

Fig.2(a) Step–index profile   Fig.2(b) Graded–index profile

(4) Mechanical and chemical properties
These include the strength, the bending property, the resistance to corrosion from water and atmosphere, the thermal expansion coefficient, etc.. The particulars of these properties should be considered very carefully. For example, if the difference in the thermal expansion coefficients between the core and the cladding is very large, serious noise caused by thermal stress will be produced when the cable is subject to fluctuating temperatures. For some materials, the thermal coefficients will increase with the quantity of dopants, for example, P_2O_5, GeO_2, B_2O_3, etc.; but for some other materials, the thermal coefficients will decrease with the quantity of dopants, a typical example is TiO_2.

The above mentioned are the four points for the consideration of the materials for making optical fibres. In the followings, a few typical materials will be discussed.

II. Pure silica system

Optical fibre made from the pure silica system is composed of pure silica and doped silica. The dopants are usually B_2O_3, P_2O_5, GeO_2 or $GeO_2 - B_2O_3$. The function of the dopants is to change the refractive—index. From the viewpoint of transparency, at a certain range of wave lengths, pure silica gives the best optical transparency while other oxides give higher loss owing to molecular absorption. Fig.3 gives the experimental absorption spectra for GeO_2, P_2O_5 and B_2O_3 — doped silica glass in the wavelength range of $0.2 - 0.5$ μm as compared to that of pure silica (see Izawa & Sudo, 1987).

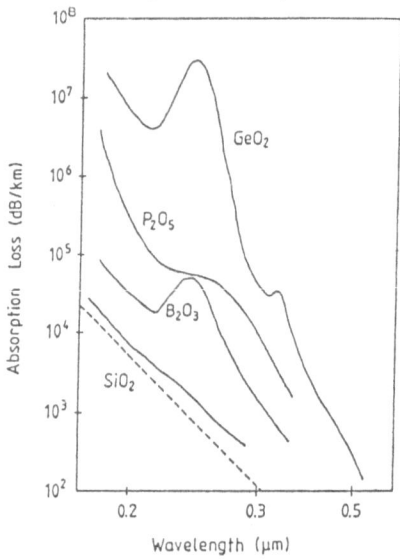

Fig.3 Absorption spectrum of GeO₂, P₂O₅ and B₂O₃

The dopants may increase or decrease the refractive—index. In the silica system, GeO_2 will increase the index whereas B_2O_3, P_2O_5 and Fluorine will decrease it. Therefore, optical fibre can be made from a core composed of $GeO_2 - SiO_2$ and a cladding of pure SiO_2 or it can be made from a core composed of pure SiO_2 and a cladding of $B_2O_3 - SiO_2$, $P_2O_5 - SiO_2$ or Silica—Fluorine compounds system. The quantity of dopants needed in the manufacture of optical fibre is usually $10 - 20$ mol% with the diameter of the core much smaller than that of the cladding. Therefore, the key issue is how to obtain pure silica.

The natural source of silica is sand or quartz. For the manufacture of optical fibre from glass however, there are special requirements. Most transition metal oxides in glass introduce absorption in the visible and the near—infrared regions. The impurity elements in glass may produce two undesirable effects: one is the absorption of light by the transition elements; the other is the scattering of light by the colloidal particles of some elements. These imply that silica of very high purity is required in the manufacture of optical fibres. Therefore, the sand must have very high silica content before it can be considered to be used as the raw material in the manufacture of optical fibre.

Most sands all around the world have brownish or reddish colour owing to the presence

204

of large concentrations of transition metals. Some metals such as copper, gold and their colloids impart a red colour while silver gives a yellow tint. CdS and CdSe give dark red or brown colour. The presence of these transition elements and colloid forming elements in the raw material are not desirable. We have found that in some area in Malaysia, the percentage of pure silica in some samples of natural silica is more than 99.4%; the main impurities are aluminium oxide (Al_2O_3) which may be less than 0.05% and iron oxide (Fe_2O_3) which may be less than 0.01%. The silica from these mines, therefore will be suitable for the manufacture of optical fibre.

The purity requirement of silica is extremely high for the manufacture of optical fibres. For examples, the amount of impurities that will give 1 dB/km loss by absorption for light of wavelength in the range of $1.0 - 1.8$ μm are: 2 ppb (two parts per billion) for Cr; 10 ppb for Mn; 20 ppb for Fe; 0.2 ppb for Co; 2 ppb for Ni and Cu. Raw materials which contain impurities of the above concentrations cannot be used directly. A special method to refine the SiO_2 is necessary.

Among the various methods, the "distillation" method is by far the most convenient method. The chemical distillation method involves the conversion of raw sand into silicon tetrachloride, $SiCl_4$, which can be easily vapourized and subsequently oxidized to pure silica.

$SiCl_4$ can be synthesized by reacting an intimate mixture of carbon and silica in a stream of chlorine, the chemical reaction is

$$SiO_2 + 2C + 2Cl_2 = SiCl_4 + 2CO$$

$SiCl_4$ is a liquid at room temperature and its boiling point is 59.6°C. The vapour of $SiCl_4$ and O_2 may react in the following way

$$SiCl_4 + O_2 = SiO_2 + 2Cl_2$$

The silica SiO_2 thus obtained is pure silica.

As a matter of fact, the pure GeO_2 is also obtained by the same method. To make a doped core, the chemical reactions of $SiCl_4 \longrightarrow SiO_2$ and $GeCl_4 \longrightarrow GeO_2$ occur at the same point. This can be done in the commercially available methods: Modified Chemical Vapour Deposition (MCVD), Outside Vapour Deposition (OVD) or Vapour–phase Axial Deposition (VAD), etc..

III. Multi–component glass

Another kind of optical fibre made from the silica system is the multi–component glass fibre which contains SiO_2, GeO_2, Na_2O, CaO, Li_2O, MgO, etc.. The advantages of using multi–component glass are low cost and lower melting temperature (750°). It can be used as a high numerical aperture fibre which has found applications in short distance image transmission. However, in view of the above four points in section 1, the transparency of multi–component glass fibre is not as good as that of pure silica. The best level is several dB/km and it is very difficult to reduce it to below 1 dB/km owing to the difficulties encountered in the purification of the raw materials. Another disadvantage lies in the weaker mechanical strength than that of the pure silica system.

IV. Plastic

In plastic fibre, the core may be made from poly–methacrylate, while the cladding may be made from vulcanized silicone. The advantages of plastic fibres are low cost, flexibility, ease to handle, etc.. The main problem here is the transparency. Although the purification process of plastic is not difficult, the strong absorption and scattering of plastic limit its transparency. The possible optical loss may be as high as 10 dB/km or more.

V. Crystalline materials

All the above mentioned materials are in the amorphous state. Crystalline materials may give very low absorption loss. Tangonan et al (1973) have tried to grow single crystals from alkali halides such as the single–crystal KCl optical fibre which is used as the CO_2 laser wave guide. The single crystal fibre they grew was about $75 - 100$ μm in diameter and 10 cm in length. The crystal was grown from an aqueous solution of the salt. The measured loss of the fibre indicated that the loss is almost entirely due to the scattering as a result of the imperfections in the crystal.

Although it is very encouraging to have crystalline fibre to further reduce the optical loss, the high cost, the difficulty in making single–mode fibre (diameter of the crystal should be $1 - 2$ μm), the difficulty to grow long crystals, etc. hinder the development at this moment.

REFERENCES

Geisler, J. , Beaven, G. and Boutouche, J. P., "Optical Fibres", Pergamon Press, (1986).

Izawa, T. and Sudo, S., "Optical Fibres: Materials and Fabrication", KTK Scientific Publishers/Tokyo, (1987).

Kapron, F. P., Keck, D. B. and Maurer, R. D., "Radiation losses in glass optical wave guides", Tech. Digest Conf. on Trunk Telecom. by Guided Waves, p148, (1970).

Senior, J. M. and Cusworth, S. D., "Spectral transmission effects concerning wavelength multiplexed optical fibre sensor systems", Optics & Laser Technology, Vol.21, No.2, p87, (1989).

Tangonan, G., Pastor, A. G. and Pastor, R. C., "Single crystal KCl fibres for 10.6 μm integrated optics", Appl. Opt. Vol.12, p1110, (1973).

Materials for Laser Spectroscopy: Rare earth doped alkaline earth fluorides

Y.L. Khong and S. Radhakrishna
Institute for Advanced Studies, University of Malaya
Kuala Lumpur, Malaysia

Alkaline earth fluorides in particular CaF_2, SrF_2 and BaF_2 are excellent hosts for doping with rare earth impurities for the purpose of optical studies. Aspects of the crystal growth and preparation will be discussed. Over the last decade or so, a wealth of laser spectroscopic information has been accumulated. Arguably, the most useful technique from the point of view of isolating and understanding the defect centres present in these crystal systems is laser selective excitation (LSE). The use of this technique for the study of some of the defect centres present is discussed. Over the last decade, high resolution lasers and pulsed sources have been increasingly used to probe the fine structures of transitions and dynamical processes of excitation belonging to the various centres. The holeburning effect studied by the high resolution lasers may provide possibility of some novel optical application.

INTRODUCTION

The spectroscopic studies of lanthanides or rare earths (RE) in crystalline environments have yielded several technological applications, most notable of which is the Nd:YAG laser (fig.1). Perhaps lesser known is the fact that alkaline earth fluoride hosts like CaF_2 doped with the RE's Nd^{3+}, Pr^{3+}, Ho^{3+}, Er^{3+} and Tm^{3+} among a few others, have also been successfully made to lase.

207

Apart from being potential laser materials, some RE doped crystal systems are used as phosphors while others present possible prospects as mass storage optical memory devices or optical masks in the frequency domain.

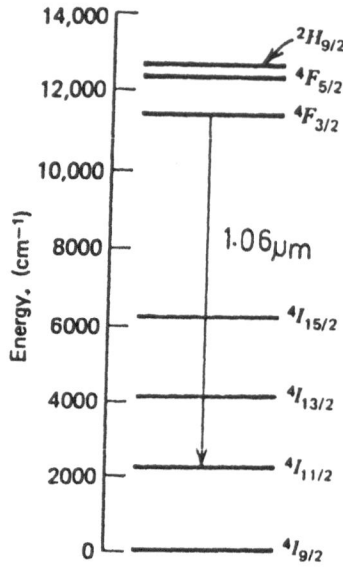

Figure 1.
The lasing transition
of the Nd:YAG laser

The purpose of this review is to discuss aspects related specifically to the spectroscopic studies of the rare earth doped alkaline earth fluorides (CaF_2, SrF_2, BaF_2) which are ideal hosts for doping with RE ions as they are mechanically and thermally robust. Further, they are also transparent over a wide spectral range from the far infrared to the UV region making them excellent hosts for optical studies.

The structure of the alkaline earth fluoride (AEF) lattice is face centred cubic[1] with the fluorine atoms forming a cubic lattice cage and the alkaline earth ions placed at the centre of every alternate cage (Fig.2). The RE's enter the alkaline earth fluoride lattice usually as tripositive ions replacing the divalent alkaline earth ions. The excess positive charge as a result of such substitutions may be compensated by negatively charged fluorines or hydride ions in the interstitials of the cubic lattice. From such substitutions arise the RE defect centres which have been the focus of most of the optical work on these crystal systems.

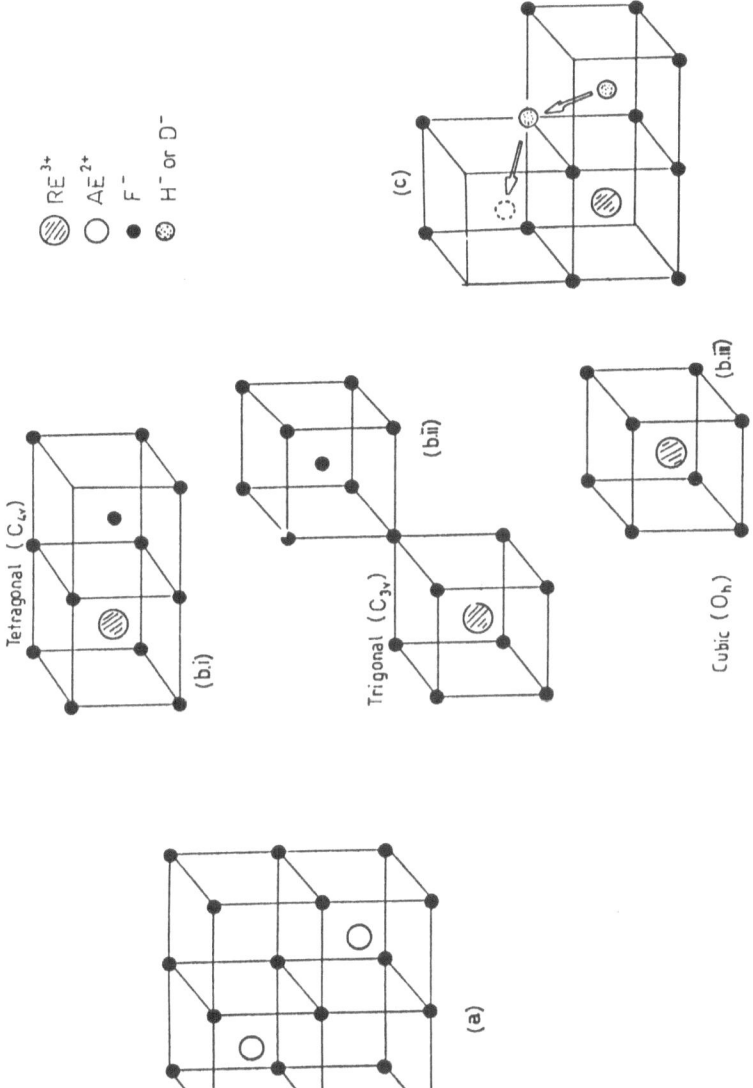

Figure 2: (a) The alkaline earth cubic lattice (b) Some of the RE symmetry centres (c) A model of the Cs(1) bleaching centre and the migration path of the hydrogenic ions.

Most RE defect centres have some form of symmetry and hence can be characterised and investigated to a certain extent using one of the 32 possible point groups. Each point group is defined by specific combinations of symmetry operations (rotations, reflections, inversions etc) and is given labels like C_{4v}, C_s, D_{3h} etc. The C_{4v} single ion defect centre is the most widely studied centre in RE doped AEF crystals because of the simplicity of its structure as well as its stronger absorption and emission strength. Energetically, this has been calculated to be the most favourable centre for RE centres in the AEF cubic lattice [2].

The RE elements are characterised by the progressive filling of the 4f shell and it is the transition between states of these 4f electrons that contribute to the absorption and emission spectra observed [3,4,5]. One interesting optical feature of the RE ions in crystals is the sharpness of the electronic transitions. This is a result of the *lanthanide contraction* whereby the 4f shell is drawn inside the $5s^2 5p^6$ shells providing the 4f electrons with a shield from external perturbations. The electronic states of the 4f electrons in a crystalline environment are governed by the "crystal field" exerted by the host ligands. The symmetry of the field exerted on the RE ions determines how the $2J+1$ degeneracy of free RE ion states are lifted, i.e how the electronic levels split. In the first approximation, the crystal field is provided by the point charges of the ligands surrounding the RE though other contributors to the crystal field including covalency or overlap of the ligand orbitals are present and could be significant.

The electronic transitions of the centres follow specific transition rules determined by group theoretical considerations [6]. For example, the requirement for a non vanishing transition probability between two states $|a\rangle$ and $|b\rangle$ with group irreps (irreducible representations) of Γ_a and Γ_b is

$$\Gamma_a \times \Gamma_D \subset \Gamma_b$$

where Γ_D is the irrep for the dipole operator. Another consequence from such group operations is that centres which have a higher symmetry than C_1 are expected to obey certain well defined polarisation selection rules.

Some of the RE dopant in the alkaline earth hosts may aggregate within the near neighbourhood of each other forming cluster centres and the presence of this type of centres appear to be an invariable feature of these systems. Short range interactions between these ions is possible. One interesting effect as a result of these interactions is *upconversion* where effectively, the emitted frequencies from these centres may be higher than the input excitation frequency, hence making possible a laser where the the output frequency is higher than the input frequency. A recent system CsCdBr doped with Er^{3+} [7] was made to lase by the upconversion process [8].

CRYSTAL GROWTH AND PREPARATION

Single crystals of AEF's doped with RE's can be grown using the Bridgman-Stockbarger method. Typically, tiny grains of AEF crystal crushed from a larger piece of stock material and the dopant in the form of rare earth fluoride powder would constitute the charge material, which would be loaded in a carbon crucible and lowered through a temperature gradient in vacuum. An RF furnace is suitable for achieving temperatures of above $1100^{\circ}C$ to melt the AEFs. For a successful growth of a single crystal, the temperature profile through which the charge material is to be lowered should consist of a gradual descent, a sharp decrease followed by another gradual descent [19]. An inappropriate temperature profile would lead to formation of polycrystals.

The single crystals so grown are relatively free from defects and they can be conveniently cut with a rotary diamond saw unless it is a double AE doped or mixed crystal system. There appear to be more defects in the mixed crystals because of the mismatch of ionic radius between the host and dopant AE and consequently, mixed crystal systems tend to chip or spontaneously crack when cut, hence they need to be handled with extra care during the cutting process [11]. The slabs of crystals obtained can be shaped and given a preliminary polish using various grades of sand paper. The final optical finish is achieved by polishing with successively finer grades of silicon carbide grains.

The host material cleaves naturally along the (111) planes and these can be used as convenient reference for orientating the crystal which is necessary for polarisation studies.

To study the hydrogenic charge compensated centres, hydrogen or its isotopes deuterium or tritium can be incorporated into the crystal following the method of Hall and Schumacher [9]. This process involves heating the crystal in the presence of molten aluminium and an atmosphere of hydrogen at between $850 - 900^{\circ}C$ for a period of typically 36 - 72 hours depending on the amount of

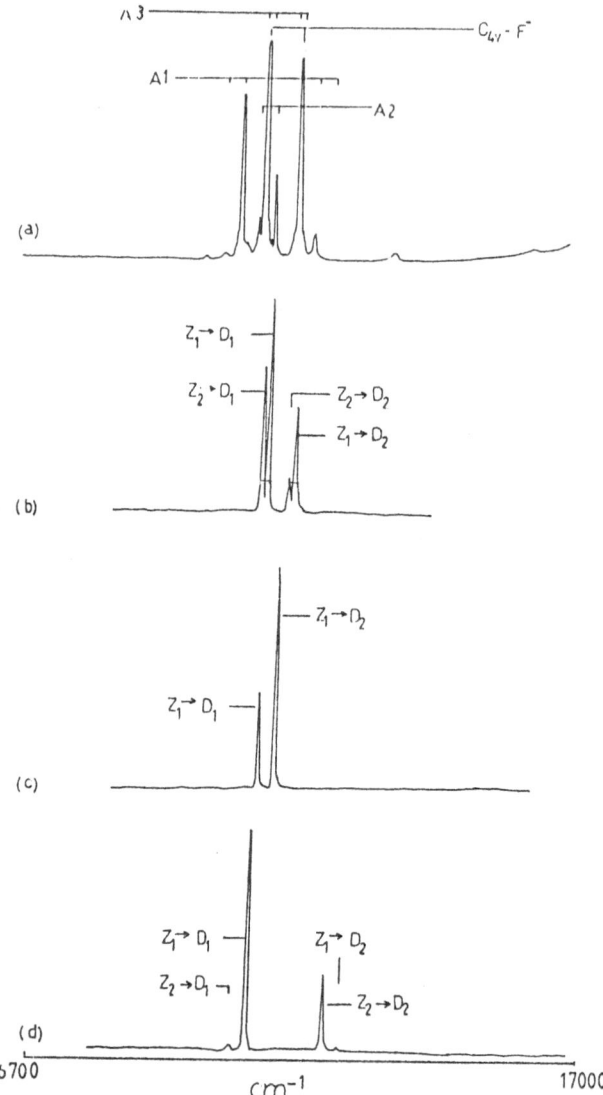

Figure 3: (a) The broadband excitation spectrum of the 1%
Sr^{2+} doped CaF_2:Pr^{3+} crystal. A1,A2,A3 are the mixed crystal
centres derived from the parent C_{4v} centre. (b)-(d) are the
selective excitation spectra monitoring a particular $^1D_2 \rightarrow {}^3H_5$
transition of the A3,A2 and A1 centres respectively thus
identifying the ground (Z_i) and excited (D_i) states belonging
to the centres.

212

hydrogen to be incorporated in the crystal. The pressure of the hydrogen in the hydrogenation chamber can also be varied to introduce the desired amount of hydrogen into the crystals.

LASER SPECTROSCOPIC TECHNIQUES

– Laser Selective Excitation

The first step in the study of the RE doped crystals is usually the identification of the defect species present and their associated electronic levels. Laser Selective Excitation (LSE) proved to be a powerful technique for this purpose. It was first employed to elucidate the electronic level structure of the centres in $CaF_2:Er^{3+}$ by Tallant and Wright [10].

Different centres in the crystal give rise to characteristic sets of energy levels. By sweeping a tunable dye laser over its frequency range and monitoring the emission from all centres present to a selected multiplet with a low resolution monochromator, the broadband excitation spectrum observed is a composite of energy transitions from all the centres present. An example of this is shown in figure 3(a) for the case of the excitation spectrum in the $^1D_2$ region of the 0.5% Sr^{2+} doped $CaF_2:Pr^{3+}$ (0.01%) [11]. If the laser is tuned to one particular frequency corresponding to the absorption energy of a specific centre, the emission spectrum observed arises out of the transitions from the excited state to the other states belonging to the same centre only. This is the basis of LSE. A variation of the experiment is to monitor a particular transition of a centre with a monochromator and sweeping the dye laser over the excitation region. In this way it is possible to determine the levels of a centre in the wavelength region of excitation (fig.3b,c,d).

The polarised emission spectra give further confidence in the assignment of a set of levels to a particular symmetry [12,13]. For these sort of measurements, the crystal must be suitably oriented with the geometry of the experiment chosen to the requirements of the results desired. For example, for the C_{4v} centre, the principal z–axis may be aligned in three orthogonal directions in an AEF cubic structure. To obtain conveniently interpreted polarised emission spectra, the geometry of the experiment may be set up as shown in figure 4 with a [100] oriented crystal. An example of the polarised spectra for the Pr^{3+} C_{4v} fluorine compensated centre in CaF_2 is shown in figure 5.

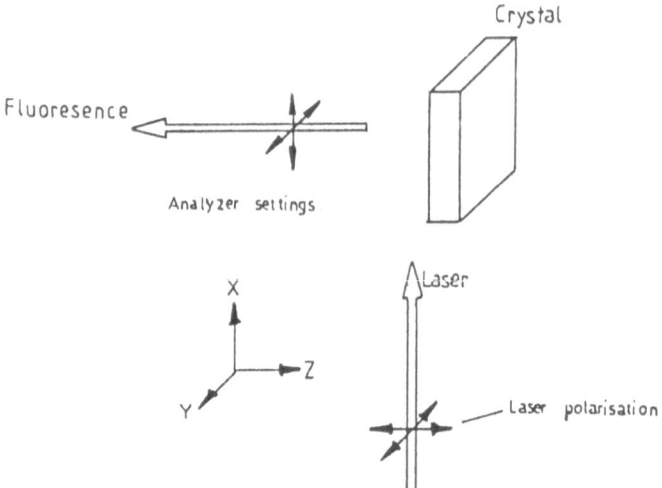

Figure 4: A convenient geometry for polarisation measurements. The laser can be polarised Y or Z while the analyzer direction can be set at either X or Y.

– Studies of dynamical processes using pulsed lasers

Using a pulsed laser source, the fluorescent lifetimes or decay rates of the centres may yield information on the nature of the centres and also on the excitation dynamics present in the crystal systems. The H^- centres invariably have much shorter lifetimes than the equivalent F^- centres (see table 1) and the reason for this is that the H^- ion being much lighter is able to to dissipate the excitation energy non-radiatively through the local modes of vibration [20]. This non-radiative dissipation of energy is also responsible for the generally weaker fluoresence intensities of the hydrogen centres hence it is common to use deuterium instead of hydrogen to study the hydrogenic centre species. Deuterium being heavier is less efficient at non-radiative relaxation.

Due to the proximity of the RE ions to each other in the cluster centres, the potential for exchange interactions exists for such centres which can be probed using pulsed laser sources [15,16]. One hallmark of the cluster centres is in fact their relatively faster lifetimes compared to the single ion centres. Transfer of energy between donor and acceptor ions has been observed in various types of mainly dimer centres though in one case, an allegedly trimer centre upconverted a red input beam to give UV output [16].

Upconversion can also occur in single ion centres by either the ETU (Energy Transfer Upconversion) or the STEP (Sequential Two Photon Excitation) processes. The lifetimes of the single ion centres were found to vary with concentration of the dopant RE ion in a fashion that suggests long range dipole-dipole interactions between the single ion centres [11,13].

– Spectroscopy of Hydrogenic Centres

The hydrogenated RE doped AEF are interesting because of the extra spectral features present together with the observation of some new phenomena in these crystal systems [21]. The most obvious spectral feature is the shift in the energy levels of centres analogous to those found in the unhydrogenated crystals. This shift is a direct result of the different crystal field experienced by the RE ion. The coupling of the vibrational local modes to the electronic states also allows observations of vibronic levels. In cases where the selection rules does not allow transition to a particular electronic level, the energy for the electronic level in question may be deduced from the identification of the vibronic level associated with it.

Possibly, the most interesting effect is the "optical bleaching" of some low symmetry hydrogenic centres [14]. These centres have hydrogenic ions substituting not only in the interstitial positions but also replacing the lattice fluorines near the RE ion. Upon optical excitation, the hydrogenic ions migrate to form a different centre (photoproduct) or the same centre in a different orientation (fig.2c). Experimentally, the intensity of the transition excited drops over time or "bleaches" up to less than 5% of the original fluoresence intensity. As far as is known, this is the only material in the solid state that exhibit this bleaching effect.

THE HOLEBURNING EFFECT AND POSSIBLE APPLICATIONS

With the advent of high resolution lasers with frequency resolution of 1 MHz or less, the forefront of spectroscopic research of RE doped materials is now in probing the fine structures within transitions. One such experiment observes the holeburning effect [17,18]. In the holeburning process, a high resolution laser is tuned to match a transition between the a ground state and an excited hyperfine level of a centre. During the de-excitation pro-

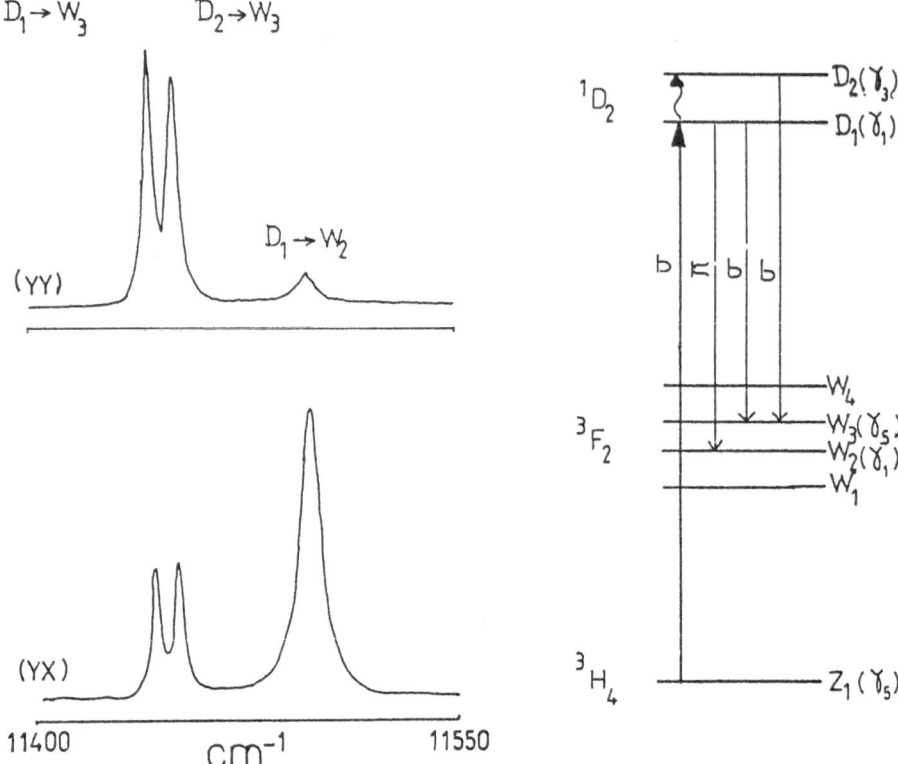

Figure 5. $^1D_2 \rightarrow {}^3F_2$ polarised emission of the A2 centre which
has a C_{4v} symmetry. The energy levels are labelled by a sub-
scripted alphabet where the subscript denotes the 1st ,2nd,3rd
etc excited level of the multiplet. The γ_i's are irrep labels
of the C_{4v} point group. A transition between two levels is
either allowed or polarised π or σ depending on the irreps of
the levels. (YY) and (YX) is the geometry of the experiment with
the first letter denoting the laser polarisation and the second
the analyzer setting.

216

cess, the population distribution of the hyperfine ground states has now changed and if another probe laser is swept through the region of excitation, a "hole" is observed in the transition band. This is illustrated in figure 6. A double resonance technique related to holeburning is ODNMR [17] (Optically Detected Nuclear Magnetic Resonance) where after a hole has been burned in the transition, an RF signal is applied to the sample and if the RF frequency matches the transition between the hyperfine state which has been depleted and another hyperfine state, the fluorescent intensity of the "hole" increases which gives the ODNMR signal.

Holeburning offers exciting and novel optical applications taking advantage of the frequency dependence of this effect. A possible application to be discussed here is in frequency domain memory storage. Conventionally, the storage element for the bits are encoded spatially i.e, one bit would occupy a position on a CD disk or the usual magnetic medium. However in frequency domain storage, all the bits that make up a basic unit of information, a byte say, can be encoded on a single spot, the size of which is limited by the focus area of the laser. Encoding and decoding information would be a matter of burning holes in the spectral line profile and probing the holes burned using the appropriate frequencies.

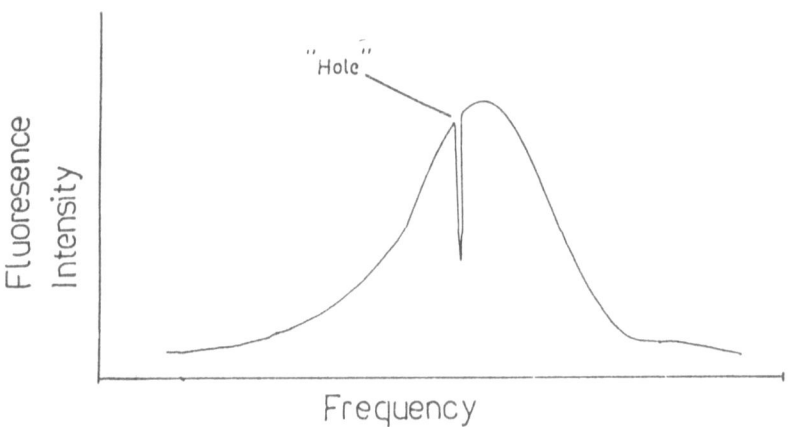

Figure 5: Illustration of an absorption profile after a "hole" has been burnt in it. Several holes may be burnt into a single profile simultaneously.

Another possible application of the holeburning effect is the fabrication of optical masks. A conventional mask would comprise regions of space where radiation is blocked or admitted to give the pattern required and one mask would give only one pattern. An example of the use of such a mask is in the photolithography technique employed to fabricate microelectronic circuits in the semiconductor industry. Applying the holeburning effect, it is possible to have a mask which has several patterns etched onto it by burning holes at a particular frequency for each pattern. Access of a particular pattern is thus by irradiating light of the correct frequency.

For these applications to be successful, it is preferable for persistent spectral holeburning (PSH) to occur, i.e the hole persist after the excitation source is removed. The search for centres in the RE doped AEF fluorides which exhibits PSH are continuing. Some low symmetry hydrogenic centres, the bleaching centres demonstrate PSH with the holes burnt persisting for several hours.

CONCLUDING REMARKS

RE doped AE fluoride crystals are convenient to work with because of its physical resilience. The defect centres which give rise to the many optical processes continue to be of interests to a broad range of researchers from spectroscopists, laser physicists, material scientists and theoreticians. The enticing possibility that out of some of these crystal materials may come new optical application or phenomena ensures that further work shall continue in the future.

Table 1: $^1D_2 \to {}^3H_4$ lifetimes of some C_{4v} Pr^{3+} centres in CaF_2 and SrF_2

Centre	Crystal	Lifetimes
$C_{4v} - F^-$	$CaF_2:Pr^{3+}(0.1)$	490 μsecs
$C_{4v} - F^-$	$CaF_2:Pr^{3+}(0.01)$	640 μsecs
$C_{4v} - F^-$	$CaF_2:Pr^{3+}(0.001)$	860 μsecs
$C_{4v} - H^-$	$CaF_2:Pr^{3+}(0.05)$	390 nsecs
$C_{4v} - D^-$	$CaF_2:Pr^{3+}(0.05)$	95 μsecs
$C_{4v} - F^-$	$SrF_2:Pr^{3+}(0.05)$	2.0 msecs
$C_{4v} - H^-$	$SrF_2:Pr^{3+}(0.05)$	270 μsecs
$C_{4v} - D^-$	$SrF_2:Pr^{3+}(0.05)$	405 μsecs

REFERENCES

1. Crystals with Fluorite Structures ed. W.Hayes.
   Oxford (Clarendon), (1974)

2. Corish J., Jacobs P.W.M. and Ong S., Phys.Rev.B 25:6425 (1982)

3. Wybourne B.G., Spectroscopic Properties of Rare Earths,
   Wiley & Sons Inc. (1965)

4. Dieke G.H., Spectra and Energy Levels of Rare Earth Ions in Crystals, Interscience Publishers (1968)

5. Hufner S, Optical Spectra of Transparent Rare Earth Compounds Academic Press (1978)

6. Tinkham, Michael, Group Theory amd Quantum Mechanics McGraw-Hill Book Company (1964)

7. Cockroft N.J., Jones G.D. and Syme R.W.G. J.Lumin. (Netherlands) 43(5):275 (1990)

8. Jones G.D., Physics Dept., Uni.of Canterbury,NZ , Private Communication

9. Hall J.L and R.T Schumacher, Phys Rev 127:1892 (1962)

10. Tallant D.R. and Wright J.C. J.Chem.Phys. 63:2074 (1974)

11. Khong Y.L., Laser Spectroscopy of Praseodymium doped Mixed Alkaline Earth Fluorides. PhD thesis(1991), Uni.of Canterbury,NZ

12. Cockroft N.J., Thompson D, Jones G.D. and Syme RWG J.Chem.Phys. 86(2):521 (1986)

13. Reeves R.J.Laser Selective Excitation of Praseodymium Ions in Hydrogenated Flourite Crsytals. PhD thesis (1987),Uni.of Canterbury,NZ

14. Reeves R.J.,Jones G.D. and Syme RWG Phys.Rev.B 40(10):6475 (1989)

15. Kliava J,Evesque P and Duran J. J.Phys.C 86:3357 (1978)

16 Arturo Lezama, Marcos Oria and Cid B. de Araujo Phys.Rev.B, 33(7):4493(1985)

17. Burum D.P., Shelby R.M. and Macfarlane R.M, Phys.Rev.B 25(5):3009 (1982)

18. Macfarlane R and Shelby R. in Modern Problems in Condensed Matter Sciences. Vol21 - Spectroscopy of Solids containing Rare Earth ions. (1987)

19. Laudise R, The Growth of Single Crystals, Prentice Hall, New Jersey (1970)

20. Reeves R.J., Jones G.D., Cockroft N.J., Han TPJ and Syme RWG J. Lumin. 38:198 (1987)

21. Jones G.D. Aust.J.Phys. 32(6):629. (1979)

Semiconductor Materials

Magneto-optics of Semiconductors at High Magnetic Fields

N. Miura, S. Sasaki, S. Takeyama and Y. Nagamune
Institute for Solid State Physics, University of Tokyo, Roppongi, Minato-ku, Tokyo 106, Japan

We present the latest results of magneto-optical investigation of the band structures and excitons in semiconductors under pulsed high magnetic fields up to a few megagauss (>100T). The first topic is the energy band structure and the Landau levels in short period superlattices of GaAs-AlGaAs. Secondly, we discuss the exciton structure in a layer-type crystal PbI_2 and PbI_2-BiI_3 superlattices.

1. INTRODUCTION

In high magnetic fields, energy levels of conduction and valence bands of semiconductors are quantized into Landau levels. Investigation of the electronic transitions between these levels by optical processes such as those shown in Fig. 1, is a powerful means for the study of the energy band structure in semiconductors. The magneto-optical measurements are also very useful for studying exciton structures, since magnetic field dependence of energy levels of excitons provides information about the reduced mass, the binding energy, the g-factors, etc. Many investigations have been undertaken on the magneto-optical spectra in semiconductors using steady high magnetic fields.

Up to now, the highest steady magnetic field is about 30T produced by hybrid magnets. For producing higher magnetic fields, pulse magnets have to be used. Recently, significant progress has been achieved in techniques for generating pulsed high magnetic fields[1]. The limitation restricting the magnitude of non-destructive magnetic fields is primarily due to the large magnetic force which can exceed the finite strength of the coil material. A great deal of effort has been put into both developing new strong wires for the windings and also the method of reinforcing the magnet[1]. At present, long pulse (duration of about several ms) high magnetic fields up to about 60T have been achieved non-destructively. Above this level, to the megagauss range, non-destructive field generation is extremely difficult, therefore, destructive methods for producing ultra-high magnetic fields have to be used. At the Institute for Solid State Physics (ISSP) of the University of Tokyo, techniques

223

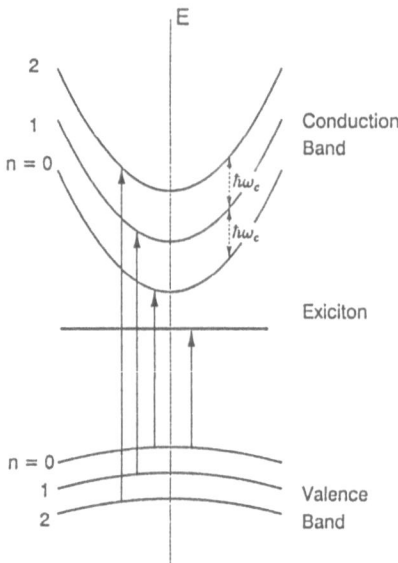

Fig.1 Electronic transitions between Landau levels and to exciton states in magneto-absorption

for generating megagauss fields have been developed by using electromagnetic flux compression and the single-turn coil technique[2], as well as the long pulse fields up to about 60T[3]. These techniques are particularly suitable for the application of ultrahigh fields to solid state physics, and have been applied in various experiments[4].

Very high magnetic fields have opened up many new possibilities. In magneto-optics of semiconductors, many new situations can be created by applying such high magnetic fields. For example, the level quatization effect is so large that we can obtain high resolution in the magneto-optical spectra. One of the interesting areas of the high field magneto-optics is the investigation of the quasi-2-dimensional electron systems in quantum wells and superlattices. Because of the quantization in the quantum potentials, additional quantization by magnetic fields allow the observation of highly quantized states. Moreover, we can investigate the magnetic field effect on the nearly 2-dimensional excitons. At ISSP, we have extensively investigated magneto-optics in GaAs-AlGaAs quantum wells. By analyzing the oscillatory magneto-optical spectra, and comparing with the exciton spectra, we have determined the binding energy and the diamagnetic shift of the nearly 2-dimensional excitons[5]. We have also found the transition from the hydrogen atom-like states at low field region to the Landau level-like states in the high field region in the quantum wells[6]. In modulation doped quantum wells, the magnetic field effects on the Fermi edge singularity have been studied[7].

Another interesting area is the study of the small radius excitons. If the radius of excitons is small, the magnetic field effects on excitons such as diamagnetic shift are usually very small. However, the application of sufficiently high magnetic fields enables us to observe the effects. We have clarified the origin of the sharp series of the exciton lines in BiI_3 in terms of the localization in stacking faults[8]. In PbI$_2$, we have obtained the assignment of the exciton line series, on which there has long been a large controversy[9]. In anthracene, magnetic field effect has been observed for the first time in the Davydov splitting of Frenkel excitons[10].

In this paper, we first describe our experimental techiques for the magneto-optical measurements in pulsed high fields, and then discuss two topics as examples of the recent high field magneto-optical investigations of semiconductors. First, we will discuss the energy band structure and the Landau levels in GaAs-AlGaAs short period superlattices, with a special focus on the Landau level broadening when the fields are applied parallel to the layers. Next, we present the study of relatively small radius Wannier excitons in layer-type crystal PbI$_2$ and its superlattices.

2. EXPERIMENTAL TECHNIQUES

Pulsed high magnetic fields up to 50-60T can be generated by using wire-wound solenoid coils with a duration of about 10ms. The coils are immersed in a liquid nitrogen bath, and we discharge a pulsed current from a condenser bank with a storing energy of 300kJ. We have recently succeeded in producing long pulse fields up to nearly 60T non-destructively, by using coils wound with strong wires containing Nb-Ti fibers embedded in copper matrix[3]. The reinforcement of the coil structure was made by impregnating ice into woven glass fibres inserted in between the windings. Impregnation is made by filling the coil container with full of water at first, and ice is formed when the magnet is cooled in a liquid nitrogen bath. The gap between the windings is automatically pressurized at the solidification of ice. The advantage of the ice impregnation is that it is much simpler than epoxy impregnation. The gap in the coil structure is completely filled without any procedures of vacuum impregnation or solidification under pressure. Moreover, the cooling time of the magnet necessary after a heavy shot is reduced to less than 15min. as compared with the case of epoxy impregnation for which the typical cooling time is more than 40min. after a 40T shot.

Higher magnetic fields up to several megagauss (>100T) are produced either by electromagnetic flux-compression or the single-turn coil technique. We have generated high fields up to 360T with a rise time of 8μs using the main capacitor bank of 5MJ[2]. Attempts to obtain even higher fields are still under way. Although the samples are destroyed every time, the fields are applied to various accurate measurements[2,4].

The second technique is the single-turn coil technique which can produce megagauss magnetic fields without causing any damage on samples or cryostat mounted inside the coil. This aspect is very convenient for experiments. The peak field depends on the coil bore. It is 150T with a bore of 10mm diameter and 200T with a bore of 6mm diameter. The pulse duration is about 7μs.

In order to measure magneto-optical spectra in non-destructive long pulsed fields, an OMA (optical multi-channel analyzer) is employed. The measuring system is schematically shown in Fig. 2[7]. The measurements can be carried out in a wide temperature range between 4.2K with liquid helium and room temperature. The gate of the OMA was opened for 1ms at the top flat part of the field pulse. Variation of the field during the aperture time was less than

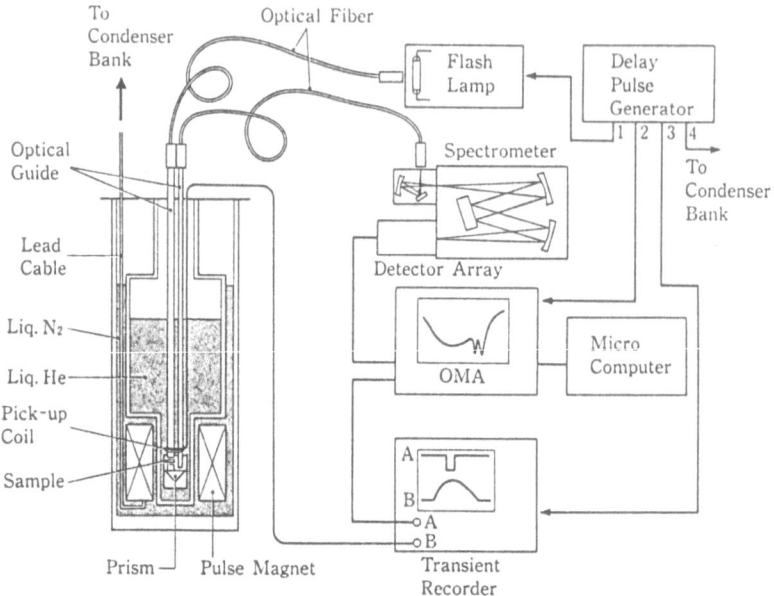

Fig.2 Block diagram of the system of high field magneto-optical measurements using a pulse magnet and an OMA

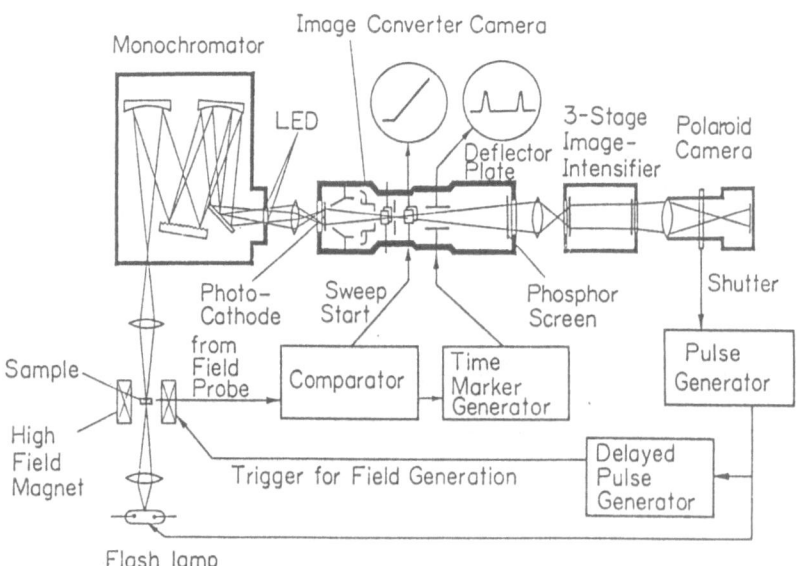

Fig.3 Block diagram of the system of magneto-optical measurements in megagauss magnetic fields using an image converter camera

3%. For the magneto-luminescence experiment, samples are excited usually by the 5145Å line of an Ar laser.

For the megagauss fields whose pulse duration is of the order of several μs, magneto-optical measurements are performed by using a spectrometer combined with an image-converter camera, as shown in Fig. 3. In this case, the optical image is swept to obtain 2-dimensional streak spectra. This is convenient because entire magneto-optical spectra can be obtained in a single shot pulse.

3. MAGNETO-OPTICS IN SHORT PERIOD GaAs-AlGaAs SUPERLATTICES

Short period $(GaAs)_m/(Al_xGa_{1-x}As)_n$ superlattices with small monolayer numbers (m, n) are of interest as new highly anisotropic crystals, as mini-bands are formed in the direction perpendicular to the superlattice layers by tunneling effect through $Al_xGa_{1-x}As$ barrier layers. In the case of $x = 1$, conduction bands originating from the X-points in AlAs layers are lower in energy than those from the Γ-points in GaAs layers. The mini-band structure in the conduction band then possesses an indirect nature both in the real and k-spaces, and the crystals are called of type II. When x is close to 0, on the other hand, the lowest mini-bands in the conduction band originate from the Γ-point in GaAs layers. These crystals have a direct nature and called of type I.

Recently, many investigations have been carried out on the energy band structure of the short period superlattices both theoretically[11] and experimentally[12]. Magneto-optical measurements, especially with high fields applied parallel to the layers, are powerful means for the study of the band structure in the direction perpendicular to the layers. High magnetic fields enable us to observe oscillatory structures and magnetic shift in the spectra even for very short period samples.

We present here the results on two kinds of samples of $(GaAs)_m/(Al_x$-$Ga_{1-x}As)_n$ short period superlattices, type I samples with Al concentration $x=0.5$ and type II samples with $x=1$. The samples were grown by Horikoshi's group at NTT Research Laboratory by MBE on (001)-oriented semi-insulating GaAs substrates. For transmission measurements, the GaAs substrates were removed by polishing and etching.

Figure 4 shows the magneto-absorption spectra in a type I sample with $(m, n)=(7,5)$ for $\boldsymbol{B}\perp$layers ($\boldsymbol{B}_\perp$) measured with the Faraday configuration and $\boldsymbol{B}/\!/$layers ($\boldsymbol{B}_{/\!/}$) with the Voigt configuration[13]. In the zero magnetic field, a very weak exciton absorption is observed as a shoulder. At about 230meV above this, the spectrum exhibits a kink corresponding to the edge of the first mini-band. As the field is increased, oscillatory structures were observed in both field directions. For $\boldsymbol{B}_\perp$, a slight difference was observed in the traces between left (σ_+) and right (σ_-) circularly polarized radiation. For $\boldsymbol{B}_{/\!/}$, the oscillation was found to disappear around the kink position at the mini-band edge, whereas it persists up to higher fields for $\boldsymbol{B}_\perp$. Figure 5 shows a plot of the photon energy as a function of magnetic field. Non-linear dependence due to the non-parabolicity was observed especially for $\boldsymbol{B}_{/\!/}$. From the linear part, the apparent reduced masses of electrons and holes were estimated to be

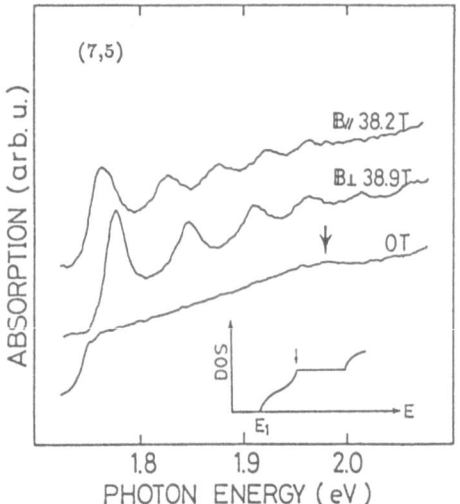

Fig.4 Magneto-absorption spectra in GaAs-AlGaAs short period super-lattices with $(m, n)=(7,5)$ for two field directions $\boldsymbol{B}_{/\!/}$ and $\boldsymbol{B}_{\perp}$. Inset shows the joint density of states. The vertical arrows indicate the miniband edge.

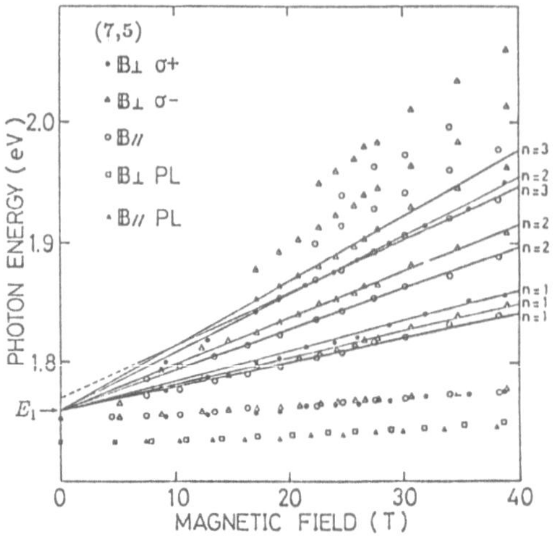

Fig.5 Plot of the photon energy of the peaks in the magneto-absorption and magneto-photoluminescence for $(m, n)=(7,5)$ for two field directions

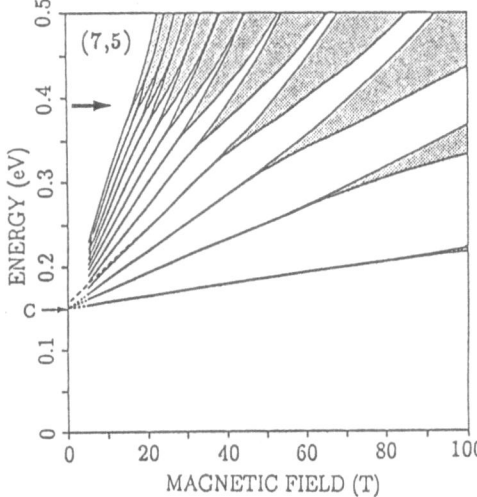

Fig.6 Calculated energy of the Landau levels in the conduction band for $(n, m)=(7,5)$ as a function of magnetic field applied parallel to the superlattice layers $(\boldsymbol{B}_{/\!/})$. Above the miniband edge indicated by the horizontal arrow, the Landau levels are broadened.

μ=0.067(n=1), 0.060(n=2) for σ_+ and μ=0.075(n=1), 0.074(n=2), 0.075(n=3) for σ_- in a unit of the free electron mass. For $\boldsymbol{B}_{//}$, a slightly heavier masses, μ=0.081(n=1), 0.085(n=2), 0.100(n=3) were obtained. Very similar values were obtained for a sample with (m,n)=(6,6). These results indicate that the anisotropy is not so large in these samples.

In order to analyze the data quantitatively and to test the validity of the effective mass approximation, the Landau levels of electrons and holes were calculated for $\boldsymbol{B}_{//}$, based on the Kronig-Penny model[14]. Taking account of both the parabolic potential of magnetic field and the periodic superlattice potential, the Schrödinger equation for electrons becomes one-dimensional equation with the center coordinate x_0 of the cyclotron motion as follows:

$$\left[\frac{\hbar^2}{2m^*}\left\{-\frac{\partial^2}{\partial x^2}+\frac{(x-x_c)^2}{l^4}\right\}+V(x)\right]\phi(x)=\epsilon\phi(x),\qquad(1)$$

where $l=\sqrt{\hbar/eB}$ is the cyclotron radius, $x_c=-l^2k_y$ is the cyclotron orbit center and $V(x)$ is the band edge profile. The equation was solved numerically. The energies in the conduction band are shown as a function of center coordinate in Fig. 6. It is evident that the Landau levels are broadened above the miniband edge.

The calculation for the valence band is more complicated because of the band degeneracy. We have to solve a 4×4 Hamiltonian matrix. If we restrict ourselves to the case k_z=0, however, the matrix is decoupled to two 2×2 matrices corresponding to $J_z=(3/2,-1/2)$ states and $J_z=(-3/2,1/2)$ states, respectively. The matrices were solved numerically by replacing all the differential operators with finite difference equations[15]. The energies of the Landau levels for holes are plotted in Fig. 7 as a function of magnetic field. The degeneracy of the light and heavy hole bands is lifted by the quantum well potential at zero field. As the hole energy becomes larger than the mini-band edge indicated by arrows, the Landau levels are broadened, due to the difference of energy depending on the center coordinate. The levels originating from the light and heavy hole bands cross each other and showed a complicated mixing at crossing points.

From such calculations, the transition energies can be calculated. Figure 8 shows the calculated results as a function of magnetic field together with the experimatal data. We can see a very good agreement between theory and experiment except for the ground state where the excitonic effect is significant. A similar good agreement was also obtained for sample with (m,n)=(6,6) and (m,n)=(3,3). This implies that the effective mass approximation is still adequate to describe the energy levels in such short period superlattices. It would be interesing to measure the shorter period samples to investigate the limit of the applicability of the effective mass theory.

As for the ground state, from the difference between theory and experiment, we can estimate the binding energy of excitons. It was about 4 meV at zero field. The magnetic field dependence of the ground state was found to be in reasonably good agreement with the theoretical calculation for the 1s bound state[16].

As for the type II superlattices, the oscillation in the magneto-absorption

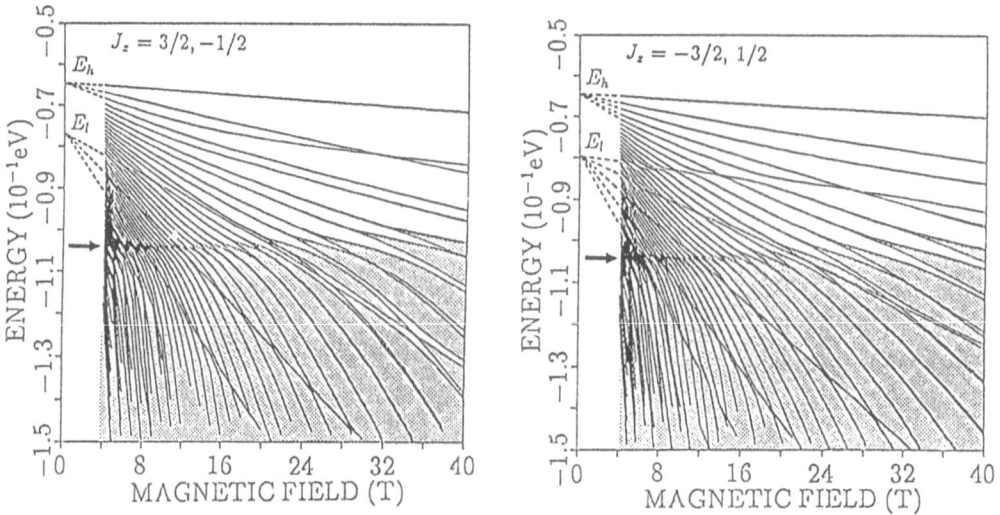

Fig.7 Caluculated energy of the Landau level in the valence band for $(n, m)=(7,5)$ as a function of magnetic field applied parallel to the layers($\boldsymbol{B}_{//}$). The horizontal arrows indicate the miniband edge.

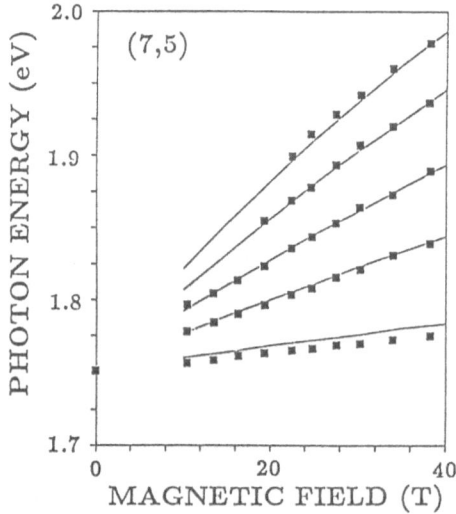

Fig.8 Plot of the photon energy of the magneto-absorption peaks for $\boldsymbol{B}_{//}$layers for $(n, m)=(7,5)$. The solid lines are the calculated results and the closed points are the experimental data.

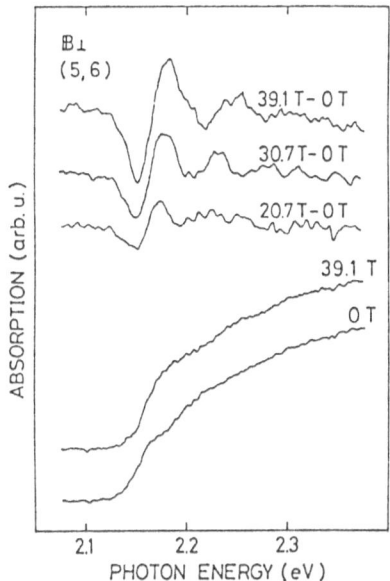

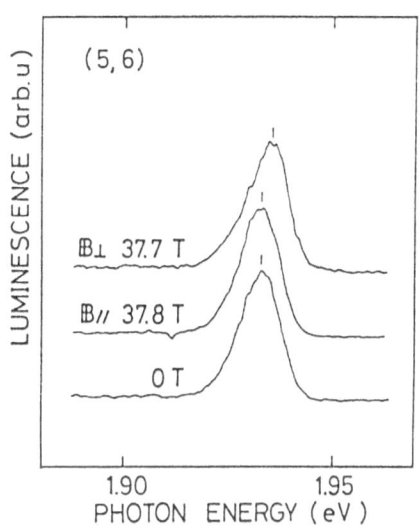

Fig.9 Magneto-absorption spectra for type Ⅱ superlattices with $(n, m)=(5,6)$. $\boldsymbol{B}\perp$layers. The upper three traces are the plot of the difference between between the spectra with and without magnetic fields.

Fig.10 Magneto-photoluminescence spectra for for type Ⅱ superlattices with $(n, m)=(5,6)$.

is very weak, and not easily descernible for samples with $(m, n)=(2,2)$, $(4,4)$, $(3,5)$. In a sample with $(m, n)=(5,6)$, a small oscillation was observed for $\boldsymbol{B}_\perp$ as shown in Fig. 9.[17] The oscillation is more visible if we substract the zero field spectra. From the slope of the $n=1$ line, the reduced mass of $0.08m_0$ was obtained.

Figure 10 shows the magneto-luminescene spectra for excitons in a $(m, n)=(5,6)$ sample for both field directions. At zero field, the peak is shifted from the band edge estimated from the fan chart of the magneto-absorption oscillation by about 230 meV. It can be seen that the luminescence peak shows a considerable shift for $\boldsymbol{B}_\perp$, which is almost proportional to the square of the field. From this dependence, we can estimate the reduced mass of excitons. The estimation is different depending on whether we take the 2D model or 3D model. It was estimated to be $0.15m_0$ and $0.26m_0$ in the 2D and 3D limit, respectively. But it is not easy to deduce whether this emission comes from X_Z(parallel to the growth direction) or X_{XY}(in-plane) from this rough estimate because of several factors. The excitons may be localized in monolayer-thick growth steps[18]. This effect leads to smaller diamagnetic shift, or enhanced reduced mass. More elaborate variational calculation that takes this effect into account would be necessary. For $B_{//}$, however, the shift is negligibly small. This large anisotropy with respect to the field configuration is consistent with the model that the lowest conduction band in type II superlattices is X_z-like[19,20]. The strucures observed in the magneto-absorption spectra, on the other hand, are considered to be those which arise from a higher lying band of Γ-like character.

4. EXCITONS IN LAYER-TYPE CRYSTALS

Next, let us turn the subject to layer-type crystals. PbI_2 is one of the layer type crystals which show distinct exciton absorption lines. Nikitine and Perny were the first who observed a series of band edge exciton absorption lines[21]. They found that the spacing between adjacent lines is not in accord with the Rydberg series. Exciton absorption spectra in $2H-PbI_2$ exhibit three main peaks, A_1, A_2 and A_3, and in some crystals a small shoulder-like peak A_x is observed in between A_2 and A_3[22], as shown in Fig. 11. If we assign the three main peaks A_1, A_2 and A_3 peaks as the $n=1$, $n=2$ and $n=3$ exciton lines as demonstrated on the left of the inset in Fig. 11(a), the binding energy of $R^*=127$meV is obtained from the spacing between A_2 and A_3. Then, the observed spacing between A_1 and A_2 is too small within the hydrogenic model. To explain the anomalous spacings, Harbeke and Tosatti proposed a negative

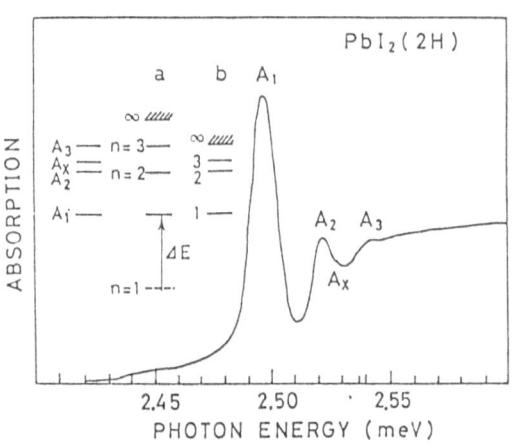

Fig.11 Exciton absorption spectra in $2H-PbI_2$. The inset shows the two models to explain the origin of the peaks A_1-A_3.

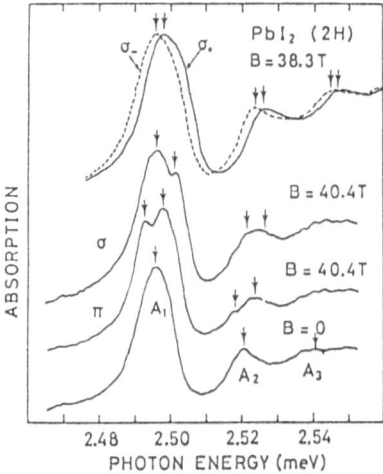

Fig.12 Magneto-absorption spectra in $2H-PbI_2$ at about $B=40$T. The top two traces (solid and dashed lines) are for $B//c$, the bottom trace is for $B = 0$ and the rest are for $B\perp c$. σ_+, σ_-, σ and π represent the polarization of the incident light. The arrows indicate the position of the peaks.

central cell correction ΔE [22], considering the cationic exciton character in PbI_2.

According to this model(model a), the binding energy $R^*=74$meV (spacing between A_1 and the continuum) is obtained after the central correction. Baldini and Franchi proposed a model in which the A_1 and A_3 are assigned as the $n=1$ and $n=2$ lines, and A_3 as a ground state of another Rydberg series[23]. Thanh et al., on the other hand, suggested that the A_1, A_2 and A_x correspond to the the the $n=1$, $n=2$ and $n=3$ lines and A_3 has some other origin [24]. This model(model b) leads to a binding energy of $R^* = 30$meV. There has been a controversy concerning these exciton models, but no definite conclusion has been obtained. High field magneto-optical measurements have been performed to solve this problem[25,26], but it was found that the magneto-optical effects are very small in PbI_2 because of the small exciton wave-function size.

Single crystals of $2H$-PbI_2 were grown by a travelling zone technique. Very thin ($\sim 0.1\mu$m) samples were cleaved for the measurements. For the measurements in very high fields produced by the single-turn coil technique, thin film samples($\sim 0.05\mu$m) of $4H$ polytype grown by the hot-wall epitaxial technique[27].

Figure 12 shows the magneto-optical spectra at about $B = 40$T obtained with the OMA system in $2H$ samples. In zero fields, three absorption peaks A_1, A_2 and A_3 were observed in this sample. Reflecting the layer type crystal structure, a large anisotropy was observed with respect to the magnetic field direction relative to the crystal axes. For $B//c$ with the Faraday configuration, a splitting was observed between the two traces for left (σ_+) and right (σ_-) cicularly polarized incident radiations, as shown in the top traces. For $B\perp c$ with the Voigt configuration, the magneto-optical spectra are different depending on the polarization of the electric field vector E against the magnetic field B. For $E//B(\pi)$ and $E\perp B(\sigma)$, a new aborption peak appeared in high magnetic fields in addition to the shift of the peak for both A_1 and A_2.

The position of the absorption peaks are plotted in Fig. 13 as a function of magnetic fields in both directions, togehter with the high field data up to 150T. For $B//c$, the splitting between the two polarizations σ_+ and σ_- for the ground state A_1 was found to be a linear function of field. From the splitting below 40T, the g-factor $g_\perp$ was obtained to be 0.89$\pm$0.09. From the midpoint between the data for σ_+ and σ_- for A_1, the coefficient of the diamagnetic shift in the expression $\Delta E_d = \sigma B^2$ was obtained as $\sigma = (9.68\pm0.39) \times 10^{-4}$meV/T^2.

If we use the hydrogen atom model, the reduced mass μ for the relative motion of the exciton can be estimated from the diamagnetic shift. Using the values $\epsilon_{\infty//} = 5.9\epsilon_0$, and $\epsilon_{\infty\perp} = 6.1\epsilon_0$ as the effective dielectric constants valid for the ground state exciton, $\mu_\perp = 0.17m_0$ is obtained. From these values of μ and ϵ, we obtain the values of the binding energy and the Bohr radius of the exciton as $R^* = 63$meV and $a_B=19$Å. The value of R^* is larger than the one obtained from model b but smaller than the one obtained from model a. Because there is a significant polaron effect in PbI_2, and it is larger for the excited states than for the ground state, the binding energy determined from only the spacing between A_1 and A_2 should be smaller than that determined from the diamagnetic shift of the ground state. Thus we can deduce that the model b is the most likely model as the assignments of the exciton absorption lines. This conclusion is supported also from the the magnitude of the diamagnetic shift of the A_2 peak which is assigned as the 2s $(n=2)$ state. The shift of

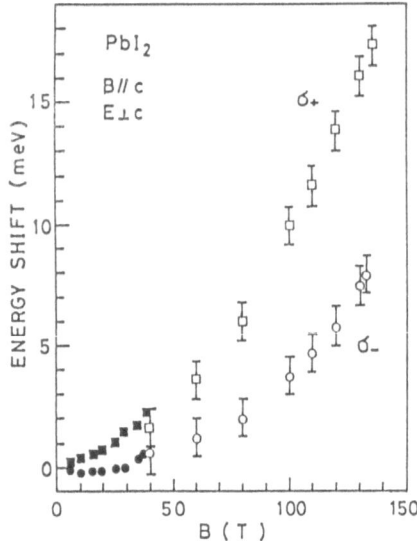

Fig.13 Plot of the photon energy of the absorption peaks in PbI$_2$ for $\boldsymbol{B}/\!/\boldsymbol{c}$. σ_+ and σ_- indicate the two circular polarizations. The solid and open points denote the data for 2H polytype in nondestructive fields and for 4H polytype in destructive fields, respectively.

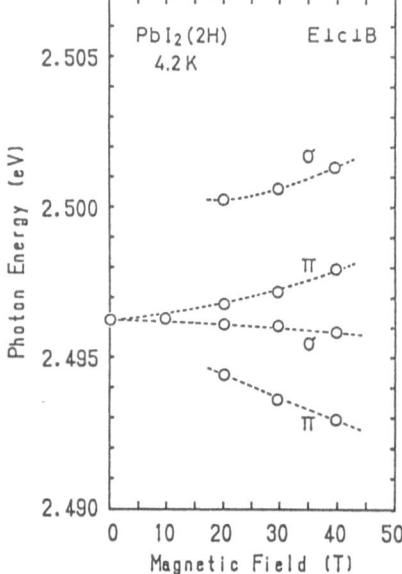

Fig.14 Plot of the photon energy of the absorption peaks in 2H-PbI$_2$ for $\boldsymbol{B}\perp\boldsymbol{c}$.

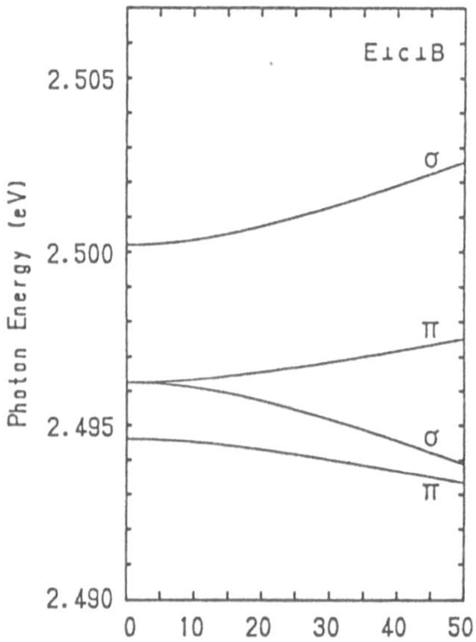

Fig.15 Magnetic field dependence of the photon energy of the absorption peaks for $\boldsymbol{B}\perp\boldsymbol{c}$, calculated based on the cationic model.

A_2 is considerably smaller than that calculated from the shift of A_1 assuming the hydrogen atom model. It is reasonable if consider again the fact that the polaron effect is larger for A_2 than for A_1.

For $\boldsymbol{B} \perp \boldsymbol{c}$, the spectra exhibit more complicated structures as well as non-linear splittings, as shown in Fig. 14. This indicates the interaction between different exciton states. In order to analyze the data we calculated the exciton levels based on the cationic exciton model[8]. The conduction band and the valence band are approximately represented by the p and s atomic orbitals of Pb. Therefore, to the zeroth order approximation, the band edge exciton is described as the excitation within the Pb atom. This is the cationic exciton model. The six p-states of the conduction band split into three two-fold degenerate levels by the crystal field and the spin orbit interaction. The two-fold lowest states of the conduction band are coupled with the two-fold s-states of the valence band by the Coulomb interaction in the exciton, resulting in four exciton states E_1, E_2 and E_3 (doubly degenerate). Here the state E_1 is a forbidden transition corresponding to a pure triplet state. E_2 is allowed for $\boldsymbol{E}/\!/\boldsymbol{z}$(c-axis) and E_3 is allowed for $\boldsymbol{E}/\!/\boldsymbol{x},\boldsymbol{y}$.

Since the wave functions of these states are represented by the atomic wave functions, the Zeeman energy can be calculated using the parameters of the crystal field Δ, the spin-orbit interaction λ and the Coulomb and exchange interactions. The energy levels were calculated using the parameters obtained from the energy differences among the spin-orbit split-off excitons, and the calculated results are shown in Fig. 15 together with the selection rule[28].

For $\boldsymbol{B} \perp \boldsymbol{c}$, the Zeeman splitting becomes nonlinear, because of the mixing of the above exciton levels by the external magnetic field, whereas a linear splitting is obtained for $\boldsymbol{B}/\!/\boldsymbol{c}$ since there is no mixing for E_3. These features well reproduce the experimental resuslts. The predicted selection rule is also in agreement with the experiment. This implies that the description of the Zeeman effect in terms of the cationic exciton model is adequate, although the model cannot lead to the diamagnetic shift. The exciton in PbI$_2$ is considered

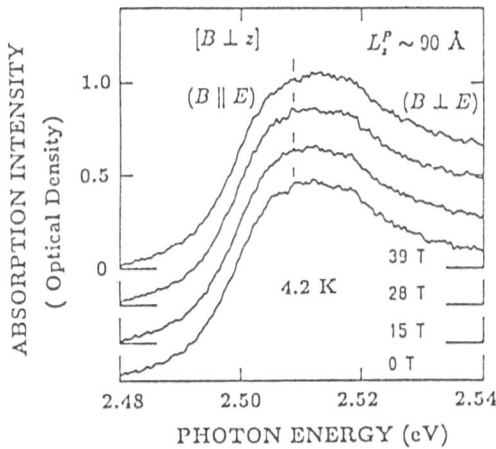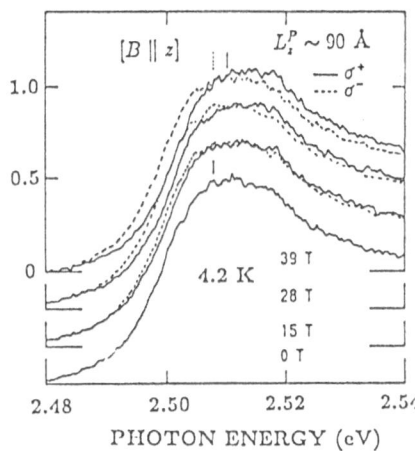

Fig.16 Magneto-absorption spectra of excitons in the 93Å thick PbI$_2$ layers in a PbI$_2$-BiI$_3$ superlattice. The left and right figures show the spectra for $\boldsymbol{B}/\!/$layer($\boldsymbol{B} \perp \boldsymbol{z}$), and $\boldsymbol{B} \perp$layer($\boldsymbol{B}/\!/\boldsymbol{z}$), respectively.

to be Wannier type excitons from the existence of considerable diamagnetic shift as well as the excitonic wave function size estimated above. However, the present results demonstrate that it has a character of cationic excitons localized in cations, at the same time, due to the band structure of PbI_2 as well as the small radius of the exciton wave function.

In PbI_2, each crystal layer is accumulated along the c-axis due to the Van der Waals force, so that the layer surfaces have no dangling bonds. Therefore, epitaxial growth in the c-direction on to a substrate of another crystal of a similar structure can easily be made[29]. Figure 13 shows high field magneto-absorption spectra for PbI_2/BiI_3 superlattices grown by hot-wall epitaxial technique[30]. A quantum confinement effect of exciton absorption lines was observed only for the PbI_2 layers and not in BiI_3 layers. In relatively wide PbI_2 layers ($L_z > 300$Å), nearly the same shift and level splitting as those for bulk PbI_2 were observed for both field directions. However, in narrower PbI_2 layers, the diamagnetic shift and level splitting were observed only for $B \perp$ layer and those for $B /\!/$ layer were negligibly small, as shown in Fig. 16. This can be explained in terms of the confinement effect of the exciton wavefunction in the z-direction; mixing of the E_3 states with the E_1 and E_2 states by the quantum well potential.

Acknowledgment

The authors are indebted to Dr. Y. Horikoshi of NTT Research Laboratories for providing the high quality GaAs-AlGaAs superlattices.

References

1. For an overview of the present status, see for example: Physica B, **155**, (1989).
2. Miura, N., Goto, T., Nakao, K., Takeyama, S., Sakakibara, T., Haruyama, T. andKikuchi, T. (1989). Physica B, **155**, pp.23-32.
3. Takeyama, S., Ochimizu, H., Sasaki, S. and Miura, N. (1991). submitted to Meas. Sci. Technol.
4. Miura, N. (1991). *Proc. Int. Conf. Application of High Magnetic Fields in Semiconductor Physics*, ed. G. Landwehr, (Springer-Verlag, to be published).
5. Tarucha, S., Okamoto, H., Iwasa, Y. and Miura, N. (1983). Solid St. Commun. **52**, pp.815-819.
6. Miura, N., Takeyama, S. and Iwasa, Y. (1987). *Proc. 18th Int. Conf. Phys. Semiconductors*, ed. O. Engström, (World Sci. Pub. Co, 1987), pp.715-718.
7. Lee, J. S., Miura, N. and Ando, T. (1990). J. Phys. Soc. Jpn. **59**, pp.2254-2273.
8. Komatsu, T., Kaifu, Y., Takeyama, S. and Miura, N. (1987). Phys. Rev. Lett. **58**, pp.2259-2262.
9. Nagamune, Y., Takeyama, S. and Miura, N. (1989). Phys. Rev. **B40**, pp.8099-8102.
10. Takeyama, S., Kobayashi, M., Matsui, A., Mizuno, K. and Miura, N. (1987). *High Magnetic Fields in Semiconductor Physics*, ed. G. Landwehr, (Springer-Verlag), pp.555-559.

11. Nakayama, T. and Kamimura, H. (1985). J. Phys. Soc. Jpn. **54**, pp.4726-4734.
12. Nakazawa, T., Fujimoto, H., Imanishi, K., Taniguchi, K., Hamaguchi, C., Hiyamizu S. and Sasa, S. (1989). J. Phys. Soc. Jpn. **58**, pp.2192-2199.
13. Sasaki, S., Miura, N. and Horikoshi, Y. (1990). J. Phys. Soc. Jpn. **59**, pp.3374-3387.
14. Belle, G., Maan, J. C. and Weimann, G. (1986). Surf. Sci. **170**, pp.611-617.
15. Fasolino A. and Altarelli, M. (1988). *Proc. 19th Int. Conf. on the Physics of Semiconductors*, ed. W. Zawadzki, (Institute of Physics, Polish Academy of Sciences, Warsaw), pp.361-364, and the reference therein by the same authors.
16. Makado, P. C. and McGill, N. C. (1986). J. Phys. C: Solid State Phys. **19**, pp.873-885.
17. Sasaki, S. and Miura, N. (1991). Appl. Phys. Lett. **59**, (in press).
18. Permogorov, S., Naumov, A., Gourdon, C. and Lavallad, P. (1990). Solid St. Commun. **74**, pp.1057-1061.
19. Brown, L. D. L., Jaros, M. and Wolford, D. F. (1989). Phys. Rev. **B40**, pp.6413-6416.
20. Minami, F., Todori, K. and Inoue, K. (1989). Semicond. Sci. Technol. **4**, pp.265-266.
21. Nikitine, S. and Perry, G. (1955). Compt. Rend. **240**, pp.64-66.
22. Harbeke G. and Tosatti, E. (1972). Phys. Rev. Lett. **28**, pp.1567-1570.
23. Baldini, G. and Franchi, S. (1971). Phys.Rev.Lett. **26**, pp.503-505.
24. Le Chi Thanh, G. Deperursinge, F. Levy and E. Mooser (1975). J. Phys. Chem. Solids, **36**, pp.699-702.
25. Miura, N., Kido, G., Katayama, H. and Chikazumi, S. (1980). J. Phys. Soc. Jpn. **49 Suppl.**, A pp.409-412.
26. Skolnick, M. S., Le Chi Thanh, Le Chi, Levy, F. and Harbeke, G. (1977). Physica **89B**, pp.143-146.
27. Nagamune, Y., Takeyama, S. and Miura, N., Minagawa, T. and Misu, A. (1987). Appl. Phys. Lett. **50**, pp.1337-1339.
28. Nagamune, Y., Takeyama, S. and Miura, N. (1991). Phys. Rev. (in press).

29. Koma, A. (1985). in *Abstracts of the 17th Conf. on Solid State Devices and Materials*, (Tokyo, 1985). pp.13.
30. Takeyama, S., Watanabe, K., Komatsu, T. and Miura, N. (1990). in *Proc. 20th Int. Conf. Phys. Semiconductors*, eds. E. M. Anastassakis and J. D. Joannopoulos, (World Sci. Pub., 1990), pp.1553-1556.

Problems of Silicon Crystal for VLSI Applications

Seigô Kishino

Department of Electronics, Faculty of Engineering
Himeji Institute of Technology, Shosha, Himeji

Problems of Silicon Wafers are reviewed from the standpoint of the needs of VLSI devices. After specifying the characteristics of VLSI devices the representative problems of silicon wafers are discussed. They are the flatness of a silicon wafer with a large diameter, the surface quality of the wafer, and a low density of heavy metal impurity. Finally a Gettering procedure, which is indispensable in the VLSI era is discussed from the viewpoint of the recent VLSI processing. As a result , it is emphasised that a clean processing is preferable to the conventional one of the normal grade in the VLSI Processes, even when the Gettering procedure is applied.

1. INTRODUCTION

The LSI industry has grown rapidly in recent years at an average annual rate of 15% (1). Especially , an MOS LSI market has increased at an even higher rate and is expected to grow by about 16 - 20% annually hereafter. The main impetus for such extraordinary market growth is the intrinsic pervasiveness of electronic products and the continued technological breakthroughs in VLSI technology. The technological breakthroughs bring the reduction of unit cost per function (2) and result in the improved performance of the VLSI.

As is well known at the present time the LSI growth is mainly owing to the exponential growth of the number of components which can be put in one IC chip (3). The growth of the number is achieved by the continued reduction of the minimum device dimension. It is reported that the annual rate of the reduction

has been 13% since 1960.   As a result, the minimum
feature length has shrunk to submicron meters in recent
years.   By 2000 year the minimum dimension is expected
to be ~0.3 micron meters.

Very large scale integrated circuits contain 10^6-
10^7 active devices, and failure of any one of them or
their deviation from the required characteristics can
greatly reduce over all circuit performance.   This
shows that a very low failure rate is required for the
individual active devices as shown in Fig. 1.
Moreover, the diameter of a silicon wafer has increased
since the development of IC's, in order to maintain the
number of producible chip, as shown in Fig. 2.

At the same time, the flatness of a silicon wafer
is strictly required because the warpage of wafers
generates catastrophic defects in the device structure
during photolithographic processes.   It is not easy
for silicon wafers to meet simultaneously the demands
of both the large diameter and the surface flatness.
Therefore, the VLSI imposes stringent demands on the
quality of silicon wafers for their fabrication.

Representative items required for silicon wafers
are as follows; (a) a low density of heavy metal impurity, (b) a flatness of the wafer, and (c) a surface
crystalline quality of the wafer.

In this note, we would like to discuss problems of
silicon wafers in connection with requirements of the
VLSI devices.

2. CHARACTERISTIC FEATURE OF LSI

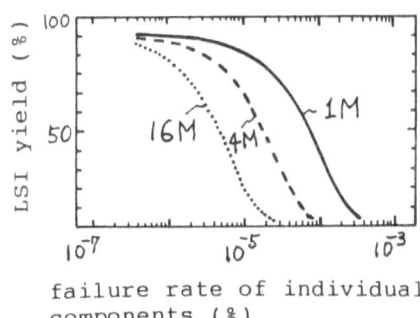

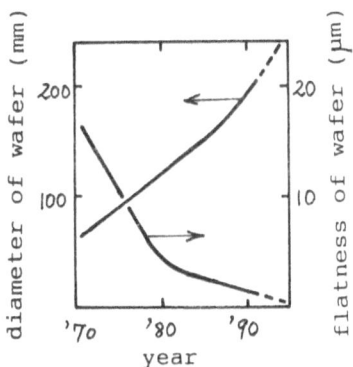

Fig.1 LSI yield versus fail-
      ure rate of device
      component

Fig.2 Wafer diameter and wa-
      fer flatness versus
      calendar year

239

VLSI devices have many characteristic points.   Of
these, following five items are closely correlated with
problems of silicon crystals.   They are as follows.
(i) VLSI comprises complicated electronic circuits composed of many devices.   This complexity is dependent
on the number of components per one IC chip.   IC complexity has advanced from small-scale integration (SSI)
to medium-scale integration (MSI), to LSI, and finally
to VLSI(or ULSI) as shown in Table 1.   These acronyms
are originated with the number of components of IC.
(ii) VLSI is provided with both the complexity and the
integration of devices.   They are due to the device
miniaturization.   The miniaturization requires micro-
fabrication with a high degree of accuracy.   For
example, the fabrication of 0.5 μm devices requires the
accuracy of 0.05 μm.

(iii) The production of VLSI is effectively performed at a very high speed.   As to the high speed
production, it may sound strange.   The high speed production offers the reasonable cost of the VLSI chip.
However, this also demands of silicon wafers severe
quality.   In the VLSI processing, many chips are designed on the surface of one wafer and are manufactured
simultaneously according to the planar technology[+).

Here, let's postulate 200 chips per one wafer.   At
the same time, a batch of wafers, which are composed of
100 sheets of silicon wafer, are assumed to be simultaneously processed in the VLSI production.   In order to
complete the VLSI processes, many process steps of 100-
200 must be conducted.   However, the processing of
VLSI is efficiently performed during a relatively short
term of period, for example, one month.   At the same
time, VLSI chips are produced with a high production
yield (for example, 50%).   This is essential in order
to reduce unit cost per function of VLSI.

Under the conditions described above, the number
of VLSI chip, which is producible per one batch,
amounts to 10,000 (=200x
100x0.5), and 43,200
(=30x24x60) minutes are
spent for the completion
of the processing.
Therefore, one VLSI chip
is produced at an aver-
age rate of 4.3 minutes.
This is very high speed
production for complexi-
ty of VLSI.   It is to

Table 1. Components per one
chip versus acronyms
of IC

acronym	components per one chip
SSI	$\sim 10^2$
MSI	$10^2 \sim 10^3$
LSI	$10^3 \sim 10^5$
VLSI	$10^5 \sim 10^7$
ULSI	$10^7 \sim$

be noted that the high speed production is only possible under the conditions that successful micro-fabrication is achieved as well as the reasonable production yield of VLSI is obtained as described above.

3. PROBLEMS OF RECENT VLSI DEVICES

Here, we would like to discuss representative problems of VLSI devices[5] inside the scope of a matter which is correlated with the crystalline quality of silicon wafers. They are (i) the short channel effect of an MOS device, (ii) a thin thermal oxide film, and (iii) the degradation of S/N ratio.

The short channel effect has occurred in the MOS transistors which are miniaturized in order to achieve both the complexity and the integration of VLSI. The effect is shown in Fig.3, where the threshold voltage V_{th} of the MOS transistor decreases as the gate length (nearly equals a channel length) L_g is reduced. This phenomenon occurs because of the departure from the long channel behavior of an MOS transistor. This departure, the short channel effect, arises as a result of a two-dimensional potential distribution.[6]

In those days when the channel length was long, the fabrication of an MOS device was not so difficult because the fluctuation of the channel length did not significantly change the threshold voltage V_{th}. However, in the short channel device, the fluctuation of the channel length can generate catastrophic failure of device characteristics through the deviation of V_{th}.

This effect shows that the flatness of a silicon wafer is very important to produce uniform MOS transistors without the fluctuation   This is because the successful micro-fabrication of the MOS transistor is not achieved without a high degree of the flatness of the silicon wafer.

As well known, the thin thermal oxide film is used as the gate insulator. Therefore, this is the key material of an MOS transistor. Especially, the thin oxide film is necessary to prevent the short channel effect from occurring. The thin oxide film is also essential in order to produce a large capacitance of an MOS memory device.

As the oxide thickness decreases,

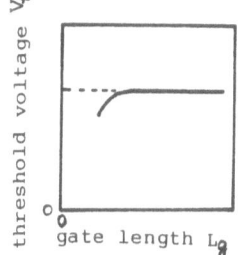

Fig.3 Short channel effect of MOS transistor

the electric field of the oxide film increases. As a result, a high break down field is required for the thin oxide film. The requirements are only acceptable upon the conditions that the thin film is formed on the crystalline surface where there are few defects and impurities. Therefore, a perfectly clean and defect-free surface is desirable as the surface of the silicon wafer.

The degradation of a signal to noise ratio (S/N) has become serious problems recently. The S/N ratio degrades by the decrease of a signal or by the increase of a noise. In the VLSI devices both the decrease of the signal and the increase of the noise may occur separately or simultaneously. This is because the device miniaturization reduces the signal charges as schematically shown in Fig. 4. For example, if 16 electronic charges are used as a signal "1" and zero charge as a signal "0", one surplus electronic charge scarecely degrades the S/N ratio as shown in Fig. 4(a) . However, if one electronic charge is used as a signal "1" as shown in Fig. 4(b), one excess electronic charge is sufficient to degrade the S/N ratio. This situation arises actually in the highly integrated VLSI. In the situation, the less the noise charges, the better.

The magnitude of the noise current is dependent on the density of the generation center in the depletion layer of a p-n junction. As to the generation center, a heavy metal impurity is most active. Therefore, it is desirable to reduce the heavy metal impurity as far as possible in the silicon wafer. The heavy metal impurity also reduces the generation lifetime and causes the refresh time failure of DRAM (Dynamic Random Access Memory)[7]. At any rate, heavy metal impurity is very harmful to the VLSI devices.

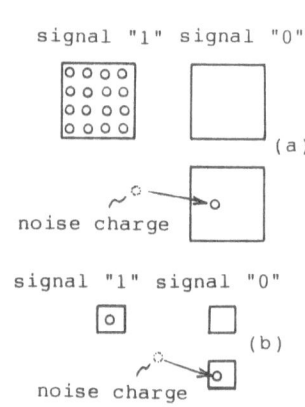

Fig.4 Decrease of S/N ratio by reducing signal charges

4. PROBLEMS OF SILICON WAFER

In the VLSI era, stringent demands are imposed on the quality of the silicon wafer in order to adjust it to the VLSI processing. Of many stringent demands, following items are especially important for the VLSI processing; they are (i) the flatness of wafer with a large diameter, (ii) the high quality of a wafer surface, (iii) a low density of heavy metal impurity, and (iv) a gettering technique of heavy metal impurities.

A surface flatness is one of the most important specifications required for the silicon wafer. The flatness must be restricted to within several microns. This restriction is due to the requirement of a lithography processing. The need of a high degree of surface flatness is not reduced, even though the diameter of the wafer increases to 200 mm and even larger than that. This is because a more miniaturized device than ever before is fabricated in the advanced ULSI age . The strictness of this specification is understandable using the analogy between the silicon wafer and the baseball ground of 150 m in diameter which is 1000 times of the 6 inch wafer. Similarly, several micron meters in the silicon wafer corresponds to several milli-meters in the ground. A height difference of several milli-meters is scarecely noticeable in the huge baseball ground[8]. A large diameter wafer is also desirable for further productivity improvement.

In highly integrated MOS devices, a very thin oxide layer is desirable for the sake of preventing the short channel effect from occurring and of obtaining a large capacitance[6]. However, the very thin oxide layer is easily affected by the surface quality of the silicon wafer. As a result, the quality of the oxide layer depends on the cleanness of the wafer surface. Microdefects in the surface region of the wafer are, at the same time, transferred to the thin oxide layer as shown in Fig. 5, and this phenomenon results in the low breakdown voltage of the MOS devices[9].

The contamination has direct effects on device properties such as leakage current and breakdown voltage. Of the contaminations, those by the heavy metal impurity could seriously hamper the VLSI fabrication process. This is because heavy metal impurities enhance the leakage current and reduce minority carrier lifetime, as shown in Fig. 6, by the increase of the deep level N_t which provides recombination center for minority carriers. Table 2 shows the correlation between the heavy

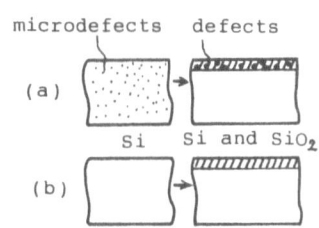

Fig.5 Effect of microdefects on oxide film quality

metal impurity, yield of discrete device, SSI, and VLSI. This Table shows that the VLSI production is only possible when the processed wafers are not densely contaminated with the heavy metal impurities.

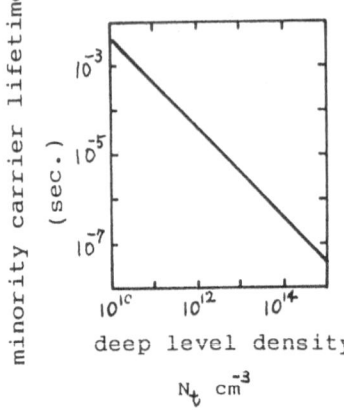

Fig.6 Minority carrier life-time versus deep level density N_t

For the sake of the contamination protection, various gettering procedures are used since the development of semiconductor devices. Of these gettering techniques, a backside damage gettering (B-G) and an intrinsic gettering (I-G) are illustrated in Figs. 7(a) and 7(b). The I-G procedure is a relatively new technique which began emerging about ten years ago. In the I-G technique, oxygen precipitates and the resultant microdefects are utilized as the gettering sites. The microdefects are composed of dislocations, stacking faults, precipitates, and their complexes. They are introduced in the interior of the silicon wafer by thermal treatments.

The I-G procedure is easily performed using a normal CZ silicon wafer. This is because the CZ silicon wafer contains an oxygen content sufficient for the I-G treatment. In the treatment, a denuded zone (d.z.) is formed besides the defective interior region as shown in Fig. 7(b). The denuded zone, which provides defect-free region for VLSI devices, is formed in the surface layer by means of the thermal annealing in a non-oxygen ambience.

5. OPTIMAL GETTERING PROCEDURE FOR VLSI PROCESSING[10]

A gettering procedure has

Table 2. Density of heavy metal impurity D_{imp} versus device yield Y_s or Y_{LSI}

D_{imp}	Y	Y_{LSI}		
		$n=10^2$	$n=10^4$	$n=10^6$
10^{15}	0.000000	0.000000	0.000000	0.000000
10^{14}	0.370000	0.000000	0.000000	0.000000
10^{13}	0.990000	0.370000	0.000000	0.000000
10^{12}	0.999900	0.990000	0.370000	0.000000
10^{11}	0.999999	0.999900	0.990000	0.370000

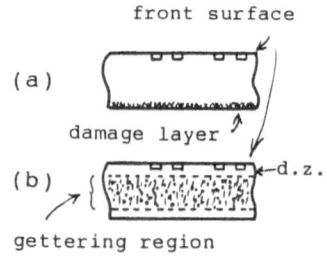

Fiig.7 Cross section of wafer showing gettering technique
(a) backside damage and
(b) intrinsic gettering

both positive and negative effects on the device characteristics. In those days when discrete devices and SSI devices were mainly produced, the negative effects were overwhelmed with the positive effects. As a result, the gettering procedures were easily conducted and were always profitable.

However, in the present VLSI era it is not easy for the gettering procedure to avoid the negative influences on the device parameters. This is because VLSI contains 10^6-10^7 active devices, and a failure of any one of them or their deviation from the required characteristics can greatly reduce over all circuit performance. This is clear from the comparison between Figs. 8 (or Fig. 9) and 10. Figures 8 and 9 show the device yield of a discrete device and of SSI as a function of gettering ability, where the impurity is of 10^{15} cm^3 and the gettering ability D_d indicates the dislocation density introduced in the gettering site region. In these figures, the sign M shows the ratio of the dislocation density between in the device area and in the gettering site region.

If a device yield scarcely depends on the magnitude of the M value, the gettering procedure is easily applicable to the processing of the device fabrication. This is because the condition of M=1 implies that the dislocation in the device area

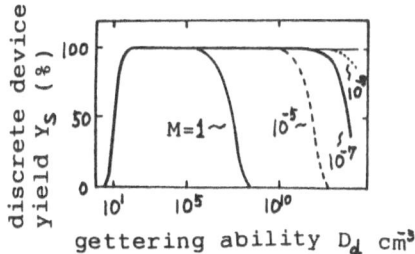

Fig.8 Discrete device yield Y_S versus gettering ability D_d in densely contaminated process (impurity density $D_{imp}=10^{15}$ cm^{-3})

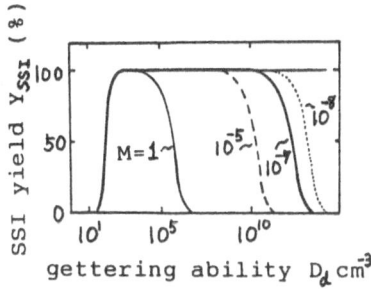

Fig.9 SSI yield Y_{SSI} versus gettering ability D_d in densely contaminated process (impurity density $D_{imp}=10^{15}$ cm^{-3})

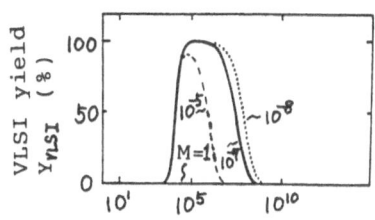

Fig.10 VLSI yield Y_{VLSI} versus gettering ability in densely contaminated process (impurity density $D_{imp}=10^{15}$ cm^{-3})

245

does not significantly affect the characteristics of devices. From Figs. 8 and 9, it is clear that the gettering procedure is profitable over a wide range of gettering ability D_d regardless of the magnitude of the M value in the processes of both the discrete device and SSI. This shows that the gettering procedure is successfully available as well as very profitable in these processes.

On the other hand, the effective gettering procedure is possible only in the narrow range of D_d in VLSI processing as shown in Fig. 10. This is due to two factors. One is the problems of the dislocation propagation from the gettering site region to the device area. This is clear from Fig. 10 where the high production yield is obtained only when $M \ll 1$. The other is the wafer warpage shown in Fig. 11[11]. As already discussed, the wafer warpage may generate catastrophic defects in the micro-structure of VLSI devices during photolithographic processes.

With the transition from SSI to VLSI, device dimensions drop significantly. Therefore, even the small wafer warpage has become harmful in the VLSI processing. The gettering procedures enhance the wafer warpage because the mechanical strength of the wafer decreases by the introduction of dislocations[12]. This shows that a large D_d value decreases the VLSI yield Y_{VLSI} significantly. On the other hand, it is also undesirable that the gettering ability is too small to eliminate the detrimental effects of impurities on device properties. Consequently, the range of the effective gettering becomes narrower in VLSI processes as shown in Fig. 10. The effective gettering

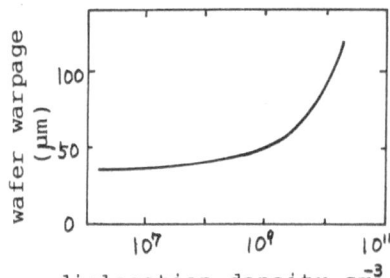

Fig.11 Wafer warpage after
thermal treatments
versus dislocation
density in bulk Si

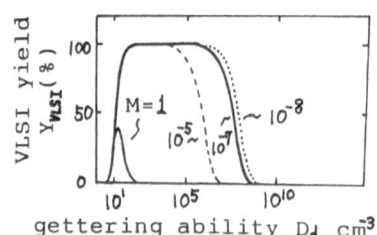

Fig.12 VLSI yield Y_{VLSI} versus
gettering ability in
less-contaminated
process
(impurity density $D_{imp} = 10^{12} cm^{-3}$)

and the good mechanical strength appear to be opposing requirements. Therefore, the trade-offs must be adjusted between two.

In the VLSI processing, strong gettering is the cause of negative effects on the device characteristics . If the rather soft gettering is applied, the profitable results may be attained as is clear from the comparison between Figs. 10 and 12. Figure 12 shows the case where the processing is not so densely contaminated with the heavy metal impurity. In this case a rather wide gettering ability is effectively available.

6. CONCLUSIVE REMARKS

Problems of a silicon crystal for VLSI applications are, as a matter of course, densely correlated with the problems of recent VLSI devices. Recent VLSI devices require the silicon wafer to provide stringent demands of VLSI. They are the flatness of a silicon wafer, the perfectness of the wafer surface, and an extremely low density of heavy metal impurities. These items correspond to the affairs of VLSI devices such as micro-fabrication of devices, a high quality of the thin oxide film, the degradation of S/N ratio, etc.
Especially, even a low density of metal impurity will generate catastrophic failure in the yield of VLSI. This is because a failure of any one of them can greatly reduce over all circuit performance of VLSI. The excess heavy metal impurities could be removed by means of the gettering procedure. However, strong gettering will adversely affect circuit performance of VLSI. Therefore, a clean processing is decidedly preferable to the conventional one of a normal grade in VLSI process even when the gettering procedure is applied during the processing.

ACKNOWLEDGMENTS

We would like to thank Dr. H. Niu of our laboratory for various discussions on problems of heavy metal impurities.

REFERENCES

1) Sze, S. M. (1983). "VLSI Technology" ed. Sze, S. M.,

McGraw-Hill International Book Company, p.1-2.

2) Noyce, R. N. (1981). Microelectronics, "The Micro-electronics Revolution" ed. Forester, T., MIT Press, p.29

3) Moore, G. (1980). VLSI, What Does The Future Held, Electron. $\underline{42}$, pp.14

4) Hoerni, J. A. (1960). Planar Silicon Transistors and Diodes, IRE Electron. Devices Meeting, Washington, D. C.

5) El-Mansy, Y. (1982). MOS Device and Technology Constraints in VLSI, IEEE J. Solid-State Circuits SC-$\underline{17}$(2), pp.197-203.

6) Dennard, R. H., et al. (1974). Design of Ion-Implanted MOS FET's with Very Small Physical Dimensions, IEEE J. Solid-State Circuit SC-$\underline{9}$(5), pp. 256-268.

7) Sun, R. C. and Clements, J. T. (1977). Characterization of Reverse-Bias Leakage Current and Their Effect on The Holding Time Characteristics of MOS Dynamic RAM Circuits, IEDM Technical Digest, p.254.

8) Kishino, S. (1985). Material Characterization for VLSI Applications, Electrochem. Soc. Proc. of The 3rd International Sympo. on VLSI Sci. and Technol. p. 399-418.

9) Itsumi, M. and Kiyosumi, F. (1982). Origin and Elimination of Defects in SiO_2 Thermally Grown on Czochralski Silicon Substrate, Appl. Phys. Lett. $\underline{40}$ (6), pp.496-498.

10) Kishino, S., Yoshida, H., and Niu, H. (1991). Shift of Gettering Condition with The Transition from IC to VLSI, Jpn. J. Appl. Phys. to bs published.

11) Shimizu, H., Watanabe, T., and Kakui, Y. (1985). Warpage of Czochralski-Grown Silicon Wafers as Affected by Oxygen Precipitation, Jpn. J. Appl. Phys. $\underline{24}$(7), pp.815-821.

12) Yonenaga, I. and Sumino, K. (1978). Dislocation Dynamics in The Plastic Deformation of Silicon Crystal, Phys. Stat. Sol. (a) $\underline{50}$, pp.685-693.

Molecular Beam Epitaxy

P.R. Vaya and K. Ponnuraju

Semiconductor Device Research Laboratory, Centre for Systems and Devices
Indian Institute of Technology, Madras-600 036, India

Molecular Beam Epitaxy (MBE) has become a well-established technique for the growth of ultra-thin films and devices with precise control of thickness, doping concentration and composition. The importance of MBE, basic growth processes, different forms of MBE and some of its recent applications are described in this review article.

1. INTRODUCTION

The importance of epitaxy in semiconductor device fabrication is a direct consequence of two critical needs; for thin, defect-free single-crystal films with precisely defined geometrical, electrical, and optical properties, and for heterojunction structures free of interfacial impurities and defects. The traditional techniques of liquid-phase epitaxy, (LPE), which is dependent on thermodynamic phase equilibria, and vapor-phase epitaxy, (VPE),which achieves growth by chemical reactions in the gas phase on a heated substrate, can certainly meet certain subsets of the foregoing requirements, but not all of them. The newer technique, molecular beam epitaxy,(MBE), achieves epitaxial growth by the reaction of one or more thermal atomic or molecular beams of the constituent elements with a crystalline substrate surface held at a suitable elevated temperature under ultrahigh vacuum (UHV). The technique, as it stands today, has come a long way since the pioneering studies of Davey and Pankey [1]. which have led to a better understanding of the growth kinetics and improved-quality of materials and devices [2-10]. Essentially confined to research and development until about 1976, MBE has now emerged as a reliable growth technique for the realization of stringent device requirements. The uniquences of the technique lies mainly in the tremendous precision in controlling layer doping, thickness, and composition; in growing modulated structures whose periods are typically less than the electron mean free path; and in achieving nearly perfect heterointerfaces and surface morphologies.

2.BASIC GROWTH PROCESS

MBE is a controlled thermal evaporation process under ultrahigh-vacuum conditions. The process is shown schematically in Fig.1 for the growth of GaAs with the possibility for dopant incorporation. The effusion cells are designed such that realistic fluxes for crystal growth at the substrate can be realized while maintaining the Knudsen effusion condition (i.e., the cell aperture is smaller than the mean free path of the vaporized effusing species within the cell). The individual cells are provided with externally controlled mechanical shutters whose movement times are less than the time taken to grow a monolayer. Therefore, very abrupt composition and doping profiles are possible. Interfaces that are one monolayer abrupt can be obtained fairly easily. The epitaxial growth process by MBE involves a series of events taking place at the heated substrate surface(11,12): (1) adsorption of the constitutent atoms and molecules; (2) surface migration and dissociation of the adsorbed molecules; and (3) fixation of the atoms to the substrate in crystallographically and energetically preferred sites. These processes result in the growth of a single crystal film with a crystallographic structure related to that of the substrate exist and complete with the growth process, so that the overall growth rate is determined by a precise balance between these various events.

From pulsed molecular beam experiments Arthur [2,13] observed that below a substrate temperature of $480^\circ C$, Ga had a unity sticking coefficient on (100) GaAs. Above this temperature the coefficient is less than unity. The adsorption and incorporation of As_2 or As_4 molecules is more complex. It was found that in general As sticks only when a Ga adatom plane is already established. Joyce and coworkers [9] have described the kinetic processes leading to the growth of GaAs from Ga and As_2 or As_4 molecules. The cation is in atomic form when it reaches the heated surface. It attaches randomly to a surface site and undergoes several kinetically controlled steps before it is finally incorporated. The As_2 or As_4 atoms is first physisorbed into a mobile, weakly bound to a precursor state. As this state moves on the surface, some loss occurs due to reevaporation; the rest is finally incorporated in paired Ga lattice sites by dissociative chemisorption.

The important fact which emerged from the early kinetic studies is that stoichiometric GaAs can be grown over a wide range of substrate temperatures by maintaining an excessive overpressure of As_2 or As_4 over the Ga beam pressure. Substrate temperatures during growth are usually close to and slightly higher than the congruent evaporation temperature of the growing compound. Under normal MBE growth conditions, where the incorporation rate of the cations is nearly 100%, the cation surface migration needs to be high. Otherwise, MBE growth will occur by a three-dimensional island mode rather the step-growth mode which is necessary for producing high-quality and abrupt interfaces.

3. SYSTEM DESCRIPTION

A typical present-day MBE growth facility consists of the UHV growth chamber into which the growth substrate is introduced through one or two sample-

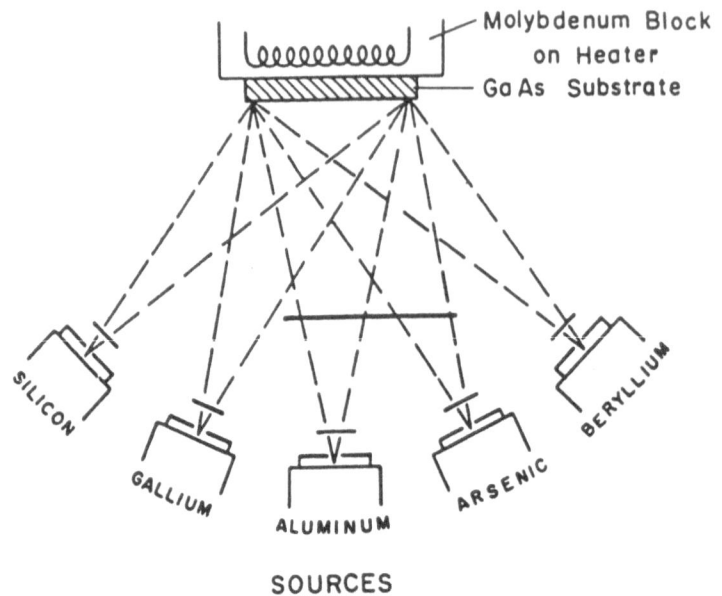

Fig.1   Schematic illustration of the molecular
        beam expitaxial process.

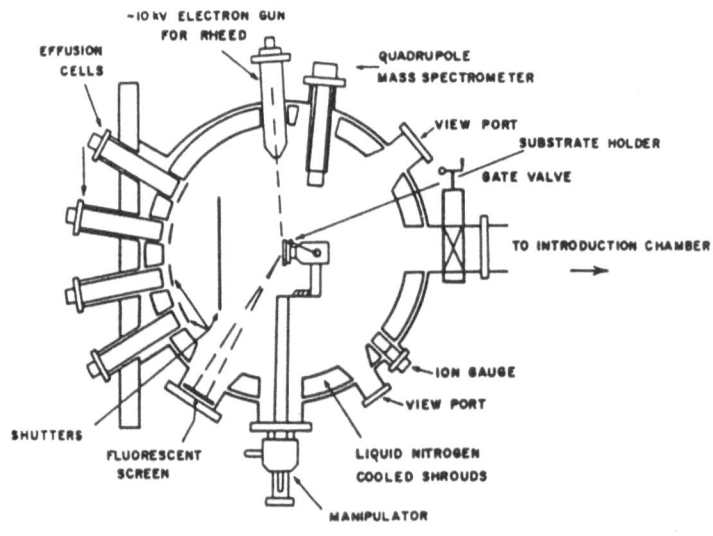

Fig.2   Schematic cross section of a typical MBE
        growth chamber equipped with in situ
        diagnostic tools.

exchange load-locks. The base pressure in the growth chamber is usually about 10^{-11} torr, while the other chambers are at about 10^{-10} torr. The schematic of the growth chamber is shown in Fig.2. The effusion cells are made of pyrolytic Bornon Nitride (BN). To obtain the dimers from the tetramer species of the group V elements, external cracking cells are incorporated. Cracking is enhanced in the presence of a loosely packed catalytic agent. Charge interlocks with auxiliary pumping is sometimes used for the more rapidly depleting group V species. The growth chamber and effusion cells are provided with liquid N_2 cryoshrouds, which are kept cold during growth.

Most growth systems are equipped with in situ surface diagnostic and analytical capabilities in the growth and auxiliary chambers. The most common facilities in the growth chambers are a quadrupole mass spectrometer (or residual gas analyzer) which gives important information regarding the ambient in the growth chamber at all times and a reflection high-energy electron diffraction (RHEED) system which gives an insight to the surface structures of crystals and the growth mechanism. Electrons from a high-energy (~10keV) electron gun strike the substrate or the growing layer surface at glancing incidence, and the diffraction pattern is monitored on a fluorescent screen. In addition to the static RHEED pattern, it is possible to explore some features of the growth dynamics of MBE by monitoring temporal variations in the intensity of various features in the RHEED pattern. It has been found [14] that damped oscillations in the intensity of both the specular and diffused beams occur immediately after initiation of growth. By observation of these oscillations, it is possible quantitatively to determine the surface diffusion rates of adatoms and to monitor the growth of monolayers [14].

4.THE IMPORTANCE OF THE SUBSTRATE TEMPERATURE

The substrate temperature, T_s, for optimum growth conditions must be sufficiently high to ensure that the atoms sticking to the surface maintain a sufficient surface mobility to allow them to settle in their equilibrium positions, and should not be too high to cause excessive re-evaporation of the impinging fluxes. When the substrate temperature is too low polycrystalline or amorphous growth results. The actual temperature for the loss of epitaxy depends critically on the cleanliness of the growing surface and on the density of background impurities in the gas phase. The use of a sufficiently high growth temperature will also be effective in displacing the equilibrium conditions. A much higher As-flux has been found to lead to an increased incorporation of deep levels, while a substantially lower As flux will give rise to Ga-rich conditions, leading to the build-up of excess gallium on the surface with ultimately results in Ga-droplet formation and unacceptable surface morphology degradation.

For a simplified view of the growth process it can be assumed that at relatively low growth temperatures all the incident group III atoms stick on the growth surface and only enough group V elements adhere as necessary in order to give stoichiometric growth. The above model holds provided the growth temperature is above the congruent evaporation temperature of the film which insures that the group V elements are preferentially desorbed; in practice therefore, it is sufficient to provide an overpressure of the group V species during deposition. In the presence of a steady state Gallium adatom population, however, the As_2 and As_4 molecules will bind to the surface and ultimatly will be incorporated in the growing layer.

The growth of ternary or quaternary alloys sharing one common group V element (examples: $Al_xGa_{1-x}As$, $Ga_xIn_{1-x}As$, $Al_xGa_yIn_{1-x-y}As$) can be treated in a manner similar to that of the growth of the binary compounds, provided that the growth temperature is sufficiently low so that the sticking coefficients of the different group III elements involved remain close to unity. Despite the excellent results that have been achieved in some laboratories with the conventional MBE technique so far, interest for the growth of ternary and quaternary alloys with III or more group V elements is shifting to the newly developed gas source MBE or CBE techniques where the beam intensities can be more readily controlled using mass flow or gas line pressure controllers.

5. VARIOUS FORMS OF THE MBE PROCESS

The various forms of the MBE process which are being used, are given below.

(1) Solid source (conventional) MBE
(2) Gas source MBE
(3) Chemical beam epitaxy
(4) Atomic layer epitaxy

5.1 Solid source (conventional) MBE

It is based on the use of solid source (for example, gallium or silicon) heated by thermal radiation or electron beam impact. Fig.2 shows schematically the principal components of a solid source MBE system.

5.2 Gas source MBE

In this system the evaporation materials are introduced in gaseous form and decomposed into their elements in a hot zone crucible (for example, the decomposition of AsH_3 into arsenic and hydrogen) or directly at the surface of the substrate (for example, decomposition of trimethylgallium and methyl radicals on the heated substrate surface).

In the Ga-In-P-As gas source system (Fig.3), the group III elements are introduced at low pressures (typically 10 to 100 torr) in the growth chamber in the form of metalorganic gaseous compounds such as, for the present example triethylgallium (TEGa) and trimethylindium (TMIn), while the group V elements are introduced as hydrides (in this case arsine and phosphine) at intermediate to high pressures (100 to 760 torr). The hydrides are decomposed in a hot zone crucible held at temperatures above $900^{\circ}C$ before reaching the substrate, while the group III elements

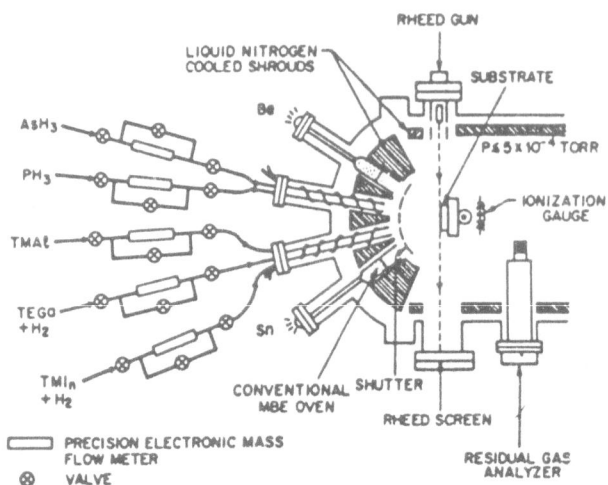

Fig.3   Schematic of a gas source MBE system for
growth of GaInPAs quaternary alloys.

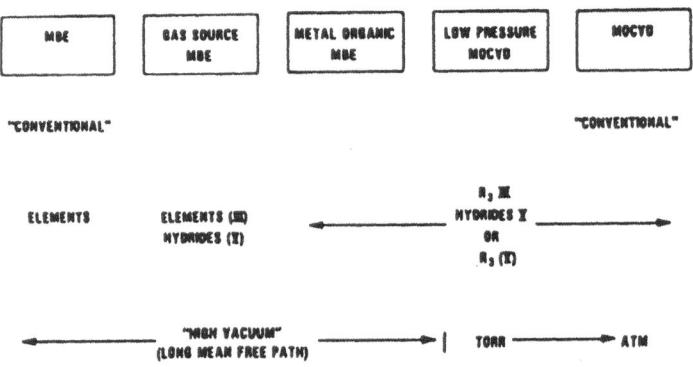

Fig.4   Typical pressures during growth for conventional MBE, gas source MBE or CBE,
low pressure MO-CVD and atmospheric
pressure MO-CVD.

reach the substrate in metalorganic compound form and decompose catalytically on the substrate surface. In this system, growth rates are controlled by the group III element beam fluxes, while the layer compositions are determined by the Ga to In and P to As ratios at the given growth temperature. Again, the group V equivalent beam pressure over the substrate should be sufficiently large to prevent layer decomposition, and the doping levels are controlled by separate thermally heated solid cells. Layer transitions are realized in this system either by gas source switching or by shuttering.

During the epitaxial deposition process, the background pressure in the growth chamber extends over the range from 10^{-7} torr in solid source MBE to 10^{-5} torr for gas source MBE due to the presence of the hydrogen and hydrocarbon decompositon products. At these pressures, the mean free paths of the molecules range from approximately 50 m down to 50 cm, and are thus much larger or at most comparable to the dimensions of the growth chamber. In this pressure range, the behavior of the particles in the growth chamber is therefore entirely determined by their surface coilisions, and particle to particle interactions in free space can be neglected. This feature constitutes the main distinction with the alternative epitaxial growth techniques for these materials such as atmospheric pressure or low-pressure (1 to 100 torr) chemical vapor deposition (CVD) where mean free paths range from 50 um down to less than 0.1 um (Fig.4).

5.3 Chemical Beam Epitaxy

There are significant differences between various gas source MBE techniques depending on whether the elements constituting the film arrive at the substrate surface in elemental form (for example, as Ga, As, As_2, As_4, etc. molecules) or as constitutent elements of a more complex molecule (for example, in the form of trimethylgallium, trimethylaluminum, etc.) In the first case, the surface reactions are identical to those found in conventional solid-source MBE, whereas in the second case, the surface reaction kinetics and surface chemistry may be significantly different. For this reason, the term chemical beam epitaxy (CBE) has been proposed for the MBE technique based on the simultaneous use of hydride and metalorganic sources.

The difference between the gas source MBE and CBE is that in CBE, group III elements are obtained from alkyls directed to the substrates unlike they are obtained by conventional evaporation methods in GSMBE whereas group V elements are obtained by thermally cracking in cracker cells in both the techniques.

5.4 Atomic Layer Epitaxy

Atomic layer epitaxy (ALE) introduced by Suntola in 1985 [15] has become a well established crystallization technique for growing compound semiconductor films. In ALE, a monolayer of each constituent atom or molecule, is layered down separately in place of having a mixed flux at the substrate. For example (see Fig. 5) a first layer of As is deposited on the substrate, and excess arsenic is swept out of the system. This layer is followed by a flux of gallium that reacts with arsenic layer, filling sites to complete a layer of gallium and the compound GaAs. Again, excess gallium is removed from the system. This is, in essence, a building of a compound or alloy as a monolayer superlattice.

Usually, a characteristic delay time called 'dead time' is introduced in ALE growth between the sequential pulses of the reactant species. During this time, the deposition process is interrupted to enable the free-reevaporation of these atoms of the deposited species, which are in excess for the first chemisorbed monolayer and, therefore, are only weakly bound to the substrate surface. ALE can be considered a 'digital' deposition process because the thickness of the epitaxial layer is determined primarily by the number of exposure cycles and is much less sensitive to parameters like growth time or fluxes of the consituent reactants. A review of the state of the art of ALE is presented in the ref [16].

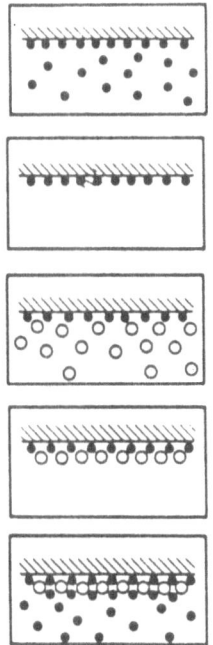

Fig.5 Atomic layer epitaxy.

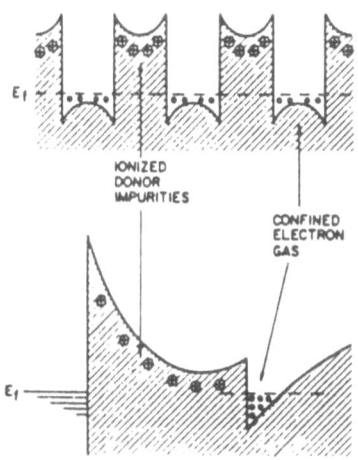

Fig.6 Modulation doping (a) in a superlattice,
(b) in a single heterojunction with an
attached Schottky surface barrier.

6.APPLICATIONS

Although the applications of MBE are large in number a few will be mentioned here. Microwave devices such as FETs, varactors, mixers and optoelectronic devices such as heterostructure lasers and photodetectors have been farbicated from MBE prepared material. MBE techniques have been employed to study the growth of metal films on III-V compounds. This includes both Schottky barriers and non-alloyed ohmic contacts. $Al_xGa_{1-x}As$ solar cells [17] and high electron mobility transistor (HEMT) with higher performance and fast speed have been fabricated by MBE [18]. Modulation doped FET's with reduced low-frequency noise and thermally stable performance have been achieved with this technique [19]. Double heterostructure LEDs and lasers have also been realized [20]. Recently MBE is used for the growth of superlattices, strained layer superlattices, single and multiquantum well structures.

A superlattice is defined as a one dimensional periodic structure consisting of alternating ultrathin layers whose period is less than the electron mean free path. When the above condition for superlattice formation is fulfilled, the entire electron system enters into a quantum regime and exhibits a number of novel and very interesting features. These superlattices have formed the basis for the study of a large family of novel physical and device concepts in semicondcutors, such as the study of transport properties in the direction perpendicular to the superlattice, resonant tunneling between superlattice wells, luminescence, light scattering, lasing action, etc. Subsequently, interst developed in the transport and optical properties of the two-dimensional electron gas system in the plane of the superlattice layer.

It is also possible to spatially separate the free carriers - electrons or holes — from their parent impurity atoms by locating the doping impurities in the regions of the potential barriers in suprlattices (called modulation doped superlattices). As a result, the free carriers suffer a much reduced ionized

260

impurity scattering, so that high mobilities and high carrier concentrations can be simultaneously achieved. This concept was first successfully implemented by Dingle [21] in modulation-doped GaAs/AlGaAs superlattices as illustrated schematically in Fig.6(a). The modulation doping concept was subsequentially applied to single-well heterostructures of the type illustrated in Fig.6(b), and applied towards the development of a new high-speed field effect transistor structure [22,23] called HEMT or TEGFET (two-dimensional electron gas field effect transistor), whose performances significantly better those achievable with traditional MESFET devices made of the same parent material. Structures of this type have also been used to demonstrate the quantized Hall effect and have led to the precise determination of the fine structure constant using the two-dimensional electron gas in a GaAs/AlGaAs single quantum well structure [24].

The superlattice and modulation doping concepts already finding increasing applications in the field of high frequency devices based on the HEMT or TEGFET principle, as well as in the area of quantum-well diode lasers, superlattice avalanche photodiodes, multiple resonant tunneling devices and oscillators, and so on. These results have opened up a new area of interdisciplinary investigations in the fields of materials science and device physics, and have provided new challenges towards the achievement of ever greater control and perfection of the MBE epitaxial technique.

REFERENCES

1. Davey, J.E. and Pankey, T.J. (1968). Journal of Applied Physics, 39, P. 1941.

2. Arthur, J.R. (1974). Surface Science, 43, P. 449.

3. Cho, A.Y. and Arthur, J.R. (1975). Progress in Solid State Chemistry, 10, P. 157.

4. Ilegems, M. (1977). Journal of Applied Physics, 48, P. 1278.

5. Robinson, J.Y. and Ilegems, M. (1978). Reviews of Scientific Instrumentation, 49, P. 205.

6. Gossard, A.C. (1982). Thin solid Films, 49, P. 205.

7. Chang, L.L., Esaki, L., Howard, W.E., Ludeke, R. and Schul, G. (1973). Journal of Vacuum Science and Technology, 10, P. 655.

8. Ploog, K. (1980). Crystal Growth Propagation Application, 3, P. 73.

9. Joyce, B.A. and Foxon, C.T. (1975). Journal of Crystal Growth, 31, P. 122.

10. Holloway, H. and Walpole, J.N. (1979). Progress in Crystal Growth and characterization, 2, P. 122.

11. Neare, J.H., Joyce, B.A., Dobson, P.J. and Norton, N. (1983). Applied Physics, A, 31, P. 1.

12. Cho, A.Y. (1980). Growth and Properties of III-V semiconductors by MBE, 27, P. 191.

13. Arthur, J.R. (1968). Journal of Applied Physics, 39, P. 4032.

14. Neare, J.H. and Dobson, P.J. (1985). Applied Physics Letters, 47, P. 100.

15. Suntola, T. and Ayvarinen, J. (1985). Annual Review of Materials Science, 15, P. 177.

16. Herman, M.A, (1991) Vacuum, 42, P. 61.

17. Chikara, A. (1985). Journal of Applied Physics, 58, P. 2780.

18. Tung, P.N. (1982). Electron Letters, 10, P. 3.

19. Peng, C.K. and Klem, J. (1987). Journal of Crystal Growth, 81, P. 359.

20. Miller, D.L. and Asbeck, P.M. (1987). Journal of Crystal Growth, 81, P. 368.

21. Dingle, R.(1975).Advances in Solid State Physics, 15, P. 21.

22. Delagebeadenf, D. and Linh, N.T. (1982). IEEE Transaction on Electron Devices, 29, P. 955.

23. Panish, M.B. (1987). Journal of Crystal Grwoth, 81, P. 249.

24. Tsni, D.C., Shur, M.S., Drummond, T.J. and Morkoc, H. (1983). Journal of Applied Physics, 54, P. 6432.

Studies on ITO/Si Junctions Prepared by Spray Pyrolysis Technique

A. Subrahmanyam and V. Vasu

Department of Physics, Indian Institute of Technology, Madras 600 036, India

This paper deals with the photovoltaic behaviour of indium tin oxide (ITO)/silicon (single crystal) heterojunctions prepared by spray pyrolysis technique. The dependence of the photovoltaic properties on process temperature (T_p) and on the oxidation time (t_{ox}) have been studied. ITO on p-Si yielded ohmic contact. A photoconversion efficiency of 9.4 % is observed (under GE-ELH illumination of 100 mW/cm^2) for both small (0.04 cm^2) and large (1.0 cm^2) areas of ITO/n-Si junctions prepared at a temperature of 380°C and for an oxidation time of 60 sec. The junctions are observed to be quite stable with time. An attempt is made to understand the interfacial oxide layer (SiO_x) and its effect on the photoconversion in these junctions.

1. INTRODUCTION

Indium Tin Oxide (ITO) based heterojunctions, particularly ITO/Si junctions [1-5], have attracted considerable attention because of their several advantages in photovoltaics. Different techniques have been employed for ITO deposition onto n and p type single crystal Si wafers. Interestingly, it is found that the performance of the device depends largely upon the method of preparation of ITO films and on the conductivity type (n or p) of the silicon substrate.

Of the various techniques studied to prepare ITO/Si junctions, spray pyrolysis technique [4,5] is found to produce high performance cells ($\eta \simeq 10$ %). The technique has viability for mass production. One of the factors that govern the pyrolytic process is the reaction temperature; reaction kinetics are interesting (but less understood) in the formation of these films [6]. The high temperatures ($\sim 300 - 400$°C) required for the pyrolytic process, develop an unintentional oxide growth on silicon (SiO_x) which is responsible for the barrier formation in

ITO/Si junctions. The interface formed between ITO and Si is rather complicated as Fe and Cr may be introduced into sprayed ITO films in addition to Cl^- [7].

With a view to understand the interfacial oxide layer, present paper reports the dependence of the photovoltaic properties of ITO/Si junctions (prepared by spray pyrolysis technique) on the process temperature (T_p) and on oxidation time (t_{ox}).

2. EXPERIMENTAL DETAILS

The p and n type silicon wafers used in the present work are of <100> orientation and of resistivity 1 – 2 Ω cm. The wafers were single side polished as procured from M/s Wacker. Standard procedures were followed for cleaning the wafers which include the final step of oxide etching by the buffered hydrofluoric acid (34.6 % NH_4F : 6.8 % HF : 58.6 % H_2O). The back side (unpolished side) of n-Si was heavily doped with phosphorous by diffusion for ohmic contact fabrication. Aluminum was evaporated on to the back of the wafers at a pressure of 3.0 x 10^{-6} torr in a commercial vacuum coating unit. In the case of p-Si the contact was annealed in nitrogen atmosphere at $550° \pm 5°C$ for 10 minutes (no annealing was required for n-Si). The back contact was made only after the deposition of ITO on the polished front side. The front contact was a very small dot of air-dry silver paste.

On the polished front side of the wafers ITO films were deposited by spray pyrolysis technique. Details of the experimental technique are given in reference 6. The substrate temperature is measured by chromel-alumel thermocouple placed on a dummy substrate kept very near to the test sample; the accuracy in the temperature is $\pm 3°C$. The substrate temperature is varied from $340°C$ to $460°C$.

Spraying liquid consists of $InCl_3$ (99.99 % purity) and anhydrous $SnCl_4$ (99.99 % purity) in the proportion 95:5 by weight dissolved in 10 cm^3 of methyl alcohol. Flow rate of the liquid is maintained at 8 – 10 cm^3/min and the substrate-nozzle distance is kept at 30 cm. The details of ITO film properties and reaction kinetics of the film formation are presented in our earlier paper[8].

The tin doping in the ITO films (5 % by weight) is estimated by Electron Probe Micro Analysis (EPMA) and Energy Dispersive

Analysis of X-ray (EDAX) measurements.

Since the formation of native oxide on Si after cleaning is quite rapid at room temperature ($27°C$), following procedure is adopted for the preparation of ITO / Si junctions: the dummy substrate kept on the substrate holder inside the reaction chamber is allowed to reach the steady value of the set temperature (called as the substrate temperature); then the test Si wafer with mask is introduced into the chamber and the spraying of liquid is started after 60 seconds (this time is termed as oxidation time). It is quite likely that the test sample (Si wafer) might not have attained the steady value of the substrate temperature. Hence, the term process temperature is used. It may be pointed out that the process temperature always is less by about $10°C$ than the substrate temperature (as measured by a chromel–alumel thermocouple). After completion of spray (one minute), the Si wafers (ITO/Si junctions) have been carefully removed from the reaction chamber and were allowed to cool under ambient. Keeping process temperature at $380°C$, the oxidation time is changed between 30 sec – 90 sec for the ITO/n–Si junctions. The time (HF etched) Si wafer exposed to the ambient in the reaction chamber (furnace) at $380°C$ prior to ITO deposition is termed as oxidation time (t_{ox}). The quality of the ITO films formed at the process temperature is ascertained by following the procedure described for Si wafers for sodalime glass plates. It is found that with process temperature the sheet resistance and transmission for these ITO films follow similar dependence as observed in the case with substrate temperature reported earlier [8].

The active area of the junction is 0.04 cm^2. Large area junctions (1.0 cm^2) have also been prepared simultaneously and are found to exhibit identical results. However, areas larger than 1.2 cm^2 have shown small deviations in photoconversion efficiency. The junctions have been characterised by the conventional I–V (dark and illuminated) and C–V measurements. EG & G model 5210 Lock in amplifier interfaced (RS 232) with a personal computer is employed for C–V measurements. Necessary software has been developed by the authors. The spectral response (with suitable interference filters) and the illuminated I–V were obtained under 100 mW/cm^2 illumination produced by a GE–ELH lamp. The lamp is calibrated by a reference (secondary standard) silicon solar cell. The temperature of the samples during measurements (I–V and C–V) was maintained strictly at 25 ± $0.5°C$ by keeping them on a water cooled metal block. The temperature was measured by placing a copper–constantan thermocouple on a dummy Si wafer kept very close to the test cell.

Degradation studies on these ITO/Si junctions also have been

266

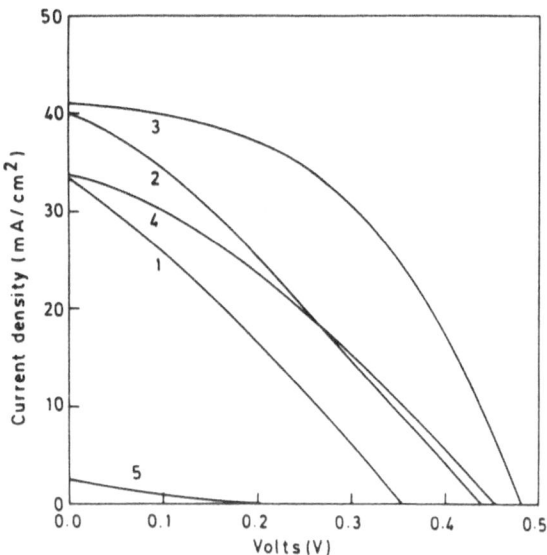

Fig. 1. I-V characteristic (under illumination) of
ITO/n-Si junctions prepared at different process
temperatures : 1. 300°C; 2. 340°C; 3. 380°C;
4. 420°C; 5. 460°C.

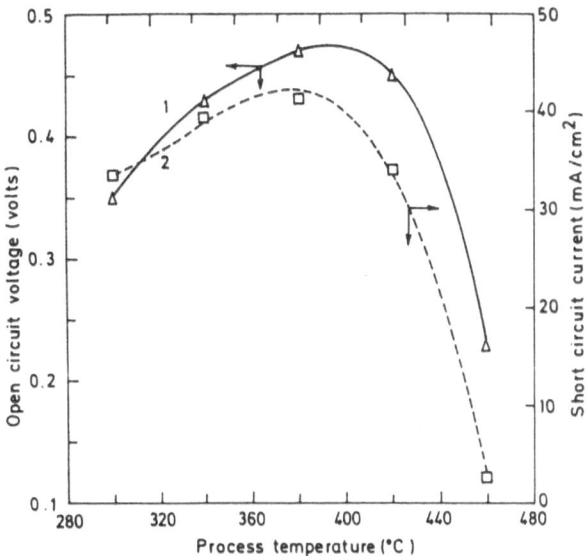

Fig. 2. Open circuit voltage and short circuit current
density of ITO/n-Si junctions as a function of
process temperature.

carried out. It is found that the junction behaviour is fairly constant (inspite of the fact that the cells are not encapsulated) and it is well within the limits of the instrument accuracy (± 2 %).

Several samples have been fabricated under nearly identical conditions and the reproducibility in the junction performance is ascertained. In each run of the experiment, 5 samples of area 0.04 cm^2 and 2 samples of area 1.0 cm^2 have been prepared. The reproducibility of the results is confirmed after repeating each run for five times. This is the reason why no specific identity (serial number or process run number) is given to the samples. The values reported in the present paper correspond to the smaller area (0.04 cm^2) cells.

3. RESULTS

3.1 Process temperature

3.1.1 I-V Characteristics

Fig. 1 shows the illuminated I-V characteristics of the ITO/n-Si junctions prepared at various process temperatures (Indium Tin Oxide on p-Si yielded ohmic contact). The related cell parameters are listed in Table 1. The series resistance (R_s) of the cells was calculated by noting the value of forward voltage (V') for which the current is the same as the short circuit current (I_{sc}) following the relation

$$R_s = (V' - V_{oc}) / I_{sc} \qquad (1)$$

The diode ideality factor n and the saturation current density J_o are determined following the variable illumination technique (in order to avoid the series resistance effects). It may be noted that both n and J_o decrease with increasing process temperature, reach minimum values around 420°C and then increase (Table 1). However, the open circuit voltage (V_{oc}) and short circuit current seem to achieve highest values around 380°C (fig. 2); the fill factor also follows the similar trend (Table 1).

It may be observed from these data that the properties of ITO/n-Si are sensitive to the ITO deposition temperature (and oxidation time). The cells prepared at 380°C exhibit a maximum efficiency of 9.4 % (under 100 mW/cm^2 illumination). It is interesting to note that though good quality ITO films (highly

268

TABLE 1
Properties of ITO/n-Si junctions (of area 0.04 cm^2) prepared at different process temperatures

T_p (°C)	V_{oc} (V)	J_{sc} (mA/cm^2)	ff Expt	ff Calc.	η (%)	n	J_o (µA/cm^2)	$\phi_{b,cv}$ (eV)	R_s Ω
300	0.35	33.51	0.29	0.57	3.40	2.42	88.30	0.69	136.4
340	0.43	39.47	0.31	0.64	5.25	2.12	15.30	0.80	82.4
380	0.47	41.25	0.48	0.68	9.40	1.93	3.86	0.81	23.8
420	0.45	34.00	0.33	0.71	5.04	1.48	0.326	0.84	68.7
460	0.23	2.68	0.18	0.56	0.11	1.68	19.20	0.72	305.6

TABLE 2
Properties of ITO/n-Si junctions (of area 0.04 cm^2) prepared at 380°C for different oxidation time (t_{ox})

t_{ox} sec	V_{oc} (V) Expt	V_{oc} (V) Calc	J_{sc} (mA/cm^2)	ff Expt	ff Calc.	η (%)	n	J_o (µA/cm^2)	$\phi_{b,cv}$ (eV)
30	0.36	0.36	32.43	0.24	0.72	2.8	1.22	0.44	0.72
45	0.41	0.41	40.77	0.32	0.70	5.4	1.45	0.47	0.99
60	0.47	0.47	41.25	0.48	0.67	9.4	1.97	4.52	0.80
75	0.40	0.38	38.16	0.27	0.76	4.2	1.07	0.048	0.79
90	0.40	0.41	25.14	0.23	0.68	2.3	1.64	1.55	0.80

Expt. : Experimental values

Calc. : calculated values

T_p : Process temperature

V_{oc} : Open circuit voltage

J_{sc} : Short circuit current

J_o : Reverse saturation current

η : Efficiency (%)

n : Diode ideality factor

ϕ_b : Barrier height (C–V)

ff : Fill factor

R_s : Series resistance

t_{ox} : oxidation time

conducting and transparent) are obtained at 460°C, the maximum photoconversion efficiency of the junction is observed at 380°C. This observation is also noticed in ITO/GaAs [9] and ITO/Si [3] junctions prepared by reactive thermal evaporation technique.

3.1.2 C–V measurements

The plot of $1/C^2$ against V_r for the ITO/n–Si junctions prepared at different process temperatures is shown in the fig. 3. The slope of the line closely corresponds to the known value of donor concentration ($\approx 2.5 \times 10^{15}$ cm^{-3}). From the voltage intercept of this line, the Schottky barrier height was evaluated [10] and tabulated (Table 1).

3.1.3 Quantum Efficiency

The quantum efficiency of the cells prepared at different process temperatures was measured in the wavelength range 400 – 1100 nm using narrow band width interference filters (Ealing), the transmittance of which are nearly equal (at peak transmission). The normalised quantum efficiency as a function of wavelength for the cells prepared at three different temperatures is given in fig. 4. The photocurrent for each of the wavelengths was normalised to the maximum value of the individual cells. Comparison of quantum efficiency among the cells prepared at different process temperatures may not be possible since the transmission of ITO films varies with temperature [8].

The peak response of the cells is observed to shift from 700 nm to 600 nm with increase of process temperature.

3.2 Oxidation time

3.2.1 I–V characteristics

Fig. 5 presents the illuminated (100 mW/cm^2) I–V characteristics of ITO/n–Si junctions prepared at 380°C as a function of oxidation time. It may be noted that V_{oc} and J_{sc} increase with oxidation time, reach maximum values at 60 seconds and then decrease (fig. 6).

It may be observed that 'n' increases with oxidation time upto 60 seconds and then decreases; J_o shows minimum value at 75 seconds (Table 2). The maximum efficiency of 9.4 % has been achieved both on large (1 cm^2) and small (0.04 cm^2) areas for a oxidation time of 60 sec.

270

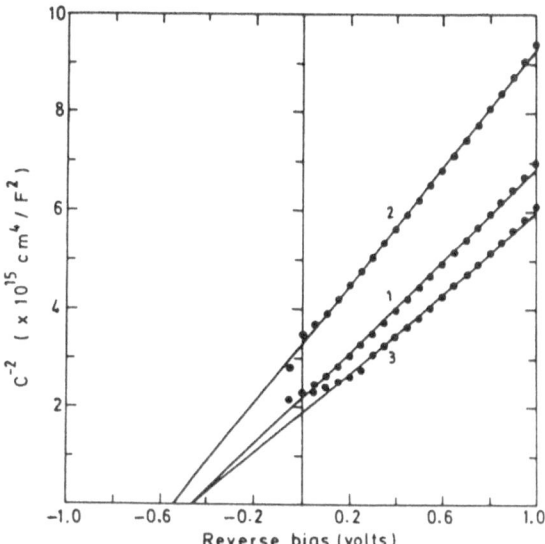

Fig. 3. C. - V characteristic of ITO/n-Si junctions prepared at different process temperatures : 1. 300°C; 2. 380°C; 3. 460°C.

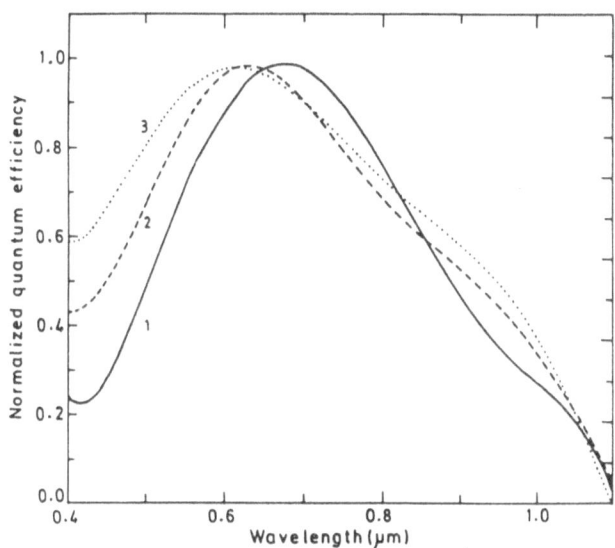

Fig. 4. Normalized quantum efficiency of ITO/n-Si junctions prepared at different process temperature: 1. 300°C; 2. 380°C; 3. 460°C.

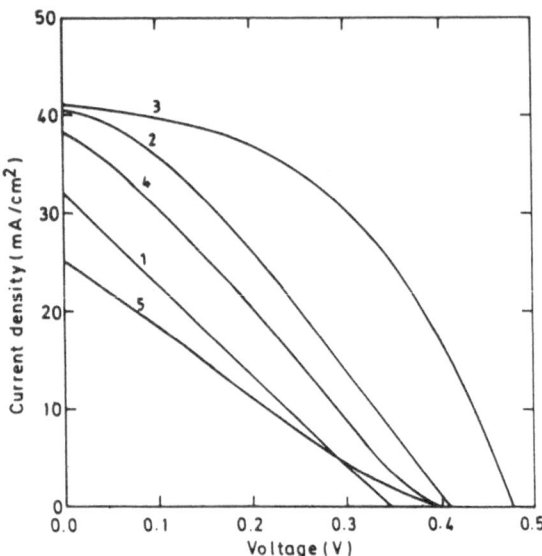

Fig. 5. I - V characteristic (under illumination) of
ITO/n-Si junctions prepared at 380°C for
different oxidation time : 1. 30 sec; 2. 45 sec;
3. 60 sec; 4. 75 sec; 5. 90 sec.

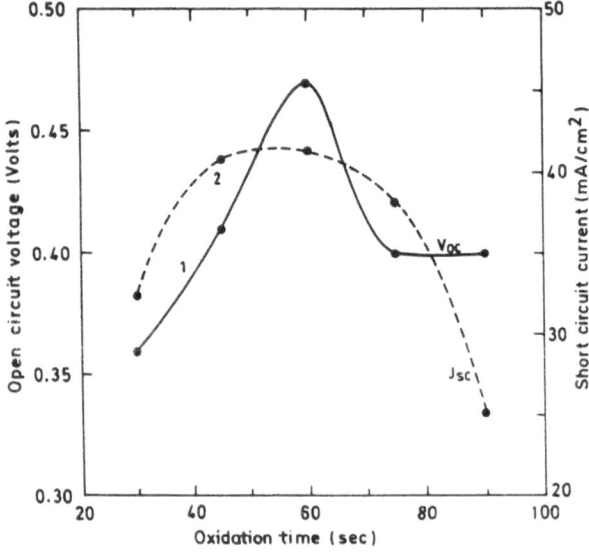

Fig. 6. Open circuit voltage and short circuit current
density of ITO/n-Si junctions as a function of
oxidation time.

The fill factor for these cells have been calculated following the relation [11]

$$ff = [\ v_{oc} - \ln(v_{oc} + 0.72)\ ]/(v_{oc} + 1) \qquad (2)$$

where v_{oc} is the normalised value of the open circuit voltage given by

$$v_{oc} = [\ V_{oc} / (nkT/q)\ ] \qquad (3)$$

and 'n' is the diode ideality factor.

3.2.2 C-V measurements

The plot of C^{-2} vs V_r for these junctions at two oxidation times (30 sec and 90 sec) have been given in fig. 7. The slope of the line closely corresponds to the known value of donar concentration ($\sim 2.5 \times 10^{15}$ cm^{-3}). The barrier height (ϕ_B) seem to attain high value (0.99 eV) at an oxidation time of 45 sec.

3.2.3 Spectral response

The normalised quantum efficiency of these cells prepared at different oxidation times is given in fig. 8. The peak response of the cells is observed to shift from 600 nm to 640 nm with increase of oxidation time. Interestingly, the short wavelength response decreases with increasing oxidation time.

4. Discussion

4.1 Process temperature

In the spray pyrolysis process, there is always a time lag between the cleaning of the wafer (substrate) and deposition; also the wafer may not attain steady state temperature immediately after introducing into the reaction chamber (This is the reason why we preferred to use the term 'process temperature' rather than the 'substrate temperature').

The $1/C^2$ versus V_r plots (fig. 3) are linear for all the devices fabricated, indicating an abrupt heterojunction, with fairly uniform donor concentration in the silicon. Since the ITO is degenerately doped, the conduction and valence bands are essentially flat in the ITO, and the entire diffusion potential is developed in silicon.

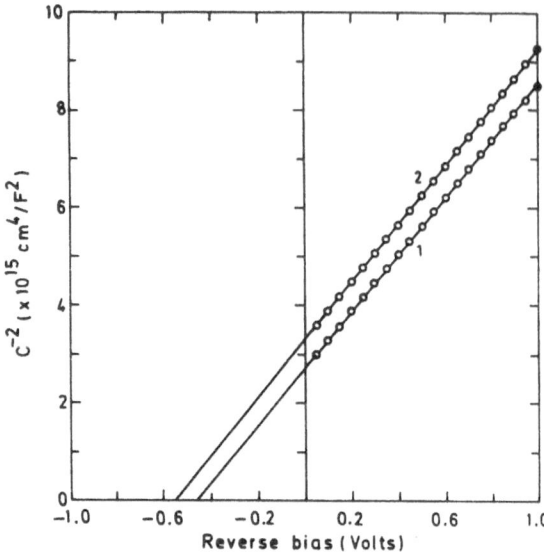

Fig. 7. C - V characteristic of ITO/n-Si junctions prepared at different oxidation time : 1. 30 sec; 2. 90 sec.

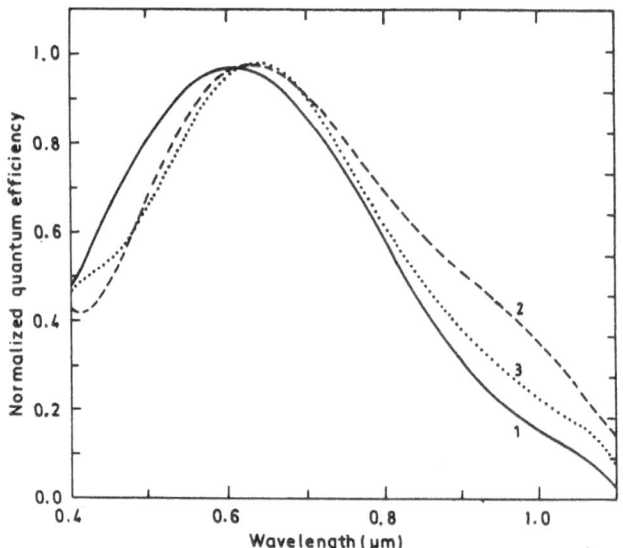

Fig. 8. Normalized quantum efficiency of ITO/ n-Si junctions prepared at different oxidation time: 1. 30 sec; 2. 60 sec; 3. 90 sec.

The increase in V_{oc} with process temperature may be attributed to increase in the oxide layer thickness and thus the barrier height as supported by the C–V measurements. However, for process temperatures above $380^{\circ}C$, the V_{oc} decreases, though the barrier height seem to attain constant value (0.8 eV), possibly due to the photocurrent suppression as indicated by the decrease in J_{sc} (fig. 2). It is known that the photocurrent suppression arises when there is an interfacial oxide layer of thickness high enough to impede the tunneling of carriers across the barrier [12,13]. The possible dominant current transport mechanism across the barrier seem to change with process temperature from 'depletion region recombination process' to 'over the barrier process' as indicated by the decrease in the ideality factor 'n' with process temperature.

The low value of fill factor observed in the present investigation may be possibly due to the high values of series resistance of the cells.

As noted earlier, the interface of ITO/Si formed by spray pyrolysis is complicated in view of the impurities, oxide formation of silicon (though C–V measurements show an abrupt interface); these complexities lead to a current transport mechanism not yet known in detail.

4.2 Oxidation time

The results of the present investigation may be understood on the basis of Semiconductor – Insulator – Semiconductor model [14,15]. With increasing oxidation time (t_{ox}) prior to ITO deposition, the thickness of the SiO_x layer increases as supported by the increase in open circuit voltage (V_{oc}) and the barrier height (ϕ_b) values. The ellipsometric data indicate that the interfacial oxide layer thickness varies from 20 Å to 30 Å with increase in oxidation time; however, the thickness of the SiO_x layer could not be confirmed accurately because of the inaccuracy involved in the assumption of the refractive index of SiO_x layer. The small decrease in V_{oc} for higher oxidation times $(t_{ox} > 75$ sec) may be attributed to the photocurrent suppression. The calculated values of V_{oc} agree well with the experimental values (Table 2).

The variation in the n and J_o values with t_{ox} are governed by the increase in oxide layer thickness and the corresponding decrease in the interface state density. The increase in n value may be understood by following the relation

$$n = 1 + \frac{q \, \delta \, D_{its}}{\varepsilon} \qquad (4)$$

where δ is oxide layer thickness, D_{its} is the interface state density and ε is the permittivity of the interfacial layer. The initial increase in n value may be due to the increase in δ. It may be noted that initial thin oxide layer growth will not affect the D_{its} value. After certain value of δ, D_{its} will be reduced which is responsible for the observed low n value at higher oxidation time. In other words, the growth of an oxide layer passivates the silicon surface and also influences the transport mechanism at the interface. At t_{ox} = 75 sec, the junction exhibits near ideal diode characteristics (n = 1.07 and J_o = 4.82 x 10^{-8} A/cm^2). However, the photoconversion is observed to be maximum (η = 9.4 %) at t_{ox} = 60 sec.

The calculated 'ff' values are higher than the experimental values (Table 2). The low values of ff observed experimentally may be due to high series resistance of the cells. The sheet resistance of the ITO window layer, and the resistance offered by the junction in the collection of the photogenerated carriers contribute to the series resistance of the cell. In the present investigation, the sheet resistance of ITO layer is fairly constant (since the deposition temperature is constant), hence, the observed changes in ff with t_{ox} may be attributed to the junction resistance (indirectly to the SiO_x insulating layer).

It is quite likely that dry oxidation of HF etched Si (<100> oriented) surface takes place at 380°C for different times (30 sec – 90 sec) prior to ITO deposition; during the time of spray the thin SiO_x surface further oxidized in presence of chlorine and water present in the reaction chamber.

At t_{ox} = 30 sec (and less), the insulating layer is so thin, barrier of enough height could not be formed (and hence V_{oc} is quite low); as oxidation time increases, the barrier height increases and reaches a maximum value (ϕ_b = 0.99 eV).

276

The photocurrent increases with t_{ox}, may be due to either to an effective decrease in the series resistance across the junction or to reduction in recombination loss at the interface (or both). At higher oxidation times ($t_{ox} > 75$ sec), the insulating property of the oxide layer (chemically tending towards stoichiometric oxide) increases affecting the photocurrents and consequently V_{oc}.

The oxidation kinetics of such thin SiO_x layers are quite complicated. Lukes [16] has observed from oxidation time versus film thickness measurements in cleaved (or HF etched) silicon (<111> orientation) that logarithmic law is followed upto thickness of about 35 - 40 Å.

It may not be possible at this stage to pass any conclusive remarks on the nature (stoichiometric) of the insulating interface layer. Further work along the lines which is in progress, particularly on the chemical analyses of the insulating layer, may give detailed understanding of the interfacial layer.

5.CONCLUSIONS

ITO/n–Si junctions prepared by spray pyrolysis technique exhibited maximum photoconversion efficiency (both on small and large areas) of 9.4 % at a process temperature of $380^{\circ}C$ and for an oxidation time of 60 sec. Higher process temperatures yield better ITO films but poor rectifying junctions. It is possible to attribute the variation in the photovoltaic properties with oxidation time as observed in the present investigation, to the series resistance offered by the interfacial oxide layer since the sheet resistance of the ITO window layer is kept constant.

The authors gratefully acknowledge Dr. Ross, University of Technology, Loughborough, UK for the EPMA measurements and Council of Scientific and Industrial Research (CSIR), India for the financial help to carryout the work.

7. REFERENCES

[1] Dubow J.B, Burk D.E and Sites J.R. 1976 "Efficient Photovoltaic heterojunction Solar Cell," Appl. Phy. Lett. 29, 494.
[2] Feng T, Ghosh A.K and Fishman C, 1979 "Efficient electron beam deposited ITO/n–Si solar cells," J. Appl. Phys. 50, 4972.
[3] Balasubramanian N, Subrahmanyam A, 1991 "Studies on evaporated indium tin oxide (ITO)/silicon junctions and an estimation of ITO work function," J. Electrohem. Soc. 138, 322.

[4] Manifacier J.C and Szepessy L, 1977 "*Efficient sprayed In$_2$O$_3$:Sn n-type silicon heterojunction solar cell*", Appl. Phys. Lett. 31, 459.

[5] Ashok S, Sharma P.P and Fonash S.J, 1980 "*Spray deposited ITO-Silicon SIS heterojunctions solar cells*", IEEE Trans. on Electron Devices, ED-27, 725.

[6] Vasu V, Subrahmanyam A, 1990 "*Electrical and optical properties of pyrolytically sprayed SnO$_2$ films – dependence on substrate temperature and substrate-nozzle distance*," Thin Solid Films 189, 217.

[7] Vossen J.L, 1977 *Physics of Thin Films*, eds.:G.Hass, et.al. (Academic Press : New York), 9, 1.

[8] Vasu V, Subrahmanyam A, 1990 "*Reaction kinetics of the formation of indium tin oxide films grown by spray pyrolysis*," Thin Solid Films 193/194, 696.

[9] Balasubramanian N and Subrahmanyam A, 1990 "*Investigations on the photovoltaic properties of indium tin oxide (ITO)/n-GaAs heterojunctions*," Solar Cells 28, 319.

[10] Henisch H.K, 1984 "*Semiconductor Contacts – An Approach to Ideas and Models*" (Oxford: Clarendon)

[11] Green M.A, 1987 "*High Efficiency Silicon Solar Cells*," TransTech Publications, pp. 96.

[12] Card H.C and Rhoderick E.H, 1971 "*Studies of tunnel MOS diodes : 1. Interface effects in silicon Schottky diodes*," J. Phys. D., 4, 1589.

[13] Anderson R.L, 1975 "*Photocurrent suppression in heterojunction solar cells*," Appl. Phys. Lett. 27, 691.

[14] Singh R, Rajkannan K, Brodie D.E and Morgan 1980 "*Optimization of Oxide-Semiconductor/Base-Semiconductor Solar Cells*," IEEE Trans. on Electron Devices ED-27, 656.

[15] Shewchun J, Burk D and Spitzer M.B 1980 "*MIS and SIS Solar Cells*," IEEE Trans. on Electron Devices ED-27, 705.

[16] Lukeš F, 1972 "*Oxidation of Si and GaAs in air at room temperature*," Surf. Sci. 30, 91.

Stoichiometry Control of Compound Semiconductor Crystals

Jun-ichi Nishizawa

Semiconductor Research Institute, Semiconductor Research Foundation,
Sendai Aoba-ku 980, Japan

Most important factor to be controlled in compound semiconductor crystals is the deviation from the stoichiometric composition. Electrical, optical and crystallographic evaluation are applied to the GaAs samples prepared by annealing under controlled As vapor pressure. It is shown that the crystal imperfection and the deviation from the stoichiometric composition are reduced to be minimum under a specific As vapor pressure (optimum As vapor pressure; $P_{As,opt}$). $P_{As,opt}$ is obtained to be similar regardless of the difference of dopant species, dopant concentration and conductivity type. Optimum vapor pressure is also obtained in other compound semiconductor crystals including GaP etc. The deviation from the stoichiometric composition affects the amphoteric manner of Si in GaAs. Vapor pressure control is also applied to the liquid phase epitaxy (LPE) in combination with the temperature difference method (TDM–CVP), and almost the same results are obtained. Temperature dependence of the optimum vapor pressure is almost the same both in annealing and LPE experiments. It confirms that the applying vapor pressure controls the composition of segregated crystals directly through the solution in LPE. Various methods including Rutherford backscattering (RBS) technique are applied to investigate the excess As atom–related defects in GaAs. The experimental results of crystal weight and X–ray anomalous transmission intensity measurements directly suggest the existence of interstitial As atoms. The RBS measurements with glazing exit angle configuration show the existence of the interstitial As atoms in As$^+$–implanted GaAs, and clarify the stable interstitial sites to be <100> split and relaxed– bond center (r–BC) interstitialcy. Formation energy of excess As atom–related defects is determined to be 0.9–1.16eV by both the lattice constant measurements and the PHCAP measurements independently. This seems likely to be the formation energy of interstitial atom rather than to that of vacancy. Vapor pressure control is effectively applied to the Czochralski (CZ) and horizontal Bridgman (HB) grown GaAs bulk crystal growth method, and enables to supply high quality substrate material with low dislocation density. High purity GaAs bulk crystals are obtained by the vapor pressure controlled floating zone (FZ) method. Liquid phase epitaxial growth by the TDM–CVP is utilized for the fabrication of GaP pure green LED without nitrogen doping, GaAlAs super bright LED and ZnSe blue LED by the pn junction etc.

Introduction

The most important factor to be controlled is the deviation from the stoichiometric composition in compound semiconductor crystals. From the investigation of iron–pyrite in 1951[1], we have carried out the annealing experiments of various III–V compound semiconductor crystals under controlled vapor pressure[2,3]. From the experimental results of the crystallographic, electrical and optical evaluation, it is shown that the nearly perfect crystals with stoichiometric composition are obtained under a specific vapor pressure of group V elements, and that the temperature dependence of the optimum vapor pressure is also obtained.

Especially, we have proposed the existence of interstitial arsenic atoms in GaAs crystals when GaAs is annealed under high arsenic vapor pressure[4]. Our results on the specific weight and the intensity of X–ray anomalous transmission show the evidence for the existence of arsenic interstitial atoms[5]. Our RBS experiments[6] have also revealed the arsenic interstitial atoms in GaAs and determined the stable interstitial sites in the regular lattices. The RBS results on the stable interstitial sites are in good accordance with those of X–ray anomalous transmission measurements[5]. The PHCAP measurements[7] under constant capacitance condition have revealed the stoichiometry–dependent deep levels and clarified the arsenic vapor pressure dependencies of

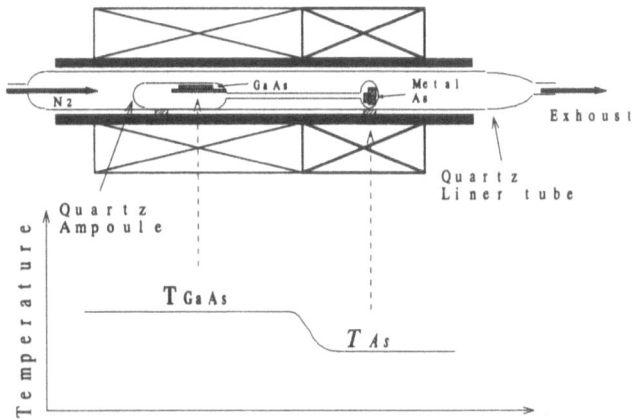

Fig.1

Schematic draw of the annealing apparatus under controlled arsenic vapor pressure.

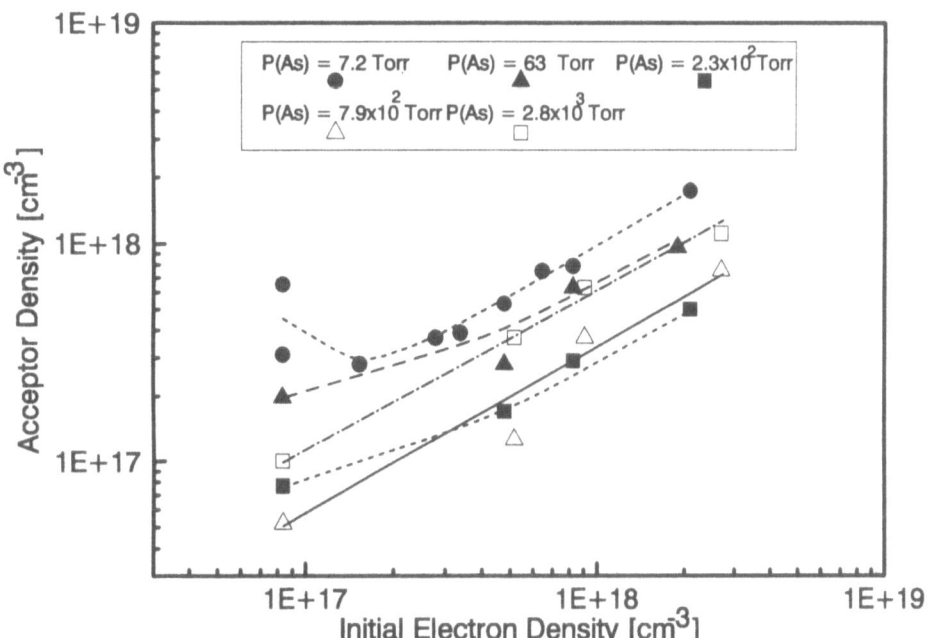

Fig.2

Intial electron density dependence of acceptor density in Te–doped GaAs after annealing.

the deep level densities. The formation energy of the defects was also obtained to be 1.16eV[8]. This relates most closely to the interstitial atoms , rather than to the vacancy.

Recently much attention has been paid to interstitial arsenic atoms in GaAs, because it is reported that the so-called EL2 level relates to the excess arsenic composition of GaAs crystals. Indeed, the recent results of X-ray quasi-forbidden reflection[9] and ESR measurements including optical detection[10] seem to justify the existence of the arsenic interstitial atoms.

By applying the vapor pressure control also on the fields of liquid phase epitaxy[11,12] and bulk crystal growth[13], it is shown that high purity GaAs crystals were obtained with stoichiometric composition and very low dislocation density. This enables to fabricate the super-bright LEDs[14], including the pure-green LEDs without nitrogen doping in GaP[15].

The vapor pressure control during crystal growth, which enables to control the stoichiometric composition, is applied extensively not only on the III-V compounds e.g. InP, but on II-VI compounds including ZnSe[16]. This should be also important in the field of the superconducting ceramics.

Annealing effects on GaAs crystals under arsenic vapor pressure

Annealing experiments were performed at 900 - 1100°C for 67h under various arsenic vapor pressure. Samples used were (100) oriented horizontal Bridgman (HB) grown GaAs with different impurity density. Defect density introduced by annealing reaches its saturating value after 67h-annealing. Figure 1 shows the schematic draw of the annealing equipment under As vapor pressure. As vapor pressure applied on GaAs crystals were obtained as

$$P_{GaAs} = P_{As} x (T_{GaAs}/T_{As})^{1/2} \hspace{3cm} [1]$$

where P_{As} is the equilibrium As vapor pressure determined from the temperature of arsenic metal (T_{As}), T_{GaAs} is the temperature of GaAs crystals. Equilibrium As vapor pressure was obtained from Honig[17]. After annealing, samples were cooled rapidly by putting them into the water in order to prevent the effect of slow-cooling. X-ray and etching inspection cannot reveal slip lines even after rapid cooling.

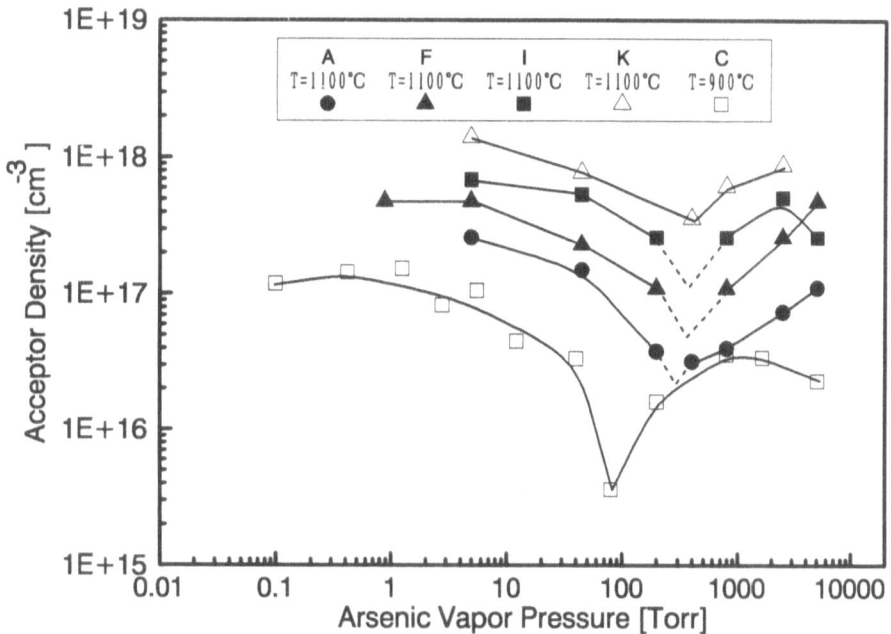

Fig.3

As vapor pressure dependence of acceptor density in annealed Te-doped GaAs.

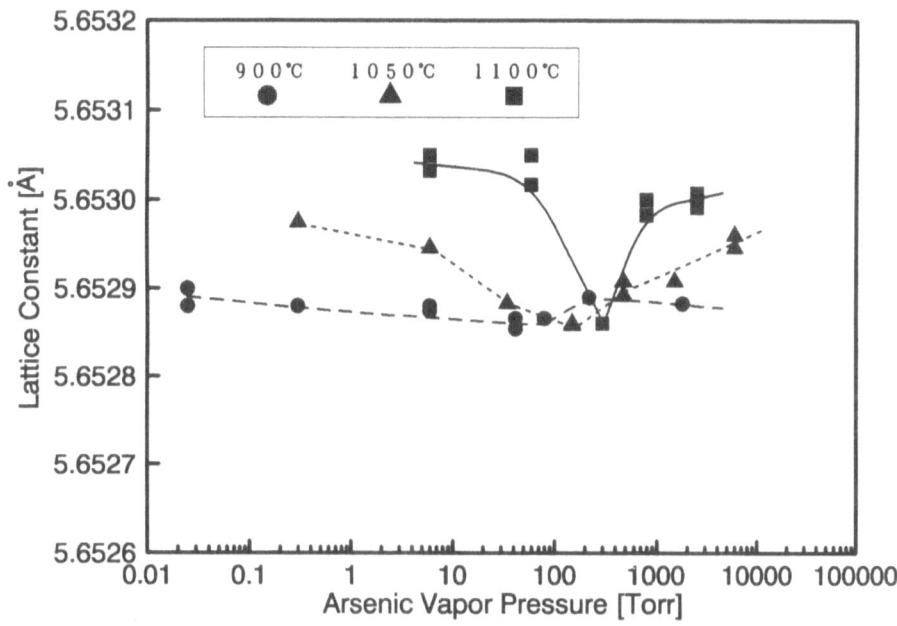

Fig.4

As vapor pressure dependence of the lattice constant.

283

Figure 2 shows the acceptor density as a function of the initial electron density in Te-doped GaAs after annealing. Acceptor density is almost proportional to the initial electron density. This shows that the acceptor-type defects relate to both the deviation from stoichiometric composition and the dopant impurity Te. Figure 3 and 4 show the As vapor pressure dependencies of the acceptor density and of the lattice constant respectively. Lattice constant was measured by using X-ray double crystal diffractmetry with (004) symmetrical configuration. Various marks in the figures denote the data obtained from different crystals with different electron density. Acceptor density shows a minimum under a specific As vapor pressure ($P_{As,opt}$). By applying almost the same As vapor pressure, lattice constant shows its minimum value. It seems that the nearly perfect crystals with stoichiometric composition could be obtained under the $P_{As,opt}$. Almost the same results were shown in Zn-doped GaAs crystals. Therefore, the $P_{As,opt}$ is independent upon the amount of the dopant and dopant species. Almost the same results were also obtained in annealing experiments of GaP and optimum phosphorous pressure ($P_{P,opt}$) was shown to improve the crystal quality.

From the experimental results, optimum vapor pressures were obtained as a function of annealing temperature for GaAs and GaP respectively to be

$$P_{GaAs,opt} = 2.6 \times 10^6 \exp [-1.05eV/(kT)] \qquad [2]$$

$$P_{GaP,opt} = 4.67 \times 10^6 \exp [-1.01eV/(kT)] \qquad [3]$$

In order to investigate the deep levels in annealed GaAs crystals, the PHCAP measurements[18] were carried out under constant capacitance conditions[19]. The PHCAP method enables to determine the accurate level density and the precise activation energy because ionization by monochromatic light irradiation at fixed very low temperature were used. Contrary to the conventional PHCAP method, depletion layer thickness is kept to be constant regardless of the change of ion density by light irradiation.

In order to obtain accurate level density and level position, fully-neutralized deep levels should be ionized at each wavelength. One method to achieve such condition is to apply forward bias injection in the dark before each photo-excitation. In n-type GaAs bulk crystals, the so-

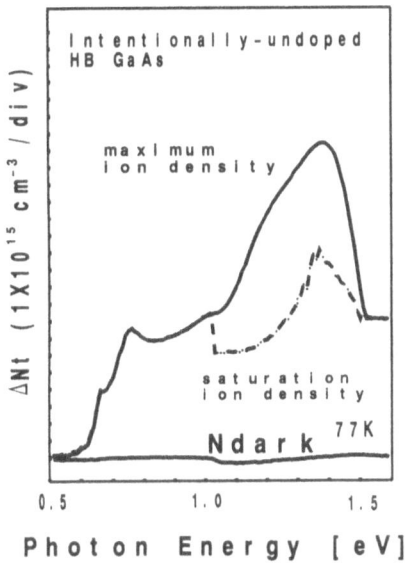

Fig.5

PHCAP spectra of intentionally-undoped GaAs

annealed under various As vapor pressure.

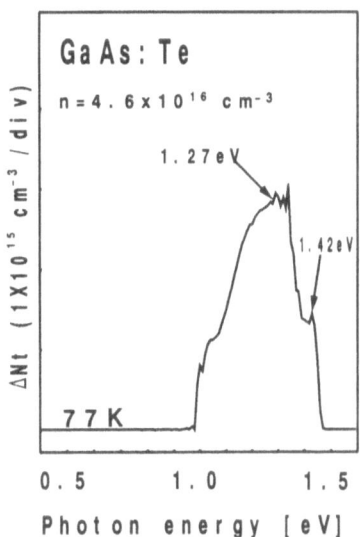

Fig.6

Deionized level density PHCAP spectrum of n-GaAs doped with Te.

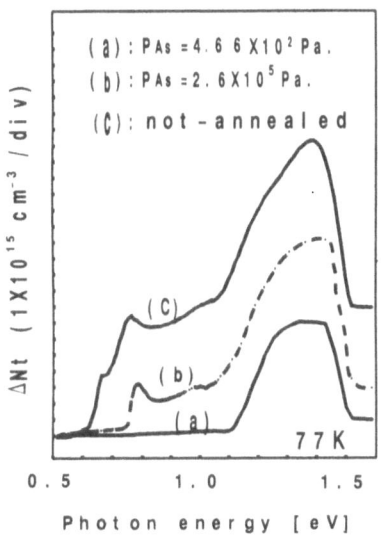

Fig.7

PHCAP spectra of intentionally-undoped HB GaAs annealed

at 900°C for 67h under various As vapor pressure.

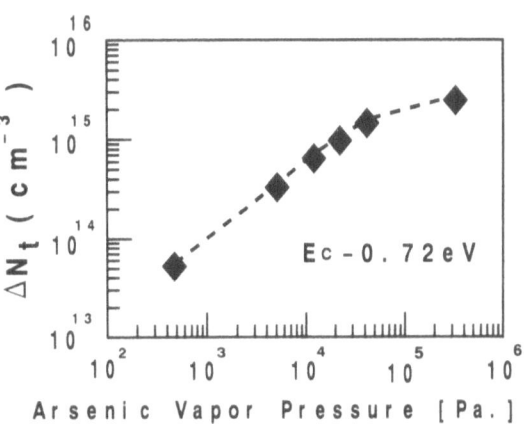

Fig.8

As vapor pressure dependence of E_c-0.72eV level density in annealed n-GaAs.

called photoquenching phenomenon[20] is observed in a specific wavelength region of about 1.0–1.5eV below about 110K. Therefore, both the maximum and the asymptotic to the saturation ion density were obtained at each wavelength.

Figure 5 shows the PHCAP spectra of the intentionally–undoped GaAs ($n=4 \times 10^{16}$ cm^{-3}) grown by HB method before annealing[21]. In the figure, (a) and (b) show the maximum (N_{max}) and the asymptotic ion density (N_{asym}) respectively. (c) represents the ion density (N_{dark}) in the dark after forward bias injection. N_{dark} corresponds to the ion density in the dark before photoexcitation. Almost constant value of N_{dark} verifies the photo–excitation of fully–neutralized deep levels at each wavelength. Figure 6 shows the PHCAP spectrum obtained from the subtraction of N_{asym} from N_{max}. This shows the deionized level density spectrum.

As shown in Fig.5, two kinds of deep donors were clearly revealed at 0.65 and 0.72eV below the conduction band. The deionized level density spectrum shows two clear peaks at 1.25 and 1.41 eV respectively. As reported previously, 1.25 1.41eV deep levels are quite different ones with different recovery temperature and ionized levels. Figure 7 shows the PHCAP maximum ion density spectra of intentionally–undoped GaAs crystals prepared by annealing under various As vapor pressure. Figure 8 shows the As vapor pressure dependencies of the E_c–0.72eV level density. It should be noted that the E_c–0.65eV level vanishes after annealing perhaps due to its thermal instability. However, E_c–0.65eV level is stable in more strained crystals and the level density increases monotonically with increase of As vapor pressure. Level density of E_c–0.72eV donor and the 1.25eV deionized level increase monotonically with increasing As vapor pressure. These deep levels are shown to be detected commonly in various GaAs bulk crystals with different dopant impurity and conductivity types[22]. Monotonical increase of level density suggests that these deep levels closely relate with the excess As composition of GaAs crystal. From the spectral correspondence among the PHCAP, DLTS and deep level PL[23], it is shown that the E_c–0.65 and 0.72eV level exhibit larger difference between optical and thermal activation energy compared with the so–called EL2 level. Whereas both our finding E_c–0.72eV level and the so–called EL2 level are closely related with the excess arsenic composition of GaAs crystals, these are quite different from each other. They show different optical excitation energies and amount of

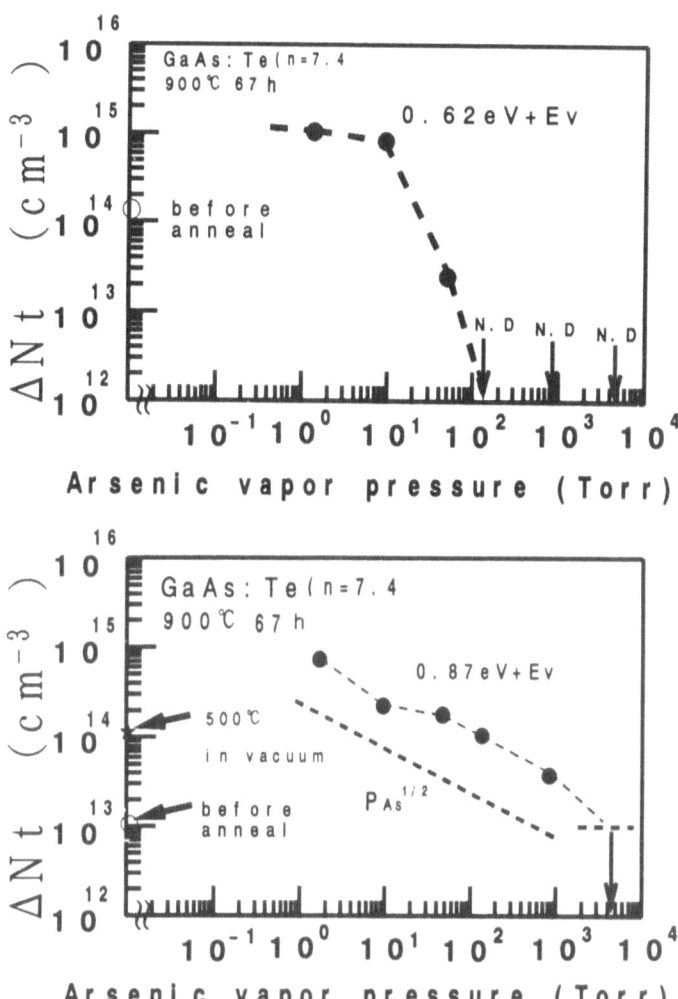

Fig.9

As vapor pressure dependencies of the 0.62 and 0.83eV+E$_v$ level in heavily Te–doped GaAs crystals.

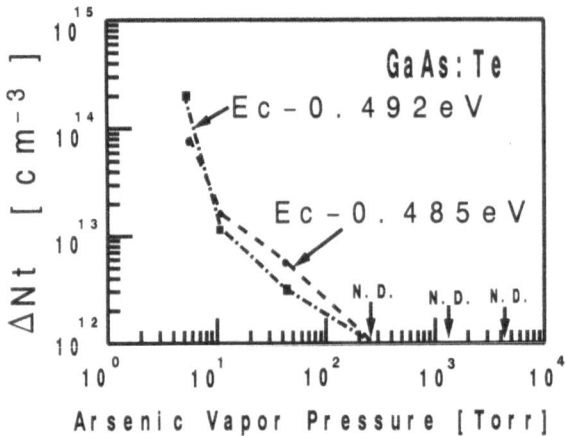

Fig.10

As vapor pressure dependence of E_c–0.48eV level density in Te–doped GaAs crystals prepared by annealing under As vapor pressure.

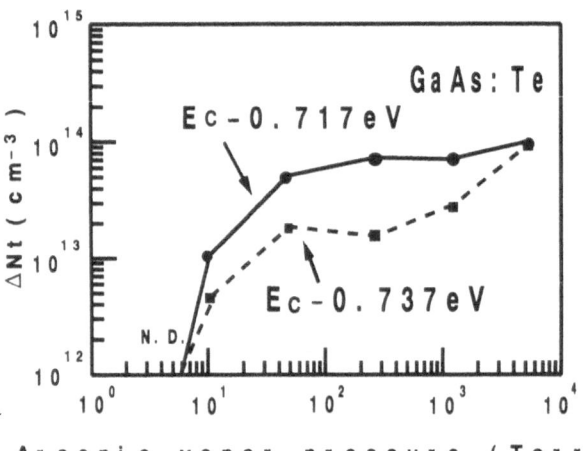

Fig.11

As vapor pressure dependence of E_c–0.72eV level density in Te–doped GaAs crystals.

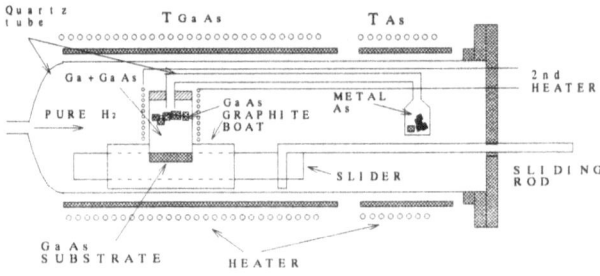

Fig.12

Schematic draw of the liquid phase epitaxial growth apparatus by our temperature difference method under controlled vapor pressure (TDM–CVP).

Frank–Condon shift (d_{FC}).

The PHCAP method also revealed the As–vacancy related deep levels in heavily Te–doped GaAs prepared by annealing under As vapor pressure[22]. Figure 9 shows the As vapor pressure dependencies of the 0.62 and 0.87eV+E_v level density, which were measured after 1.44eV–monochromatic light irradiation at each wavelength to emptize the deep levels. The level density decreases monotonically with increasing As vapor pressure. These deep levels could be also detected in vacuum–annealed samples (500°C for 50h) but not be detected in the virgin samples. This confirms the close relation between these deep levels and the As vacancies.

The precise PHCAP measurements at 20K revealed the stoichiometry–dependent deep levels at around E_c–0.48eV in Te–doped horizontal gradient freeze (HGF) grown GaAs prepared by annealing under As vapor pressure (Fig.10)[23]. Level density decreases monotonically with increasing As vapor pressure. E_c–0.48eV level was detected in Te–doped GaAs, but not be observed in intentionally–undoped nor Si doped GaAs crystals. Therefore, this should relate with at least the dopant impurity Te and the As vacancies. E_c–0.72eV deep donor was also detected in the same samples, and the As vapor pressure dependence of the level density was almost the same as that in intentionally–undoped HB GaAs crystals. This suggests that the E_c–0.72eV level is originated from the excess As atom–related intrinsic defects but not from some impurities.

Liquid phase epitaxial growth by the temperature difference method under controlled vapor pressure (TDM–CVP)

Vapor pressure control technique was successfully applied to the LPE growth, and this enables to supply epitaxial layers with good crystal quality. Figure 12 shows the schematic draw of the LPE apparatus of TDM–CVP. Crystal growth proceeds at a fixed temperature due to the difference of solubility and the kinetic energy, which is originated from the temperature difference at the upper and lower part of the solution. Controlled vapor pressure of group V elements

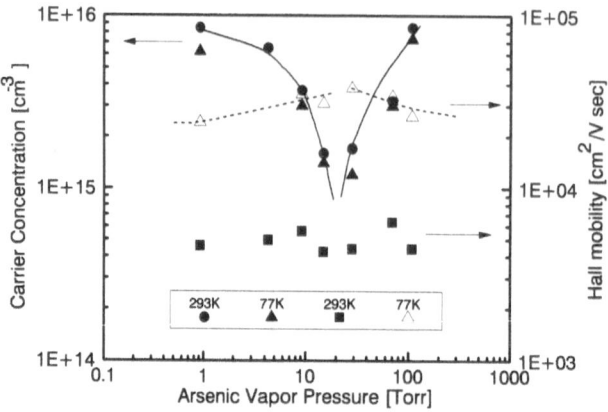

Fig.13

As vapor pressure dependencies of electron density and Hall mobility in LPE GaAs grown by the TDM-CVP.

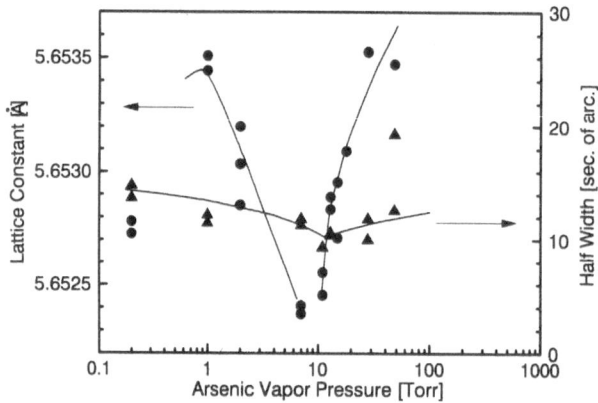

Fig.14

As vapor pressure dependencies of the lattice constant and the FWHM of X-ray rocking curve in LPE GaAs.

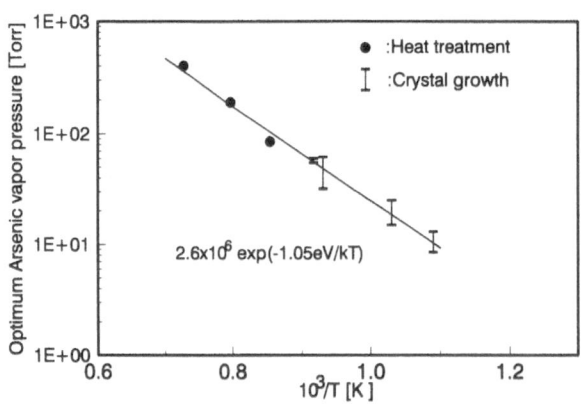

Fig.15

Temperature dependence of the optimum As vapor pressure obtained from both the annealing and LPE growth results.

290

is also applied in order to control the stoichiometric composition of segregated crystals through the Ga–As solution.

Figure 13 shows the As vapor pressure dependencies of carrier concentration and the electron mobility. Figure 14 shows the As vapor pressure dependencies of the lattice constant and the half width of the X–ray rocking curve of LPE GaAs. Lattice constant was measured by using GaAs crystal for the first crystal with the (004) symmetric configuration. Therefore, the half width of the X–ray rocking curve is dependent on the perfection of specimen crystals. Under a specific As vapor pressure, carrier concentration shows its minimum value, and the Hall mobility shows a maximum. The lattice constant and the half width of X–ray rocking curve also show minimum values under almost the same As vapor pressure. This leads to a crucial conclusion that the high purity LPE crystals with good perfection were obtained. Similar results were also obtained in LPE GaP crystals using TDM–CVP. Especially, photoluminescence (PL) measurements revealed the existence of excitons bound to shallow impurity even at room temperature. This also confirms the high crystal quality of LPE GaP grown by the TDM–CVP[24].

Vapor pressure dependencies of these crystal characteristics seem to be similar to those obtained from the annealing experiments. Figure 15 shows the temperature dependencies of the optimum As vapor pressures obtained from both the LPE growth by TDM–CVP and the annealing experiments under As vapor pressure[25]. These seem to be identical in both experiments.

The deviation from the stoichiometric composition of GaAs influences the amphoteric manner of group IV elements in GaAs. As shown in Fig.16, the diffusion phenomenon of Si in GaAs is strongly affected by the deviation from the stoichiometric composition[25]. Similar results on Si–diffusion in GaAs was also reported by Omura[26].

Interstitial As atoms in GaAs

Figure 17 shows the As vapor pressure dependencies of the crystal weight of GaAs prepared by annealing. It is shown that the crystal weight increases monotonically with increasing As vapor pressure. Figure 18 shows the results of X–ray anomalous transmission intensity

291

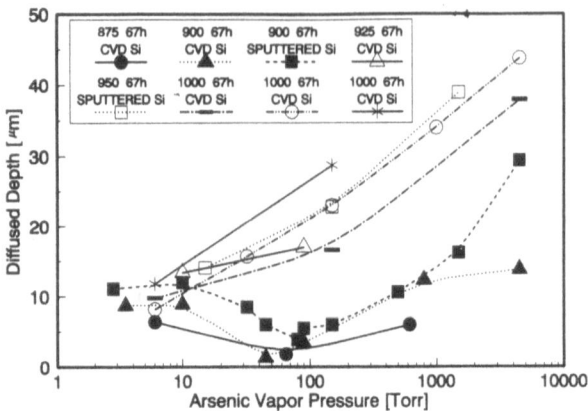

Fig.16

As vapor pressure dependencies of the diffusion depth of Si into GaAs. Inset of the figure shows the diffusion conditions of temperature, diffusion time and diffusion source.

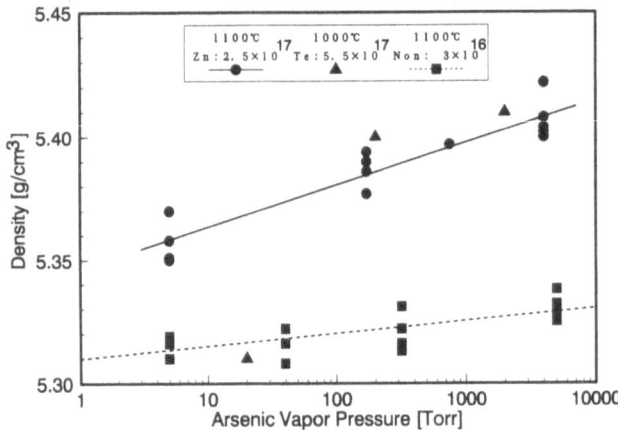

Fig.17

As vapor pressure dependence of the crystal weight of annealed GaAs.

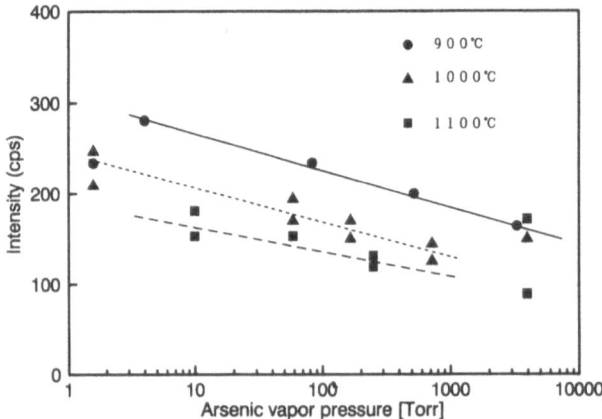

Fig.18

As vapor pressure dependence of the X-ray anomalous transmission intensity of annealed GaAs crystals.

292

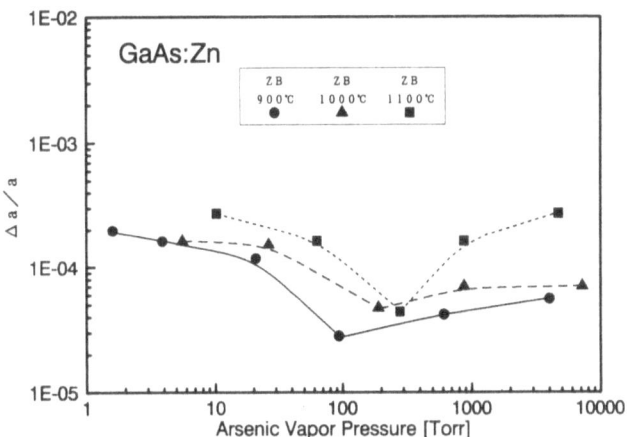

Fig.19

As vapor pressure dependence of the lattice constant in Zn–doped GaAs.

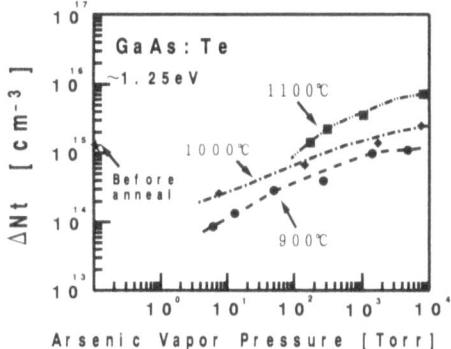

Fig.20

As vapor pressure dependence of the 1.25eV level density in Te–doped GaAs.

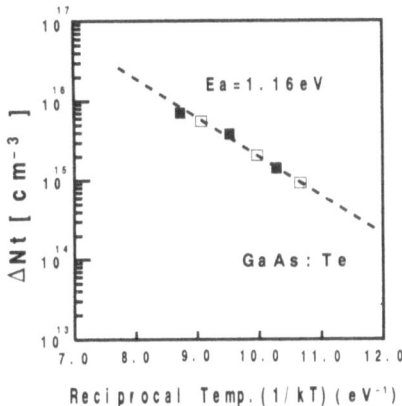

Fig.21

Temperature dependence of the saturating 1.25eV level density in Te–doped GaAs.

measurements as a function of As vapor pressure. The intensity of X-ray anomalous transmission decreases with increasing As vapor pressure. X-ray anomalous transmission is strongly influenced by the existence of interstitial atoms, because this phenomenon is caused by the propagation of Poynting vector of X-ray along the diffracting lattice plane. These results seem to confirm the existence of interstitial type defects when GaAs crystals are annealed under extremely high As vapor pressure.

Figure 19 shows the As vapor pressure dependence of the lattice constant. Samples used were Zn-doped HB grown GaAs prepared by annealing at 900°C for 67h under As vapor pressure. As already shown, the lattice constant shows minimum value at the optimum As vapor pressure. In higher As vapor pressure region, the lattice constant increases monotonically with increasing As vapor pressure and shows saturating manner. From the temperature dependence of the saturating lattice constant in high As vapor pressure region, the formation energy of the defect was obtained to be about 0.9eV.

Figure 20 shows the As vapor pressure dependence of the 1.25eV level density obtained from the PHCAP measurements. This level is followed by the so-called photoquenching phenomenon. The level density increases monotonically with increase of As vapor pressure, and saturates under high As vapor pressure. From the temperature dependence of the saturating level density, the formation energy of the defect is determined to be 1.16eV (Fig.21). This formation energy is very similar to that obtained from the lattice constant measurements.

In view of the theoretical calculations of formation energy by Bennemann[27] and Swalin[28], the defect formation energy of 1.16eV is rather close to that of interstitial atoms than that of vacancy. Therefore, we consider that the interstitial atoms should be introduced at the primary stage of defect formation when GaAs crystals are annealed under high As vapor pressure.

In order to investigate the stable interstitial site, we applied the RBS technique to the As$^+$ implanted GaAs crystals. Depth resolution was enhanced by using the grazing exit angle configuration of silicon surface barrier (SSB) detector. From the results of multi-directional and high depth resolution RBS measurements, the stable interstitial site was assumed to be <100> split

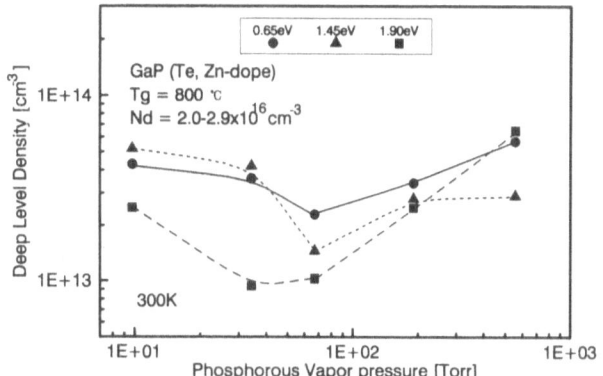

Fig.22

Phosphorous vapor pressure dependence of deep level density in LPE GaP crystals prepared by the TDM-CVP.

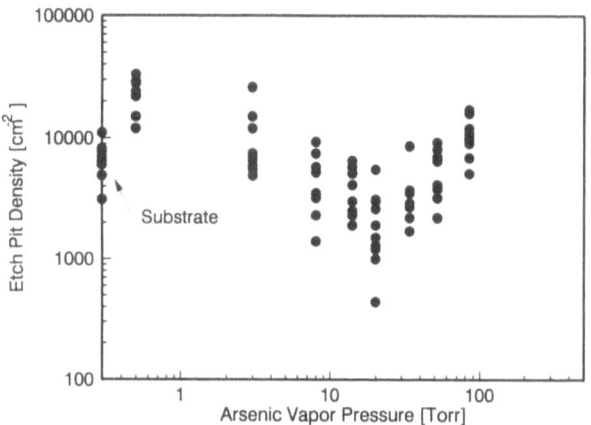

Fig.23

As vapor pressure dependence of the etch pit density (EPD) in LPE AlGaAs layers prepared by the TDM-CVP.

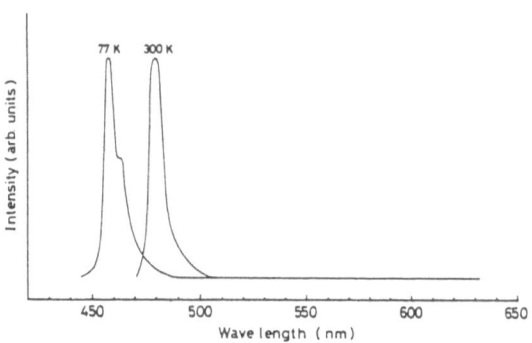

Fig.24

Electro-luminescence (EL) spectra of the blue light emission from ZnSe pn-junction prepared by the vapor pressure control method.

and relaxed bond center (r–BC) interstitialcy. This result corresponds to that obtained from the X–ray anomalous transmission measurements.

Application of vapor pressure control method to the fabrication of semiconductor devices

GaP pure green LED without nitrogen doping

Luminous efficiency of LED is strongly influenced by a small amount of defects. Therefore, the characteristics of LED is very sensitive to the crystal perfection.

It was shown that the TDM–CVP enables the LPE GaP to be almost free from dislocation, and GaP pure green LED was fabricated by this method. At that time, it has been believed that the nitrogen should be doped to improve the luminous efficiency, because GaP has indirect transition. But the emission wavelength shifts to a longer one by N doping, and the color of GaP:N LED is close to yellow rather than green. The GaP LED grown by the TDM–CVP exhibits a pure green emission (λ = 550nm) with high luminous efficiency without N doping. Wavelength of this emission light corresponds to the direct energy gap of GaP crystal.

Figure 22 shows the phosphorous vapor pressure dependence of deep level density measured by the conventional PHCAP method. Deep level density shows minimum under the optimum phosphorous vapor pressure. These levels were shown to be stoichiometry–dependent and act as a non–radiative recombination center. It is shown that the TDM–CVP controls the introduction of these deep levels caused by the deviation from the stoichiometric composition.

GaAlAs super bright LED

The TDM–CVP has also been applied to the LPE of AlGaAs, InGaP and AlGaAsP ternary and quoternary compound semiconductor crystals. Super bright LED was already fabricated with the quantum efficiency up to 30%. Temperature difference method is suitable for the mass-production because of its constancy of growth temperature.

Figure 23 shows the As vapor pressure dependence of etch pit density (EPD) in $Al_xGa_{1-x}As$ (x=0.3). EPD shows minimum value under a specific As vapor pressure. Optimum As vapor pressure should be dependent on the composition x. Precise control of the deviation from the stoichiometric composition is expected to improve luminous efficiency.

ZnSe blue LED

Vapor pressure control technique has been successfully applied to the II–VI compounds to achieve the precise control of stoichiometric composition.

In II–VI compounds, control of wide variety of conductivity type has been very difficult. Only a p–type crystal is grown preferentially in case of ZnTe. On the other hand, it is quite easy to grow n–type ZnSe, but it's difficult to obtain p–type ZnSe perhaps due to the preferential introduction of stoichiometry–dependent defects.

As already reported, p–type ZnSe crystals were grown using Se solution under controlled Zn vapor pressure. In Zn–Se system, both elements show high vapor pressure. Therefore, stoichiometric composition of ZnSe was independently controlled under maximum Se vapor pressure by use of Se solution and under Zn vapor pressure. Using the experimental procedure mentioned above, blue light emission was obtained at 480nm with a half width of 7nm from the ZnSe pn junctions. Figure 24 shows the EL spectra from ZnSe pn junctions.

InP bulk crystal growth by the vapor pressure controlled zone melting method

Vapor pressure control technique has also been applied to the InP bulk crystal growth in combination with the zone melting method. Figure 25 shows the phosphorous vapor pressure dependence of the carrier concentration and the Hall mobility. The carrier concentration and the Hall mobility at 77K show minimum and maximum value under a specific phosphorous pressure of about 22.5 atm at the melting point. The present value of mobility is greater than that obtained

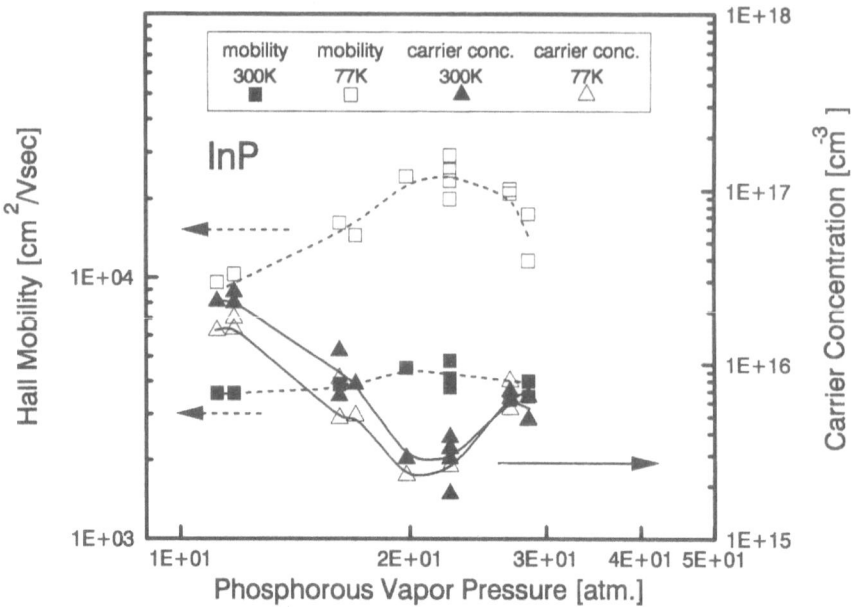

Fig.25

Phosphorous vapor pressure dependencies of the carrier concentration and the Hall mobility in InP bulk crystals prepared by the vapor pressure controlled zone melting method.

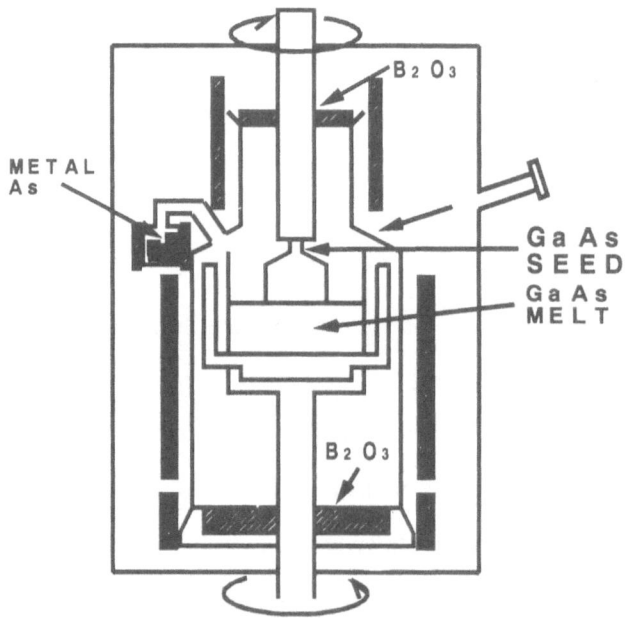

Fig.26

Schematic draw of the GaAs bulk crystal growth apparatus using the vapor pressure controlled Czochralski (PCZ) method.

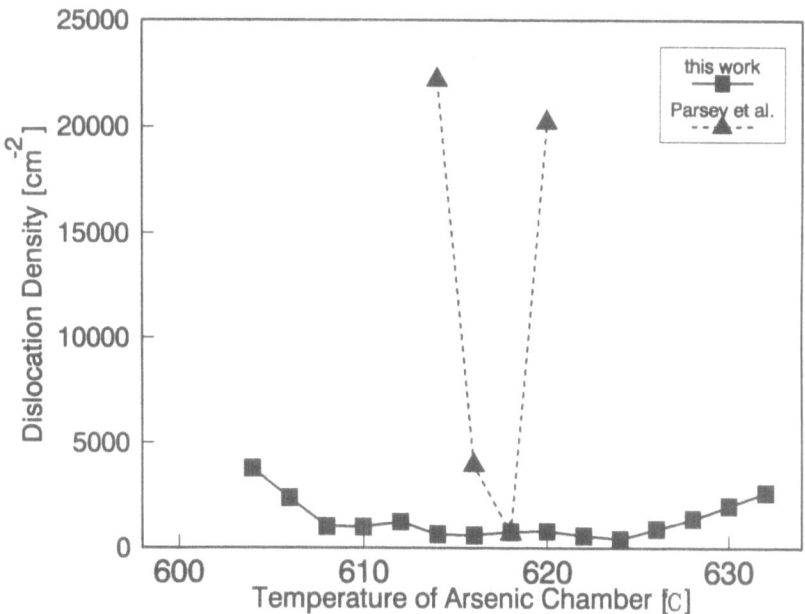

Fig.27

As temperature dependence of the dislocation density in PCZ grown GaAs bulk crystals. The results of HB grown GaAs crystals were also shown in the figure.

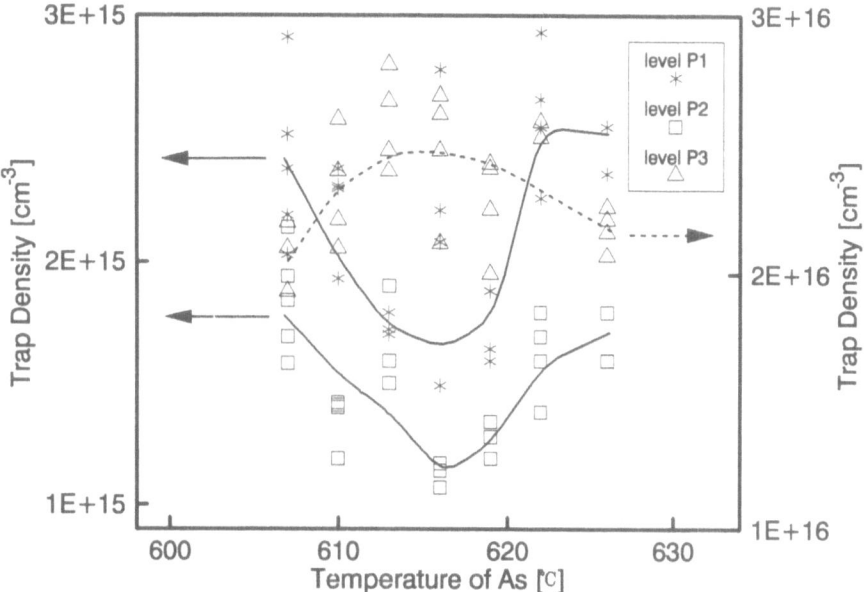

Fig.28

As temperatue dependence of the trap density in PCZ grown GaAs.

299

from the conventional LEC method.

GaAs bulk crystal growth by the vapor pressure controlled Czochralski method

The vapor pressure control of the GaAs solution growth was carried out at first by Suzuki and Akai[29] in the field of the horizontal Bridgman method. High quality GaAs bulk crystals with very low dislocation density were grown under our finding optimum As vapor pressure of 830 Torr, where the temperature of metal As is 617°C. These crystals are supplied all over the world. This work was verified by Gatos and Nanishi[30], and it was shown that the dislocation density and the defect density decreased abruptly at a specific metal As temperature of 617°C within 1°C. This specific As vapor pressure is nearly equal to that obtained from our equation [2] concerning on the temperature dependence of the optimum As vapor pressure. Optimum As vapor pressure at melting point of GaAs is obtained to be 830 Torr from the eqn. [2]. The temperature of the metal As corresponds to be 617°C according to the eqn.[1].

Conventional CZ method enables to grow GaAs bulk crystals with large diameter compared to HB method. But the dislocation density has been greater than that of HB crystals due to its large thermal strain. By applying the vapor pressure control method to the CZ growth (PCZ), high quality CZ GaAs bulk crystals with very low dislocation density were obtained. Figure 26 shows the schematic draw of the CZ growth apparatus with As vapor chamber. As shown in Fig.27, 2–inch wafers with the dislocation density as low as 2000 cm^{-2} were grown without impurity doping under the optimum As vapor pressure. The EPD of the intentionally–undoped LEC GaAs is about 10^5 cm^{-2}. This is higher at least one order of magnitude than that of our PCZ GaAs crystals. DLTS measurements revealed three kinds of deep donors. Figure 28 shows the As vapor pressure dependence of each deep level density. It is shown that the deep level density also shows a minimum under optimum As vapor pressure. Similar results were also obtained from the PHCAP measurements. Figure 29 shows the photographs of the PCZ and conventional CZ grown GaAs ingots. Surface of the PCZ GaAs crystals is shown to be very brilliant compared with that of the conventional LEC crystals.

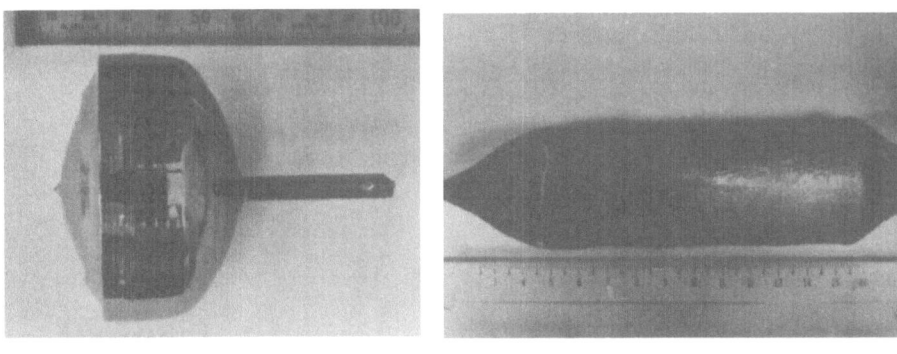

(a)   (b)

Fig.29

Photographs of the GaAs ingot grown by (a) the VPC–CZ and (b) the conventional LEC method.

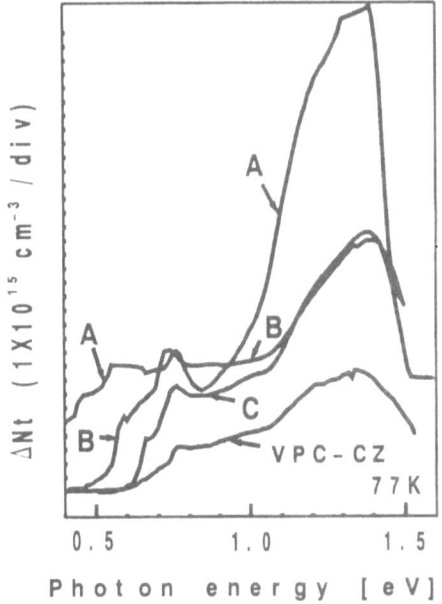

Fig.30

PHCAP maximum ion–density spectra of various n–GaAs bulk crystals with the carrier concen-

tration of about 1×10^{16} cm^{-3}.

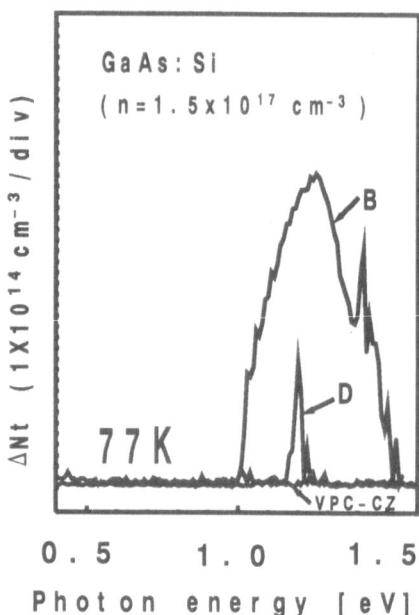

Fig.31

PHCAP deionized level-density spectra of various n-GaAs bulk crystals with the carrier concentration of 1.5×10^{17} cm^{-3}.

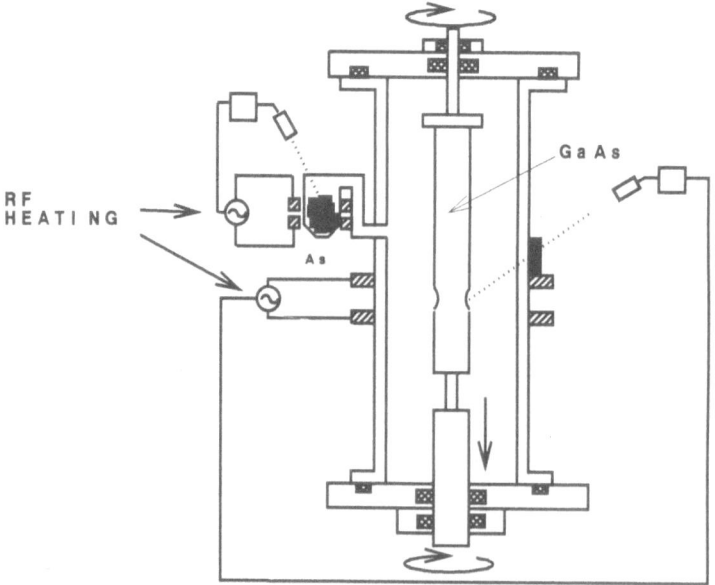

Fig.32

Schematic draw of the GaAs bulk crystal growth apparatus using the vapor pressure controlled floating zone (FZ) method.

In order to evaluate deep levels in various GaAs bulk crystals, the PHCAP measurements were carried out under constant capacitance condition. Figure 30 shows the PHCAP spectra of various intentionally–undoped n–GaAs crystals with the carrier concentration of about 4×10^{16} cm^{-3}. As reported previously, the so–called photoquenching phenomenon is observed in the spectral region of 1.0 – 1.50eV below about 110K. After irradiation of monochromatic light, ion density increases rapidly and then decreases gradually. Figure 30 represents the maximum ion density spectra in each samples. PCZ crystal shows minimum deep level density compared with that of other crystals. This confirms the possibility for perfect crystal growth by the PCZ method. Figure 31 shows the PHCAP spectra obtained from the subtraction between maximum and saturating ion density at each wavelength. Samples used were various Si–doped GaAs bulk crystals with the carrier concentration of 1.5×10^{17} cm^{-3} grown by various methods. PCZ crystal shows extremely low deep level density, which shows the so–called photoquenching phenomenon.

GaAs bulk crystal grown by the vapor pressure controlled FZ (VPC–FZ) method

In order to obtain high purity bulk crystals with stoichiometric composition, vapor pressure controlled FZ method (VPC–FZ) has been applied to GaAs bulk crystal growth[31]. Figure 32 shows the schematic draw of the VPC–FZ apparatus. Whereas the growth condition has not been optimized and the As vapor pressure could not be strictly controlled yet, high purity p–GaAs crystals were obtained with the carrier concentration as low as about 1×10^{15} cm^{-3}. Figure 33 shows the PHCAP spectrum of the VPC–FZ grown p–GaAs crystal. Stoichiometry–dependent deep acceptors were detected at 0.53, 0.71, 0.90 and 1.0eV above the valence band. These deep levels were commonly detected regardless of the variation of dopant impurities and growth method. As vapor pressure dependencies of the level density were also clarified by our PHCAP measurements. However, the level density in VPC–FZ GaAs is much lower than that in LEC and HB grown samples by three order in magnitude. This confirms the ability of VPC–FZ method to supply high purity GaAs crystals with stoichiometric composition.

303

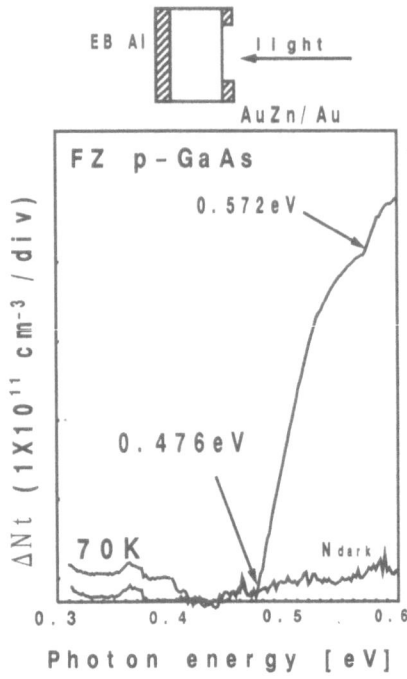

Fig.33

PHCAP spectra of the vapor pressure controlled FZ GaAs crystal in the spectral region of 0.30–0.60eV at 70K.

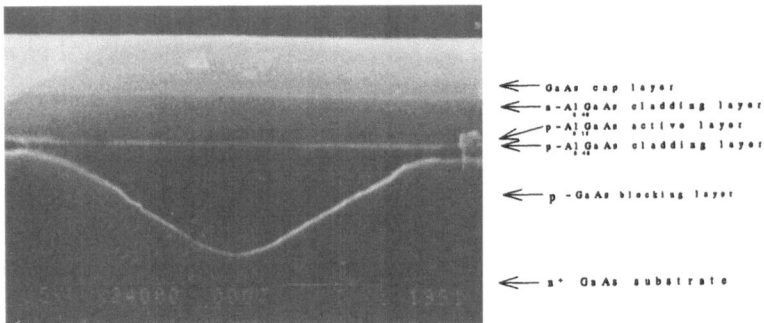

Fig.34

Cross sectional SEM micrograph of the visible semiconductor laser diode prepared by the TDM–CVP.

Visible semiconductor laser

The TDM–CVP has also been applied to the fabrication of AlGaAs visible semiconductor laser diodes. This LPE method enables to be a low temperature and successive epitaxial growth. In combination with the CVP method, the low temperature growth is effective to reduce the defect density introduced by the deviation from the stoichiometric composition.

Figure 34 shows the SEM microphotograph of the visible laser diode with the V–grooved substrate inner stripe (VSIS) structure. The structure is not optimized yet. But these CW operated VSIS laser diodes show quantum efficiency up to about 45% at present. More precise control of the stoichiometric composition leads to improve LD characteristics.

Semiconductor Ramann laser

Semiconductor Ramann laser[32], which enables to operate at over the tera Herz (10^{12} Herz), has also been fabricated by the TDM–CVP with the GaP active layer and GaAlP cladding layers. Lasing operation were obtained by low pumping light power of 500mW at present[33].

Summary

Evaluation and control of the deviation from stoichiometry has been increasingly important in the field of compound semiconductor crystals. Each device process including crystal growth and diffusion etc. cannot be completed without consideration of the stoichiometry, because the compound semiconductor device needs precise control of electrical and optical feature. Then more precise control of stoichiometric composition should be required. Stoichiometry–control should be also important in the field of the superconducting ceramics.

305

References

1)   Y.Watanabe, J.Nishizawa and I.Sunagawa, Kagaku, vol21(3), p.p.140–141, Iwanami (1951)

2)   H.Otsuka, K.Ishida and J.Nishizawa, Jpn. J. Appl. Phys., 8, 632 (1969)

3)   Y.Okuno, K.Suto and J.Nishizawa, J. Appl. Phys., 44, 832 (1973)

4)   J.Nishizawa, H.Otsuka, S.Yamakoshi and K.Ishida, Jpn. J. Appl. Phys., 13, 46 (1974)

5)   S.Yamakoshi, Doctor Thesis, Tohoku University (1975) supervised by Prof. Nishizawa

      J.Nishizawa, N.Toyama, Y.Oyama and K.Inokuchi, Proc. of the 3rd International School on Semiconductor Optoelectronics, Cetniewo, ed. by Marian A. Herman, PWN–Polish Scientific Publishers (Warszawa, 1980) p.p. 27–77 (1980)

6)   J.Nishizawa, I.Shiota and Y.Oyama, J. Phys. C: Solid State Phys., 9, 1 (1986)

7)   J.Nishizawa, Y.Oyama and K.Dezaki, J. Appl. Phys., 67(4), 1884 (1990)

8)   J.Nishizawa, Y.Oyama and K.Dezaki, Phys. Rev. Lett., 65, 2555 (1990)

9)   I.Fujimoto, Jpn. J. Appl. Phys., 23, L287 (1984)

10)   H.J. von Bardeleben, J.C.Bourgoin, D.Steivenard and M.Lannoo, Proc. of International Symp. GaAs and Related Compounds, Heraklion, Greece, 399 (1987)

      J.–M Spaeth, D.M.Hofmann, M.Heinemann and B.K.Meyer, ibid, 391 (1987)

11)   J.Nishizawa, Y.Okuno and H.Tadano, J. Crystal Growth, 31, 215 (1975)

12)   J.Nishizawa and Y.Okuno, IEEE Trans. Electron Device, ED–22, 716 (1975)

13)   K.Tomizawa, K.Sassa, Y.Shimanuki and J.Nishizawa, J. Electrochem. Soc., 131, 2394 (1984)

14)   J.Nishizawa, K.Itoh, Y.Okuno, M.Koike and T.Teshima, Proc. IEEE International Electron Devices Meeting (IEDM), p.p. 311–314 (1983)

15)   J.Nishizawa, Y.Okuno, M.Koike and F.Sakurai, Jpn. J. Appl. Phys., 19, 377 (1980)

16)   J.Nishizawa, K.Itoh, Y.Okuno and F.Sakurai, J. Appl. Phys., 57, 2210 (1985)

17)   R.E.Honig, RCA Rev., 30, 285 (1969)

18)   A.Itoh, T.Sukegawa and J.Nishizawa, Tech. Report of Transistor Specialist Committee,

IEE Japan (Jan. 1967)

A.Itoh, T.Sukegawa and J.Nishizawa, Tech. Report of Research Institute of Electrical Communication, Tohoku University, TR-32 (Feb. 1969)

19)   Y.Oyama, Denshi Tokyo (IEEE Tokyo), 28, 130 (1989)

20)   A.L.Lin, E.Omelianovski and R.H.Bube, J. Appl. Phys., 47, 1852 (1976)

21)   J.Nishizawa, Y.Oyama and K.Dezaki, J. Appl. Phys., 69(3), 1446 (1991)

22)   J.Nishizawa, Y.Oyama and K.Dezaki, to be published in J. Appl. Phys., (1991)

23)   J.Nishizawa, Y.Oyama and K.Dezaki, to be published in J. Phys. Condensed Matter, (1991)

24)   K.Suto and J.Nishizawa, J. Appl. Phys., 67(1), 459 (1990)

25)   J.Nishizawa and Y.Okuno, Proc. of 2nd International School on Semiconductor Optoelectronics, Cetniewo, ed. by Marian A. Herman, PWN-Polish Scientific Publishers (Warszawa, 1978), p.p.101-130

26)   E.Omura, X.X.Wu, G.A.Vawter, L.Coldren, E.Hu and J.L.Merz, Electron. Lett., 22, 496 (1986)

27)   K.H.Bennemann, Phys. Rev., 137, A1497 (1961)

28)   R.A.Swalin, J. Phys. Chem. solids, 18, 290 (1961)

29)   T.Suzuki and S.Akai, Bussei, 144, 12 (1971)

30)   J.M.Parsey Jr, Y.Nanishi, J.Lagowski and H.C.Gatos, J. Electrochem. Soc., 128, 937 (1981)

31)   FZ grown GaAs crystals were supplied from the Tohoku Steel Co'Ltd, Sendai Japan

32)   J.Nishizawa and K.Suto, J. Appl. Phys., 51(5), 2429 (1980)

33)   K.Suto, S.Ogasawara, T.Kimura and J.Nishizawa, J. Appl. Phys., 66(11), 5151 (1989)

Superconducting Materials

Physics of Superconducting Devices and Recent Topics on High T$_c$ Superconducting Devices

Seigô Kishino

Department of Electronics, Faculty of Engineering, Himeji Institute of Technology, Shosha, Himeji

Superconducting phenomena and their applications to electronic devices are reviewed with an emphasis on a high Tc superconductor. After a short mention of fundamentals of superconductivity, characteristics of Josephson junction (J-J) are discussed where current-voltage (I-V) and differential conductance-voltage (dI/dV-V) curves are involved as well as the indispensable conditions for J-J. Subsequently, recent developments on high Tc superconducting devices are followed. Of these, an all high Tc Josephson junction and a proximity effect with an extraordinary long penetration depth are included, besides the I-V & dI/dV-V curves, Shapiro steps, SQUID, etc. Lastly, a simple technique preparing high Tc films is shortly mentioned.

1. INTRODUCTION

The transition temperature Tc from normal conductivity to superconductivity had been confined to beneath 30 K for a long time since the discovery of superconductivity by Kamerling Onnes in 1911. However, in 1987 an oxide superconducting material has been discovered by Bednorz and Müller[1], and then, the transition temperature Tc has been increased to above a liquid nitrogen temperature at one effort by the successive researches of many groups[2].

It is expected that such a high Tc material would bring about radical changes in the field of electrical engineering. This is because device operation is possible at a liquid nitrogen temperature in the cryodevices such as a Josephson device, a high frequency microwave detector, and a highly sensitive magnetic detector, viz. SQUID (Superconducting QUantum Interference Device). With the use of a high Tc material, an improvement of device characteristics such as a switching speed is feasible besides the device operation at a higher temperature.

However, the high Tc materials have some difficulties as follows. i) Very short coherence length ξ. For example, the coherence length ξ_{ab} in a or b plane is about 16.4 Å and ξ_c in c plane, about 3.0 Å in YBaCuO[3,4] Tc of which is about 90 K. Such a short coherence length makes the fabrication of the tunnel junction difficult as described later. ii) The device operation at a high temperature also induces the increase of thermal noise. In order to reduce the disadvantages induced by thermal noise, a high current density as much as 10^5 A cm^{-2} is necessary during device operation. This requires us high Tc materials provided with a high critical current density. In addition, iii) the mechanism of the superconductivity has not been solved for the high Tc materials as yet. This also brings about large difficulties in the field of the device application. This is because it occurs that we must develop a new device without any reliable guidance.

2. SUPERCONDUCTING PHENOMENA

As to the mechanism of superconductivity, we shall confine ourselves to the conventional treatment supported by the BCS theory[5]. Firstly, we would like to consider the correlation between London equation[6] and the current equation based on the quantum mechanics. According to the London eq., the supercurrent density J_S is given by the equation,

$$J_S = - \frac{ne^2}{mc} A, \qquad (1)$$

where A is a vector potential for the magnetic field. On the other hand, with the use of the quantum mechanics, we can obtain the expression for the normal current density J(x) by the equation,

$$J(x) = - \frac{i\hbar e}{2m} \left\{ \phi^*(x) \, \text{grad} \, \phi(x) - \phi(x) \, \text{grad} \, \phi^*(x) \right\} - \frac{e^2}{mc} \left| \phi(x) \right|^2 A \qquad (2)$$

where $\phi(x)$ is an electron wave function. The resulting equation is in agreement with that of Ginzburg-Landau equation[6], if the electron wave function $\phi(x)$ is replaced by a complex-order parameter $\psi(x)$.   This is the first point to be noted.

If there is no magnetic field and, therefore, no vector potential A either, then the current density will, of course, vanish.   This shows that the first term of eq.(2) becomes zero. If the first term remains zero in the presence of a magnetic field, the current density is in agreement with eq.(1).   It might appear at first sight as though every conductor could, therefore, become a superconductor. But, this is by no means the case. The appearance of a vector potential has an effect on the electron wave function $\phi(x)$. Then, the first term of eq.(2) is compensated by the last expression in eq.(2) which is proportional to a vector potential A in the normal state.

However, the vector potential has no effect on the wave function of paired electrons in the superconductting state. As a result, the current density is given, in the absence of an electric field, by the equation,

$$J(x)=0, \qquad \text{(in the normal state)} \qquad (2a)$$
$$J(x)\neq 0, \qquad \text{(in the superconducting}$$
$$=-\frac{e_s^2}{m_s c}\left|\psi(x)\right|^2 A \qquad\qquad \text{state)} \qquad (2b)$$

Why does the first term of eq.(2) remain zero in a magnetic field and then, the second term come to dominate ? This would be the case if the wave function had some rigidity and, therefore, remained unchanged despite the application of a magnetic field. To tell the truth, it is the energy gap what brings the rigidity of the wave function. The energy gap 2Δ , which is introduced by Bardeen, Cooper, and Schrieffer (BCS)[5], is temperature-dependent.   A schematic figure of the energy gap is shown in Fig. 1.

3. CHARACTERISTICS OF JOSEPHSON JUNCTION

3.1 Josephson tunneling current

Here, we assume the structure of a tunneling device as shown in Fig.2, where S_1 , S_2 , and I are superconductor 1, 2, and an insulator, respectively. Signs φ_1 and φ_2 show respective phases of paired electrons present in respective superconductor 1 and 2. According to Josephson eq[7] , the tunneling current J of paired electrons is given by

$$J=J_c \sin\theta, \qquad (3)$$

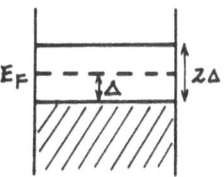

Fig.1 Energy diagram of superconductor

where $\theta = \varphi_2 - \varphi_1$,   (4)
and J_c shows the supercurrent at $\theta = \pi/2$ which flows between two superconductors through the insulator I.

3.2 Current vs. voltage curve and differential conductance curve

As described in the preceding section, a superconductor has an energy gap. This imposes some restrictions on the tunneling phenomenon of electron. For example, there is no restriction for the electron tunneling through the insulator between two normal metals (N-I-N). However, the circumstance changes between a normal and a superconductor (N-I-S) or between two superconductors (S-I-S).

This is because there is no state for electrons at the Fermi level in the superconductor as shown in Fig. 3 and then, no tunneling current flows at the zero bias between the normal metal and the superconductor shown in Fig. 3. For the occurrence of the electron tunneling through the insulator between a metal and a superconductor, positive or negative bias as large as Δ is necessary as shown in Fig. 4. The resulting current-voltage (I-V) characteristics come to that shown in Fig. 5(a) where the tunneling current is existing in the voltage region above Δ/e or below $-\Delta/e$ and is zero between them. These tunneling electrons are called quasi-particles.

In the S-I-S junction, the tunneling of paired electrons, viz. super-

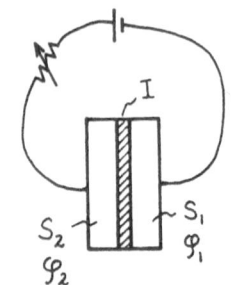

Fig.2 Josephson tunneling junction

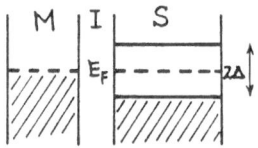

Fig.3 Energy diagram of N-I-S structure

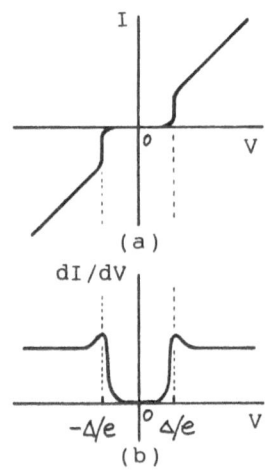

Fig.5 Current versus voltage(a) and differential conductance(b) curve in N-I-S junction

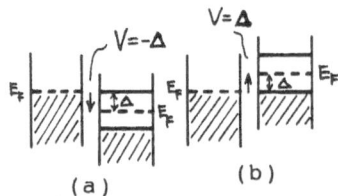

(a) (b)

Fig.4 Tunneling of quasi-particle by adding a bias voltage $-\Delta$ (a) or $+\Delta$ (b)

314

current flow, occurs besides the
tunneling of the quasi-particles.
The supercurrent is tunneling at
the zero bias condition. The resultant I-V characteristics become
a profile shown in Fig. 6(a). In
the I-V curve of S-I-S junction,
no currents flow between the voltages of $2\Delta/e$ and $-2\Delta/e$. Accord-
ingly, we can measure the energy
gap 2Δ of the superconductor from
the I-V curve. Note that the region, which the current flow is
profibited in the S-I-S junction,
is double of the N-I-S junction.

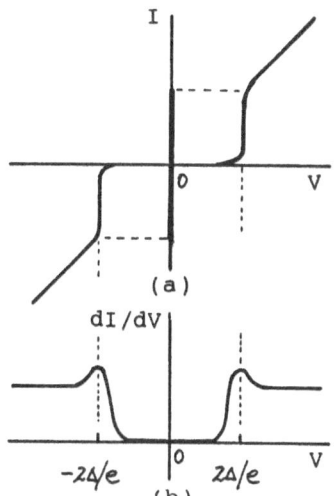

Fig.6 Current versus volt-
age(a) and differen-
tial conductance(b)
curve in S-I-S junc-
tion

The characteristics of differential conductance (dI/dV-V) curve
are shown in Fig.5(b) and Fig.6(b)
for the N-I-S and the S-I-S junction, respectively. The dI/dV-V
curve brings information on the
detailed structure of the energy gap and then, this
measurement is indispensable for the study of the
superconducting mechanism. In fact, the dI/dV-V curves
of the high Tc materials show different shape from
those of the conventional S-I-S or(N-I-S)junction shown
in Fig. 6.

The shape of dI/dV-V curve in the high Tc materials
can be approximated to that shown in Fig.7. Based on
the shape, Anderson[8] proposed a new model on the superconducting mechanism for the high Tc materials. However, the validity of the model has not been made clear.

4. NECESSARY AND SUFFICIENT CONDITIONS FOR JOSEPHSON
   JUNCTION

In this section, we confine ourselves to discussing the Josephson
junction (J-J) in a narrow sense[7].
Under the condition, we consider the
necessary and sufficient conditions
for J-J. Consequently, the following
four phenomena must be observed; (i)
supercurrent, (ii) energy gap, (iii)
Shapiro step, and (iv) Fraunhofer diffraction. As to the supercurrent and
the energy gap, we explained them

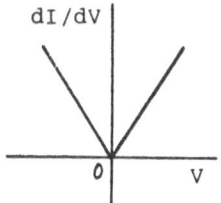

Fig. 7 Sketch of dif-
ferential conduct-
ance curve in S-I-S
(or N-I-S) junction
using high Tc mate-
rials

already in the preceding section.

Before describing the Shapiro step, we must know about ac Josephson effect. In the presence of the potential difference V across a J-J, the phase differ-ence θ between the both sides satisfies the equation[7],

$$\frac{\partial \theta}{\partial t} = \frac{2e}{\hbar} V. \qquad (5)$$

If V=const. (V=V_0), then the phase differnce is given by

$$\theta = \theta_0 + \frac{2eV_0}{\hbar} t = \theta_0 + \omega_0 t, \qquad (6a)$$

where $\quad \omega_0 = \frac{2e}{\hbar} V_0 = 2\pi f. \qquad (6b)$

With the use of eq.(6a), the following equation is obtained;

$$J = J_c \sin(\theta_0 + \frac{2eV_0}{\hbar} t). \qquad (7)$$

This eq. shows that an alternative current will flow across the junction. This is the ac Josephson effect.

Associated with the ac Josephson effect, another peculiar phenomenon, Shapiro step, is discovered in which voltage steps apear in the V-I characteristics when a varying potential difference $v_1 \cos\omega_1 t$ is applied to the junction. Resultant potential difference V(t) across the J-J is given by

$$V(t) = V_0 + v_1 \cos\omega_1 t. \qquad (8)$$

Using eqs.(3),(5),and (8), the Josephson current is equal to

$$J = J_0 \sum_{n=-\infty}^{\infty} (-1)^n J_n(\frac{2ev_1}{\hbar\omega_1}) \sin\left\{(\omega_c - n\omega_1)t + \theta_0\right\}. \qquad (9)$$

This equation shows that the dc current flows at the condition of $\omega_c = n\omega_1$. In the experiments using micro-wave irradiation, the resulting potential difference is observed at a given current as shown in Fig. 8.

Fraunhofer diffraction is the Josephson effect in presence of magnetic field. In the limited space available, we would like to show only the out-line of the phenomenon. When the field flux is applied across the junction, the maximal super-current through the junction I_{max} is given by

$$I_{max} = J_J \left| \frac{\sin(\pi\Phi/\Phi_0)}{\pi\Phi/\Phi_0} \right|, \qquad (10)$$

where Φ_0 is the fluxoid quantum which is defined as $\Phi_0 = hc/2e$. Equation (10) predicts "diffrac-tion pattern". This is called

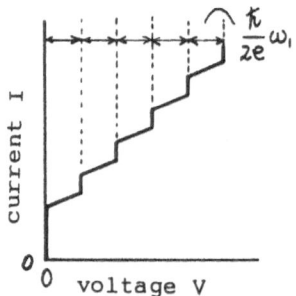

Fig.8 Theoretical estima-tion of Shapiro step

Fraunhofer diffraction pattern, which is also verified
experimentally.

5. VARIOUS WEAKLY COUPLED JUNCTIONS AND SQUID

5.1 Various weakly coupled junctions

There are various weakly coupled junctions as shown
in Figs.9(a),9(b), and 9(c). Figure 9(a) shows the
point contact junction and Fig. 9(b) shows the standard
Josephson tunneling junction. On the other hand,
Fig.9(c) shows the special junction using an irregular-
ity of crystal. In this case, a grain boundary is
used as a junction. A crack junction will be one of
the alternatives of the grain boundary junction.

5.2 SQUID

Much higher resolution in a magnetic field than that
implied by eq.(10) can be obtained by increasing the
area in which flux is effective in causing a phase
change. This can be performed readily by using two
separate J-J's in a superconducting circuit as illustrated in Fig. 10. This is called de-SQUID (Superconducting QUantum Interference Device). In the SQUID
shown in Fig. 10,
we obtain an
oscillation of
the supercurrent
as illustrated
in Fig. 11.
The maximal
supercurrent $J_{\gamma 0}$
can be expressed
by the equation,

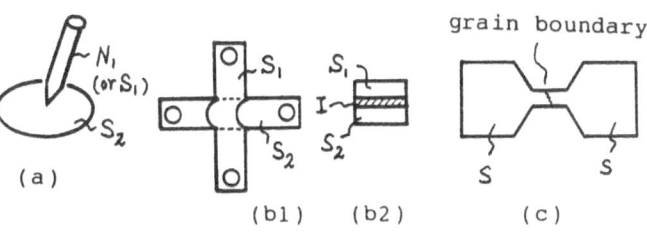

(a)

(b1) (b2)   (c)

grain boundary

S

S

Fig.9 Several kinds of weakly coupled
junctions (J)
(a) point contact J, (b) Josephson J,
(c) special J using crystal defects

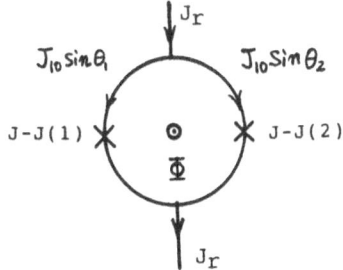

Fig.10 Schematic diagram of
dc-SQUID

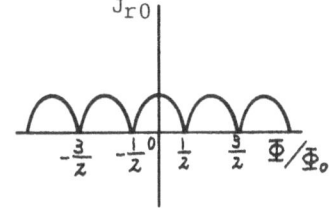

Fig.11 Maximal supercurrent $J_{\gamma 0}$
versus magnetic flux Φ

$$J_{r0} = 2J_{10} \, |\cos(\pi \Phi / \Phi_0)| \, , \qquad\qquad (14)$$

if an equal critical current J_{10} flows through two J-J's as shown in Fig. 10.

By the way, a magnetic field B is shown as $B = \Phi/s$, where s is the area of the magnetmeter loop. If we can read to about thousandth of a fringe shown in Fig.11, we can resolve the field increment of $\sim 10^{-9}$G. Then, we can measure a very weak magnetic flux, for example, that of a brain wave ($10^{-9} \sim 10^{-10}$ G).

6. HIGH Tc SUPERCONDUCTOR AND THEIR DEVICES

6.1 Current-voltage characteristics of Josephson junction

Here, we deal with a J-J in a wide sense[7] in which a weakly coupled junction is looked upon as a J-J. Several studies on the various S-I-S or N-I-S junctions have been reported. Recently, an all-high Tc J-J was reported as shown in Fig.12 using BaKBiO (Tc=30 K)[9]. Although the authors called a BaKBiO compound a high Tc material, the properties of BaKBiO are a little different from those of the cuprate high Tc materials, Tc of which is actually higher than 77 K, such as YBaCuO. This is because the crystal structure is cubic, a coherence length is rather large (50-75 Å), and the differential conductance curve can be explained by BCS theory[10].

As to the J-J's using the high Tc materials such as YBaCuO, ideal J-J's have not been attained so far. This is attributed to the extremely short coherence length or to the imperfection of the materials. Combinations of high and low Tc materials are also used as an S-I-S junction. As the low Tc material, Pb and Nb are in majority. In this combination, nor has an ideal J-J been obtained.

6.2 Differential conductance curve

Associated with the superconducting mechanism, many people have attacked the dI/dV-V characteristics of S-I-S or N-I-S junction. No one has succeeded in obtaining the ideal curve of dI/dV-V. In the early stage, Crommie et al.[11] reported the dI/dV-V curve using the point contact on YBaCuO as shown in Fig.13. This falls in the category of N-I-S junction.

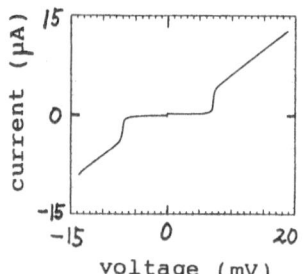

Fig.12 I-V characteristics of all-high Tc Josephson junction using BaKBiO (Tc=30 K)

The shape of the curve is, on the whole, similar to that shown in Fig. 7. From the curve, they obtained the following characteristic constants; $2\Delta/k_B Tc \sim 3.9$ and $\Delta \sim 15$ meV

After that, Gurvitch et al.[12] obtained the dI/dV-V curve with the use of YBaCuO-I-Pb junction. Using the nearly same structure, Iguchi et al.[13] reported a different curve from that of Gurvitch et al. As schematically illustrated in Figs. 14(a) and 14(b), the former has one valley and the latter two valleys. According to Iguchi[14], the discrepancy is due to the fabrication technique of the insulator. Namely, the insulator is of a natural oxide and an artificial insulating film in the former and the latter junction, respectively.

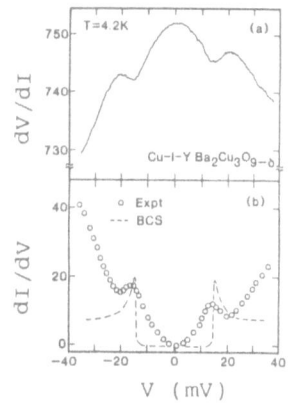

Fig.13 Differential conductance curve of S-I-N junction using YBaCuO

6.3 Shapiro step

Many studies on Shapiro steps are reported in the weakly coupled junction using high Tc materials. As the junctions used for the measurement, there are point contact, grain boundary, and crack junctions. The Shapiro steps in the case of the grain boundary junction[15] is representatively shown in Fig. 15.

6.4 A kind of "proximity effect"

Near the contact between a normal metal and a superconductor, there is the possibility of superconductivity in this boundary layers of the normal metal side. This is so-called proximity effect[16] which is

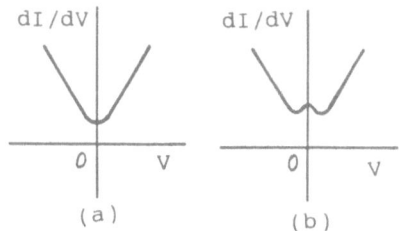

Fig.14 Sketches of differential conductance curves actually obtained in tunneling junctions using high Tc materials (a) natural oxide, (b) artificial oxide

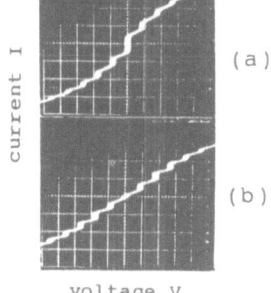

Fig.15 Shapiro steps obtained using grain boundary junction of YBaCuO material

319

due to the mutual influence of a superconductor and a
normal metal in contact with it.

In the junction of a normal metal and a high Tc material, the proximity effect is also expected. However,
even if the effect could be observed, a very short
penetration depth of the supercurrent would be estimated by the equation[7] derived from the conventional
theory[6]. The following "proximity effect" has an extraordinary long penetration depth.

Recently, Kasai et al.[8] investigated the characteristics of the superconductivity on the YBaCuO/LaCaMnO/
YBaCuO junction shown in Fig. 16 where the LaCaMnO (LCM
O) is, of course, non-superconducting material. To our
astonishment, they observed supercurrent between the
upper and the lower YBaCuO (YBCO) film at the condition
that the thickness of the LaCaMnO extends to 5000Å.

They explained the phenomenon in terms of "proximity
effect". Namely, they insist that the non-superconducting LCMO film becomes superconductive by the influence of YBCO films which sandwich the interfacial LCMO
film. However, the distance from the contact is too
long, compared to the penetration depth of the superconducting carriers which is estimated by the conventional theory as already discussed.

Similar results are separately reported by other
groups[19,20]. At any rate, if the validity of this phenomenon could be verified, the phenomenon offers a very
promising prospect for the device application of High
Tc materials.

6.5 SQUID
A SQUID is one of the most promising devices which
are producible by high Tc materials. This is probably
because the weakly coupled junction of the SQUID is acceptable even if it is not necessarily to be the J-J in
a narrow sense. The characteristics of the SQUID
might be good enough if the
property of the weakly coupled junction is only reliable.

| YBCO |
| LCMO |
| YBCO |

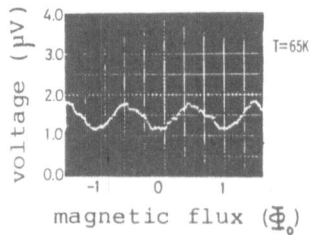

magnetic flux (Φ_0)

Fig.16 Junction structure of
YBCO/LCMO/YBCO showing
special proximity effect

FIg.17 Signals of dc-SQUID actually observed using high Tc
materials

On the evidence of this prospect, we have several reports on the actual operation of a SQUID. One of them is representatively shown in Fig. 17[21].

7. SIMPLE TECHNIQUE OBTAINING HIGH Tc FILM WITH THE USE OF ELECTROPHORETIC DEPOSITION

Of many techniques[22] preparing a high Tc film, the film coating by·electrophoretic deposition[23] is not only unique but also inexpensive technique. This technique is schematically shown in Fig. 18. In this technique, corpuscles of superconducting materials such as YBaCuO powder are dispersed in acetone solution, and a MgO substrate covered with silver film, which operates as a negative electrode, is dipped in the solution with a positive electrode of a Pt plate. The coating of YBaCuO film occurs by adding the dc voltage (several hundreds voltage) between two electrodes.

As-deposited film is not superconductive. Then, deposited film is firstly prebaked for two hours at a rather low temperature ($\sim 500°$C) in air or O_2 ambience. After that, a firing heat-treatment is normally carried out in air or O_2 ambience in order to make the film superconductive. However, we fired it in He ambience. As a result, the film showed a transition from a normal state to a superconducting state as shown with a broken line in Fig. 19.

When the O_2 ambience is used, a two-step R-T curve

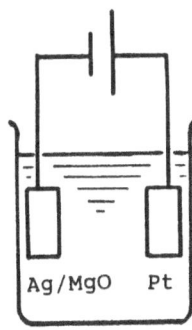

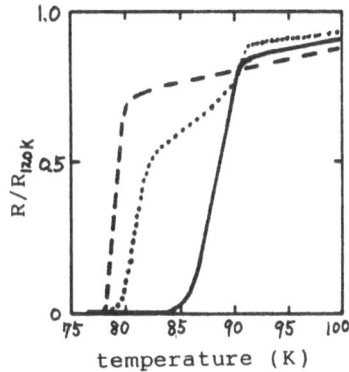

temperature (K)

Fig.18 Schematic diagram of electrophoretic deposition technique

Fig.19 Resistance versus temperature curve of YBCO film coated by electrophoretic deposition technique

is obtained as shown in Fig. 19 with a dotted line.
With the use of the He ambience, the single step R-T
curve was obtained as shown with a broken line. But,
the transition temperature was rather low with the use
of this temperature firing. By reducing the temperature of firing in He, we was able to attain a single
step R-T curve with a high transition temperature as
shown with a solid line in Fig. 19[24].

The effect of the He ambient firing on the film characteristics has not been obvious as yet. One plausible
reason is that the He ambient firing heat-tretment reduces the melting point of YBaCuO materials[25]. The low
temperature during the firing treatment probably pro-
hibits the non-superconducting particles from being
formed inside the YBaCuO film.

8. CONCLUSIVE REMARKS

Superconducting devices and their physics are reviewed upon those of high Tc materials. As to the
development of all high Tc devices, there are many
problems unsolved. As a result, following items seem
to be noticeable points.
(i)   Superconducting devices using high Tc materials
       are now in an early stage of development.
(ii)   Josephson junction using all high Tc materials
       has not been developed except the case of BaKBiO
       compound, Tc of which is about 30 K.
(iii) A SQUID is one of the most promising superconduc-
       ting devices using high Tc materials.
(iv)   A special kind of proximity effect, in which the
       penetration depth of the supercurrent is extraor-
       dinary long, is recently observed in several com-
       binations of a normal material and a high Tc
       material.
(v)   It is quite necessary for the development of the
       superconducting devices to clear up the super-
       conducting mechanism of high Tc materials.

ACKNOWLEDGMENTS

We would like to thank Dr. H. Niu of our laboratory
for various discussions and comments on the contents of
the manuscript.

REFERENCES

1) Bednortz, J., et al.(1986). Possible High Tc Super-conductivity in Ba-Ca-Cu-O System, Z. Phys. B64, pp.189

2) Jorgensen, J. D., et al. (1987). Oxygen Odering and Orthorhombic-to-Tetragonal Phase Transition on $YBa_2Cu_3O_{7-x}$, Phys. Rev. B36(7),pp.3608-3616.

3) Welp, U., et al. (1989). Magnetic Measurements of The Upper Critical Field of YBaCuO Single Crystal, Phys. Rev. Lett. 62(16), pp.1908-1911.

4) Deutscher, G. (1988). Superconducting Glass and Related Properties, Physica C153-155, pp.15-20.

5) Bardeen, J., et al. (1957). Theory of Superconductivity, Phys. Rev. 108(5), pp.1175-1204.

6) Tinkham, M., (1975). "Introduction to Superconductivity" (McGraw-Hill Inc.).

7) Likharev, K. K. (1979). Superconducting Weak Links, Rev. Mod. Phys. 51(11), pp.101-159.

8) Anderson, P. W. (1987). The Resonating Valence Bond State in LaCuO and Superconductivity, Science 235, pp.1196-1198.

9) Pargellis, A. N., et al. (1991). All-High Tc Josephson Tunnel Junction: $Ba_{1-x}KBiO_3/Ba_{1-x}KBiO_3$ Junctions, Appl. Phys. Lett. 58(1), pp.95-96.

10) Huang, Q., et al. (1990). Ideal Tunneling Characteristics in BaKBiO Point-Contact Junctions with Au and Nb Tips, Appl. Phys. Lett. 57(22), pp. 2356-2358.

11) Crommie, M. F., et al. (1987). Tunneling Measurements of The Energy Gap in Y-Ba-Cu-O, Phys. Rev. B 35(16), pp.8853-8855.

12) Gurvitch, M., et al. (1989). Reproducible Tunneling Data on Chemically Etched Single Crystals of YBaCuO , Phys. Rev. Lett. 63(9), pp.1008-1011.

13) Iguchi, I., et al. (1987). Tunneling Spectroscopy of Y-Ba-Cu-O Compound, Jpn. J. Appl. Phys. 26(5), pp.L645-L646.

14) Iguchi, I. (1991). Private Communication.

15) Yamashita, T., et al. (1988). Josephson Effects at 77K in Grain Boundary Bridge Made of Thick Films, Jpn. J. Appl. Phys. 27(6), pp.L1107-L1109.

16) De Gennes, P. G. (1964). Boundary Effects in Superconductors, Rev. Mod. Phys. 36(1), pp.225-237.

17) Seto, J., et al. (1974). Theory and Measurements of Lead-Tellurium-Lead Supercurrent Junctions, in Low Temperature Physics-LT-13, vol. 3, N. Y., Plenum, pp. 328-333.

18) Kasai, M., et al. (1990). Current-Voltage Charac-
    teristics of YBaCuO/LaCaMnO/YBaCuO Trilayered-Type
    Junctions, Jpn. J. Appl. Phys. 29(12), pp.L2219-
    L2222.
19) Rogers, C. T., et al. (1989). Fabrication of
    Heteroepitaxial YBaCuO-PrBaCuO-YBaCuO Josephson De-
    vices Grown by Laser Deposition, Appl. Phys. Lett.
    55(19), pp. 2032-2034.
20) Mizuno, K., et al. (1990). Fabrication of Thin-Film
    -Type Josephson Junctions Using A Bi-Sr-Ca-Cu-O/Bi-
    Sr-Cu-O/Bi-Ca-Cu-O Structure, Appl. Phys. Lett. 56
    (15), pp. 1469-1471.
21) Nakane, H., et al. (1987). DC-SQUID with High-Crit-
    ical -Temperature Oxide-Superconductor Film, Jpn.
    J. Appl. Phys. 26(11), pp. L1925-L1926.
22) Humphreys, R. G., et al. (1990). Physical Vapor De-
    position Techniques for The Growth of YBaCuO Thin
    Films, Supercond. Sci. Technol. 3, pp. 38-52.
23) Koura, N. (1988). Preparation of Various Shape
    Superconducting Ceramics Using The Electrophoretic
    Deposition Method, Denki Kagaku 56(3), pp. 208-209.
24) Niu, H., et al. (1991). Fabrication of Y-Ba-Cu-O
    Superconducting Film by Electrophoresis with The
    Use of Firing in Helium Ambience, Supercond. Sci.
    Technol., Submitted.
25) Idemoto, Y., et al. (1990). Melting Point of Super-
    conducting Oxides as A Function of Oxygen Partial
    Pressure, Jpn. J. Appl. Phys. 29(12),pp.2729-2731.

Synthesis of High T_c Superconductors

B. Srinivas and G.V. Subba Rao

Materials Science Research Centre, Indian Institute of Technology, Madras-600 036, India

A brief account of the known high T_c ceramic oxide superconductors followed by the optimized procedures for their synthesis in polycrystalline powder/pellet form is presented.

1. Introduction

Superconductivity is the phenomenon whereby a solid substance exhibits zero electrical resistivity at and below a critical temperature, T_c. Above T_c, the substance behaves like a normal metal. In addition to zero resistance, a superconductor exhibits many other interesting physical properties below T_c. The discovery of superconductivity in 1987 with a T_c above the liquid nitrogen temperature ($\geq 77K$) in the mixed copper oxide compound, $YBa_2Cu_3O_7$ has evoked worldwide interest in the study and exploitation of these high temperature superconductors (HTSC). There are now available four types of HTSC: (i) The T_c = 90K rare earth (Ln) containing oxide materials isostructural to $YBa_2Cu_3O_7$ (popularly called 123 or YBCO), (ii) The 80K and 110K superconductors based on the Bi-Sr-Ca-Cu-O system, (iii) The 105K and 125K materials in the Tl-Ba-Ca-Cu-O system and (iv) The 77K superconductors based on the Pb-Sr-(Ca,Ln)-Cu-O system. In addition, phases in the Ln-Ba-Cu-O system with the formula $YBa_2Cu_4O_{8+y}$ (124) and $Y_2Ba_4Cu_7O_{14+y}$ (247) have also been isolated exhibiting T_c varying from 40-80K and these phases are structurally related to the 123 system. Crystal structure and T_c data of the various HTSC oxide materials are given in Table I. In addition, there are related compounds which exhibit T_c less than 77K. They are: (i) 40K compounds based on Sr-doped La_2CuO_4; (ii) 25K compounds based on Ce or Th doped Nd_2CuO_4; (iii) 13-30K compounds based on the bismuth oxides, $BaPb_{1-x}Bi_xO_3$ ($x = 0.25$) and $K_xBa_{1-x}BiO_3$ ($x = 0.4$) and (iv) $LiTi_2O_4$ with a T_c of 13K. Excellent review articles, book series and Conference proceedings are now available in the area of HTSC [1-10].

2. Structural Aspects and T_c Behavior

All the above HTSC systems are perovskite-based copper containing layer type oxides. The principal feature of all the phases is the presence of corner shared square planar CuO_2 units forming infinite 'sheets' perpendicular to c-axis. These 2D layers are responsible for the observed HTSC. The crystalline symmetry is either orthorhombic or tetragonal in all these HTSCs. In the 123 compound with the orthorhombic structure, in addition to 2D-layers, there exist O-Cu-O linear chains along the b-axis in the basal plane. The oxygen in these chains is very labile and the normal state and superconducting properties depend crucially on 'δ' in $YBa_2Cu_3O_{7-\delta}$. HTSC (90K) is encountered only for $\delta = 0.05-0.2$; for $\delta = 0.2-0.4$, the T_c drops to about 60K whereas for $\delta > 0.6$, no superconductivity has been observed. Hence, optimization of oxygen content in the final stages of synthesis of 123 compound to obtain low values of δ is an essential step in order to obtain HTSC property.

The bismuth (Bi) and thallium (Tl) containing mixed copper oxide superconductors have a more complex crystal structure compared to 123. They form a series of isostructural tetragonal or orthorhombic layer-type phases (polytypes) with large c parameters (24-36 Å depending on the number of layers) arising out of stacking of metal oxygen layers (Table I). Each unit

Table I . Structure and T_c data of the known high T_c oxides

Compound	T_c, K	Crystal structure	Lattice param., Å			Layer sequence
			a	b	c	
$YBa_2Cu_3O_7$	90	Orthorhom.	3.823	3.886	11.681	-CuO-BaO-CuO_2- Y-VuO_2-BaO-
$Bi_2Sr_2CaCu_2O_8$	80	Orthorhom.; Tetra.	5.408 3.812	5.413 --	30.871 30.66	-BiO-SrO-CuO_2-Ca- CuO_2-SrO-BiO
$Bi_2Sr_2Ca_2Cu_3O_{10}$	110	Tetra.	5.39	--	37.10	-Bio-SrO-CuO_2-Ca- CuO_2-Ca-CuO_2- SrO-BiO-
$Tl_2Ba_2CaCu_2O_8$	105	Tetra.	3.85	--	29.30	-TlO-BaO-CuO_2-Ca- CuO_2-BaO-TlO-
$Tl_2Ba_2Ca_2Cu_3O_{10}$	125	Tetra.	3.85		35.58	-TlO-BaO-CuO_2-Ca- CuO_2-Ca-CuO_2- BaO-TlO-
$Pb_2Sr_2(Ca,Ln)Cu_3O_8$	70	Tetra.; Orthorhom.	3.813 5.402	5.433	15.76 15.74	-PbO-Cu-PbO-SrO- CuO_2-Y/Ca-CuO_2- SrO-

cell consists of two such layer sequences with a general formula of the type $A_2B_2Ca_{n-1}Cu_nO_{2n+4}$ (A = Bi,Tl; B = Sr,Ba; n = 1-4). The compounds with n = 3, viz., $Bi_2Sr_2Ca_2Cu_3O_{10}$ and $Tl_2Ba_2Ca_2Cu_3O_{10}$ (2223) exhibit the highest T_c values (Table I). Interestingly enough, single layer thallium cuprates of the general formula, $Tl_1Ba_2Ca_{n-1}Cu_nO_{2n+3}$ also do form and exhibit the HTSC behavior. Again, highest T_c is noted when n = 3, viz., 1223 and these Tl_1-series have one formula unit per unit cell with the tetragonal axis ranging from 9-15 Å depending on the value of n.

The layer-like nature of the Bi- and Tl-cuprates gives rise to intergrowth phenomenon. That is, coexistence of two or more compositionally controlled polytypes; e.g., Bi-2212 phase coexisting with 2201 (n = 1) and 2223 (n = 3) phases. Positional disorder (Sr and/or Bi partly replacing Ca etc.) also can occur. An additional feature in the Bi-system is the large van der Waals type separation (~3 Å) between consecutive Bi-0 layers and this can lead to superstructures with modulation and possible intercalation of 0-atoms in-between the layers. The net result is that during synthesis of poly-crystalline materials by high temperature solid state reaction technique, a mixture of phases (with differing n; sometimes with n ≥ 4 but are either unstable or only meta stable) is always obtained.

Thus, while $Bi_2Sr_2Ca_1Cu_2O_8$ may be somewhat easy to synthesize as a pure phase, the compounds 2201 and 2223 are difficult to synthesize as single phase materials. It is now known that partial substitution of Bi-by lead (Pb) can stabilize and facilitate the formation of Bi-2223 (T_c = 110K) phase. Similarly, it is possible to stabilize the Sr-containing phases (instead of Ba)in the Tl-cuprates by substituting partly at the Tl and/or Ca sites by Pb or rare earth (Ln or Y) and these also exhibit the HTSC behaviour. Thus chemical substitution at a crystallographic site other than the Cu-site has been found to be extremely useful in the synthesis of the HTSC's based on Bi- and Tl- cuprates. It must be pointed out that the above idealized formulae (2212, 2223 etc.) in the Bi- and Tl- cuprates have been realized in small single crystal samples. As synthesised polycrystalline materials always tend to be multiphasic (not single phase) and exhibit less-than-the ideal superconducting properties.

The HTSC cuprates have been found to exhibit slightly anomalous normal state (T>T_c) but outstanding superconducting properties. The 90K superconductor compound, $YBa_2Cu_3O_7$ has been the one extensively studied in polycrystalline, single crystal, thin and thick film, wire and tape forms. The Bi- cuprates are receiving increased attention whereas the Tl- cuprates, though exhibit the highest known T_c till todate, are more difficult to study due to the toxicity problems and inherent difficulties in the realization of single phase materials.

The HTSC cuprates are oxide ceramics, black in color and are brittle and fragile. Suitable technologies need to be developed to fabricate them in the form of flexible wires and tapes for large scale practical applications. They possess relatively high (for a metal) room temperataure resistivity (ρ_{300K} = 1-5 mΩcm), and large anisotropy in the normal state and superconducting physical parameters, a linear ρ-T behaviour in the range T_c-600K, hole-type carriers (Seebeck Coefficient and Hall effect data), low carrier density (~10^{21}/cc) and slightly anomalous optical properties. Very high critical magnetic fields (Type II superconductor; H_{c2} ~150-200 Tesla), short superconducting coherence lengths (15-20 Å), fairly large critical currents (J_c ~10^6 amp/cm^2 for thin films), good values of penetration depths (700-1400 Å), large superconducting energy gaps (20-35 meV) are the superconducting state properties exhibited by HTSC's.

Presently, we describe the methods of synthesis of the HTSC cuprates in the form of polycrystalline powder or pellets. These are pre-requisites for the good chemical and physical studies. Due to the wide varieties of chemical substitutions possible and drastic variation in the superconducting properties with the change in the oxygen content or stoichiometry or chemical substitution, optimized procedures for the synthesis are required. These have now become available through phase diagram studies and trial and error procedures. Procedures for the crystal growth, thin and thick film deposition and fabrication into wires/tapes will not be dealt with here. The interested reader may refer to work reported in the literature [1-5,7].

3. Synthesis of $YBa_2Cu_3O_{7-\delta}$ (T_c = 90K)

3.1 Solid state reaction method

Phase diagram studies have shown [11] that $YBa_2Cu_3O_{7-\delta}$ is a line phase (w.r.t. metals) and melts above 1030°C (Fig. 1). However, the oxygen content can vary from δ = 0.05-0.60; using special methods (gettering at 400°C with Ti and Zr metal; heating in vacuum or argon gas) δ can be increased to 1.0. Optimum superconducting properties are exhibited only when δ = 0.05-0.20.

The 123 phase can be synthesized most conveniently by the conventional ceramic route viz., direct solid state reaction at high temperature of the

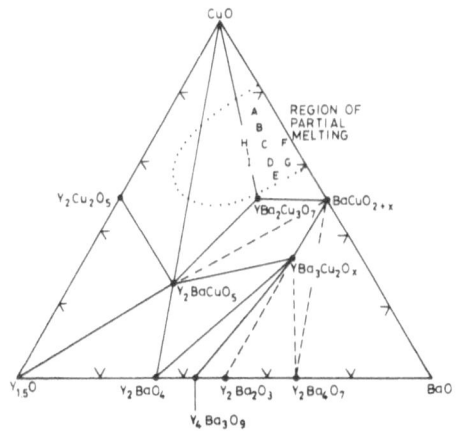

Fig.1. Y_2O_3-BaO-CuO ternary phase diagram (~950-1000°C). Dotted line represents region of partial melting (after ref. 11).

Fig. 2. SYNTHESIS OF POLYCRYSTALLINE YBa$_2$Cu$_3$O$_{7-\delta}$

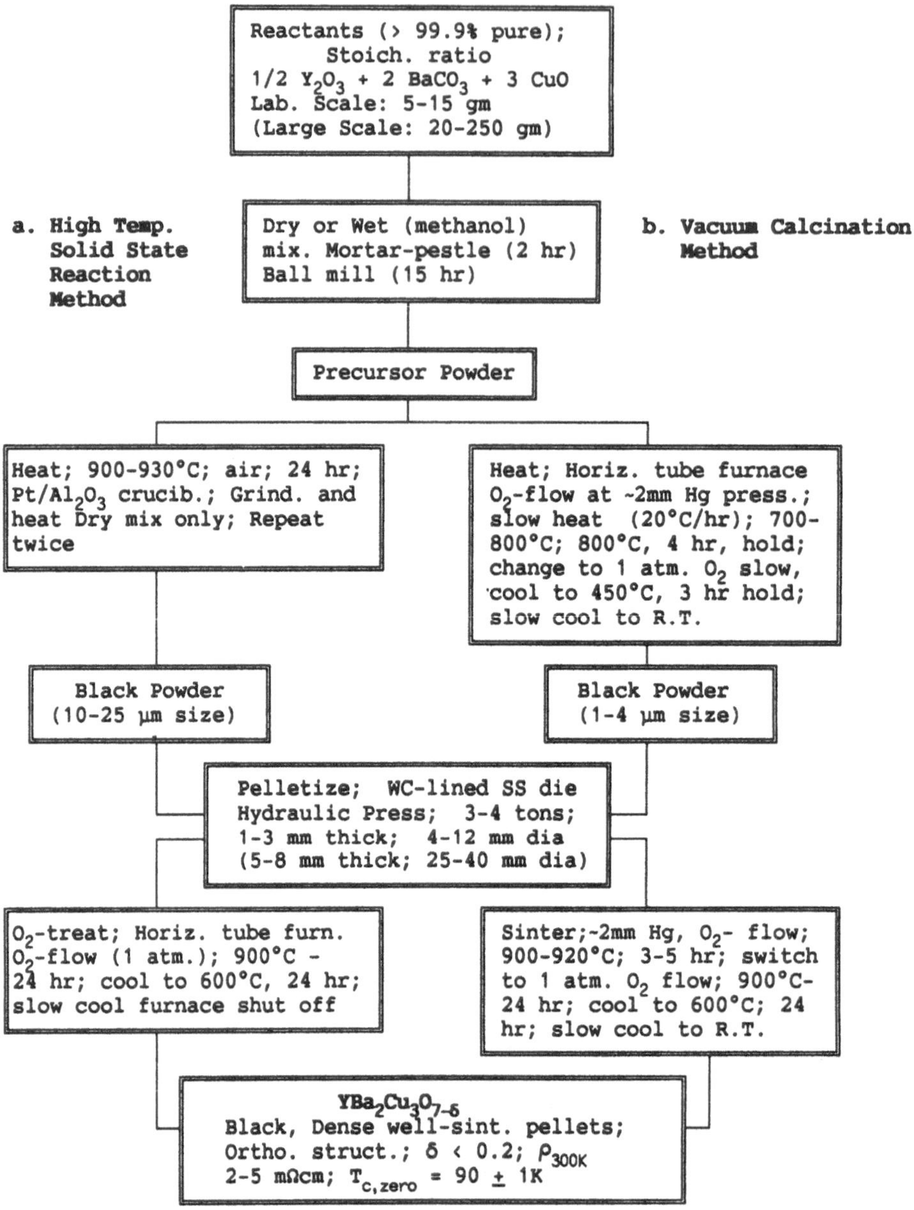

```
Reactants (> 99.9% pure);
        Stoich. ratio
1/2 Y₂O₃ + 2 BaCO₃ + 3 CuO
Lab. Scale: 5-15 gm
(Large Scale: 20-250 gm)
```

a. High Temp.   Dry or Wet (methanol)   **b. Vacuum Calcination**
   Solid State   mix. Mortar-pestle (2 hr)   **Method**
   Reaction   Ball mill (15 hr)
   Method

Precursor Powder

| Heat; 900-930°C; air; 24 hr; Pt/Al₂O₃ crucib.; Grind. and heat Dry mix only; Repeat twice | Heat; Horiz. tube furnace O₂-flow at ~2mm Hg press.; slow heat (20°C/hr); 700-800°C; 800°C, 4 hr, hold; change to 1 atm. O₂ slow, cool to 450°C, 3 hr hold; slow cool to R.T. |

| Black Powder (10-25 µm size) | Black Powder (1-4 µm size) |

```
Pelletize;  WC-lined SS die
Hydraulic Press;  3-4 tons;
1-3 mm thick;  4-12 mm dia
(5-8 mm thick; 25-40 mm dia)
```

| O₂-treat; Horiz. tube furn. O₂-flow (1 atm.); 900°C - 24 hr; cool to 600°C, 24 hr; slow cool furnace shut off | Sinter;~2mm Hg, O₂- flow; 900-920°C; 3-5 hr; switch to 1 atm. O₂ flow; 900°C-24 hr; cool to 600°C; 24 hr; slow cool to R.T. |

```
              YBa₂Cu₃O₇₋δ
Black, Dense well-sint. pellets;
Ortho. struct.; δ < 0.2; ρ₃₀₀ₖ
2-5 mΩcm; T_c,zero = 90 ± 1K
```

constituent oxides or carbonates in stoichiometric proportions of purity better than 99.9%. Being a heterogeneous solid state reaction between three different oxides (Y$_2$O$_3$, BaCO$_3$ and CuO), repeated grindings and heatings are necessary. The method can also be used for scaling up, to produce large quantities of powders (0.5-1.0 kg). The detailed procedure is given in Fig.2a. The following points should be noted: (i) Heating temp. T < 800°C will lead to the formation of impurity phases such as BaCuO$_2$, Y$_2$BaCuO$_5$ (211). Heating at T > 970°C results in the decomposition of 123 phase already formed, to a mixture of phases including (211); hence heating at T = 900-930°C is advised. (ii) Incomplete decomposition of $BaCO_3$ or unreacted BaO (formed by the decomposition of $BaCO_3$) can lead to complications and deterioration in the properties of the 123. Hence repeated grindings and heatings are necessary to ensure complete decomposition and reaction with the Y_2O_3 and CuO. (iii) Heating in O_2 atmosphere at 900°C will enable sintering of the material and cooling to 600°C and further annealing for 24hr in oxygen flow will enable oxygen uptake and formation of the orthorhombic phase with $\delta \leq 0.2$. Even though studies have shown that oxygen uptake can occur in the temp. range 400-500°C, it is our experience that 600°C is optimum for the oxygen enrichment of the 123 compound.

The above method has some disadvantages : (i) The CO_2 gas released by the decomposition of $BaCO_3$ during the solid state reaction, can recombine with BaO to re-form $BaCO_3$ if sufficient ventilation is not provided or when dealing with large quantities of the reactant material. (ii) The evolved CO_2 gas can react with the already formed 123 to produce impurity phases like 211 or $BaCuO_2$. To overcome the above difficulties the novel method of vacuum calcination was evolved recently.

3.2 Vacuum Calcination Method

In this method the constituent oxides and $BaCO_3$ are heated at T~800°C in controlled oxygen atmosphere at reduced pressure (2-20 mm Hg) in a horizontal tubular furnace. In Fig.2b is shown the procedure suggested by Balachandran [12] and others [13]. The reduced pressure will enable the decomposition of $BaCO_3$ at a lower temperature (800°C compared to 900-950°C in air) and also allows the evolved CO_2 gas to be flushed out effectively, thereby avoiding the formation of impurity phases. At this stage, switching over to oxygen flow (1 atm) is necessary to prevent the decomposition of the formed 123 phase and also Cu_2O formation. Cooling to 450°C and a hold for 3hr. at that temperature and slow cool to room temperature will enable the oxygen enrichment. The 123 compound prepared by vacuum calcination method yields very small particles (size 1-4 μm) and also exhibits $T_{c,zero}$ of 90K (Flux expulsion method). However, due to the large surface area, the compound can lose oxygen easily thereby deteriorating the HTSC properties. Hence, subsequent pelletization and sintering followed by O_2 treatment are desirable to realize the optimum superconducting properties (Fig.3).

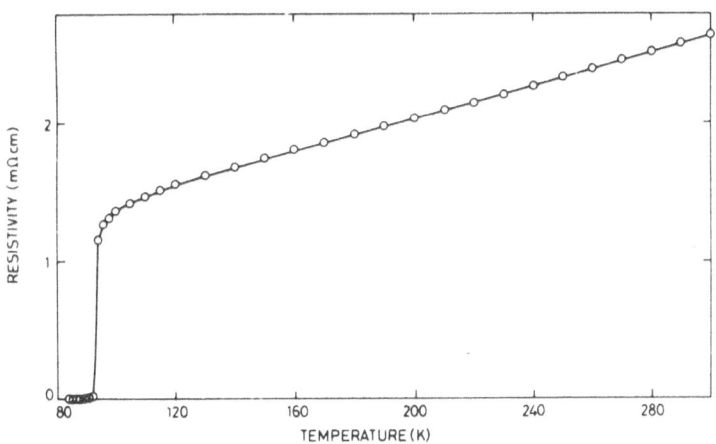

Fig.3. Resistivity (ρ) vs Temperature plot of $YBa_2Cu_3O_{7-\delta}$ pellet synthesized by method (b) of Fig.2. As can be seen, low value of ρ_{300K}, linear ρ-T behavior and a $T_{c,zero}$ of 90 $\pm$ 1K are characteristic of good 123 phase (B.Srinivas et al, to be published).

4. Bismuth Cuprate HTSC Compounds

As mentioned earlier, these phases have an idealized general formula, $Bi_2Sr_2Ca_{n-1}Cu_nO_{2n+4}$ with n = 1-4. Phases with n ≥ 2 exhibit the HTSC (T_c > 77K) behavior. Since their discovery in 1988, a variety of synthetic methods have been employed and a large number of partial phase diagrams established. Though the oxygen stoichiometry does not appear to play a significant role in the optimization of T_c, realization of the pure single phase materials in bulk form has been one of the difficult tasks in the case of Bi-cuprates. Due to the possible substitutional disorder and perhaps, vacancies at sites other than copper, intergrowth phenomenon and complexity of the phase diagrams, nominal stoichiometric starting compositions rarely yield the desired single phase materials. On the other hand, any nominal starting compositions, viz., Bi-Sr-Ca-Cu = 1112, 1112.5, 2212, 4334 or 4336 can yield significant fractions of the 2212 and 2223 phases. It should be noted that pure single phase 2223, 2234 and 2245 compounds have not yet been realized in the Bi-Sr-Ca-Cu-O system [14].

4.1 Synthesis of the phase, $Bi_2Sr_2Ca_1Cu_2O_8$ (2212; T_c = 80K)

Direct solid state reaction of Bi_2O_3, $SrCO_3$, $CaCO_3$ and CuO in stoichiometric proportions in the range 840-870°C is known to yield the 2212 phase though not in phase-pure form. Nitrate, oxalate and citrate precursors have been employed with some success. Another method is the matrix-reaction whereby a nominal $Sr_2Ca_1Cu_2O_y$ oxide is presynthesized, then mixed with Bi_2O_3 followed by heat treatment. Glassy route (melt quench of the nominal 2212 composition, followed by crystallization at a lower temperature) and partial rare earth or Y substitution for Ca (e.g., $Ca_{0.8}Y_{0.2}$) have also been tried with some success. In many cases, a $T_{c,zero}$ ranging from 54-65K is seen in the above 2212 phases; Vacuum or inert gas annealing (500°C; 1-3 hr) has been suggested to improve the T_c to 80K. However, the latter method is not always reproducible.

Recently, Ono [15] studied the phase relationships in the 2212 system and found that the 80K phase can be stabilized by starting with slight Bi excess and Sr and Ca deficient composition, viz., $Bi_{2+x}(Sr_{2-y}Ca_{1+y})_{1-x}Cu_2O_y$, 0.0 ≤ x ≤ 0.2 and 0.0 ≤ y ≤ 0.5 (Fig.4). Thus it appears that Bi at the Sr-site stabilizes the 80K-2212 phase. The recommended procedure is as follows: Starting composition: $(Bi_2O_3)_{1.05}$ + $(SrCO_3)_{1.93}$ + $(CaCO_3)_{0.97}$ + 2.0 CuO; thorough mixing and heating as powder in Al_2O_3 crucible; heating in air to 800-820°C; Pelletize, slow heat in air to 840-870°C for 24 hr and slow cool by furnace shut-off; repeat the procedure twice. Partial melting of the sample may occur above 860°C and strict control is necessary to avoid complete melting.

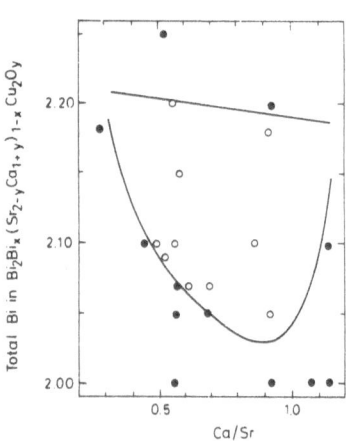

Fig.4. Single phase region (open circles) for the 80K HTSC phase in Bi-Sr-Ca-Cu-O system (see text) (after ref. 15).

4.2 Synthesis of the phase, $(Bi-Pb)_2Sr_2Ca_2Cu_3O_{10}$ (2223; T_c=110K)

Earlier studies indicated that heating of the nominal compositions (after pre-reaction of the powder) to temperatures very near the melting (875-885°C) or long time (>100 hr), low temperature (850°C) annealing can induce and enhance the content of the 2223 phase. It was also soon recognized that partial lead (Pb) substitution for bismuth can stabilize, more easily, the 110K phase. As expected, starting with off-stoichiometric compositions, the rates of heating and cooling and temperatures and times of annealing are the crucial factors for the realization of almost phase pure 2223. Sasakura et al [16] reported the partial phase diagram, which is reproduced in Fig.5, which shows that single phase samples can form over a fairly wide range of compositions in the formula, $Bi_{1.68}Pb_vSr_{1.73}Ca_yCu_zO_w$ with v = 0.28 or 0.32; 1.75 $\leq$ y $\leq$ 1.85; and 2.65 $\leq$ z $\leq$ 2.85. The recommended procedure is as follows: Prereact the powders in air at 795°C for 15 hr, cool, grind, pelletize, heat in air at 858°C for 90 hr; grind, repelletize followed by heating in air again at 858°C for 65 hr, followed by cooling at the rate of 1.0-2.5°C/min. to room temperature. Compounds prepared in this manner are found to show $T_{c,zero}$ above 105K and substantial Meissner fraction.

Synthesis by the mixed nitrate route at a reduced oxygen partial pressure (1/13 atm. O_2) was tried by Koyama et al [17] and a slightly different phase diagram was established (Fig. 6). Again, the starting compositions are off-stoichiometric and time and temperatures of heat treatments are important. The suggested procedure is as follows: Starting comp., $Bi_{1.84}Pb_{0.34}Sr_x$ $Ca_yCu_zO_w$, (1.87 $\leq$ x $\leq$ 2.05; 1.95 $\leq$ y $\leq$ 2.1 and 3.05 $\leq$ z $\leq$ 3.2): Dissolve in nitric acid and evaporate to dryness followed by decomposition at 800°C for 30 min.; pelletize and heat treat at 828-843°C for 36-130 hr and slow cool (all reactions in oxygen flow at reduced pressure (1/13 atm.)). $T_{c,zero}$ greater than 105K is exhibited by the phases prepared in the above manner.

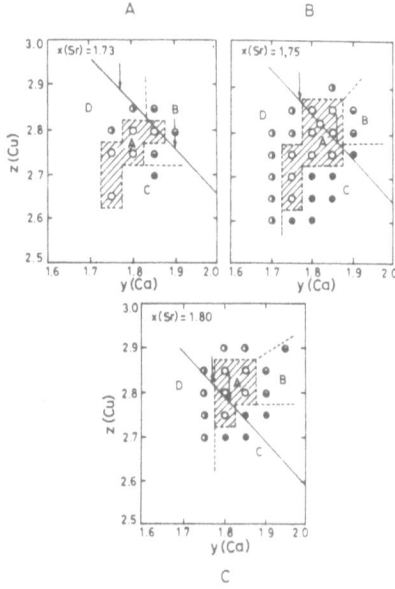

Fig.5. Composition diagrams for samples of nominal composition $Bi_{1.68}Pb_vSr_xCa_yCu_zO_w$ In (A), v = 2.8 and x = 1.73; for (B), v = 0.32 and x = 1.75 and for (C), v = 0.32 and x = 1.80. The dashed area (region A) indicates single phase - 110K region (after ref. 16).

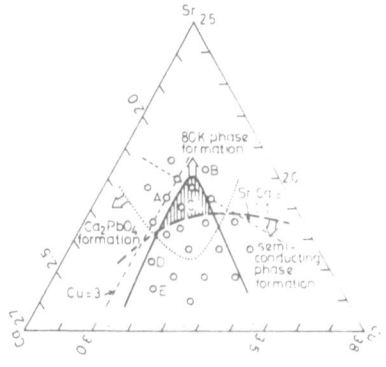

Fig.6. Partial phase diagram for phases of composition $Bi_{1.84}Pb_{0.34}Sr_xCa_yCu_zO_w$. The dashed area represents the single-phase region for the 2223, T_c = 110K phase. The samples were prepared in 1/13 atm. O_2 (after ref. 17).

5. Thallium Cuprate HTSC Compounds

In the system, $Tl_mBa_2Ca_{n-1}Cu_nO_{2n+2+m}$, m = 1 or 2; n = 1-4, the phase with n = 1 to 3 and m = 2 and those with n = 2 to 3 and m = 1 exhibit $T_c >$ 77K. As mentioned earlier the bilayer (m = 2) thallium cuprates are akin to the corresponding Bi-analogues; as can be expected, difficulties in the synthesis are also similar. Additional complications are: (i) Volatility of Tl_2O_3 and its decomposition (to Tl_2O and O_2) at T > 600°C (complete at 875°C in air). Hence the solid state reactions need to be carried out either in closed systems or for a short time at high temperatures in open systems. (ii) Extreme toxicity of thallium compounds. It is always advantageous to use the oxides or peroxides of Ba and Ca rather than the carbonates, since the latter decompose only at high temperatures and produce CO_2 which creates a reducing atmosphere, both of which are detrimental to the formation of the thallium cuprates. Synthesis under flowing oxygen gas at 1 atm. pressure is also helpful in avoiding the reduction of Tl^{3+} ion.

Detailed phase equilibrium data are not yet available but a variety of preparative methods were employed. However, by trial and error, somewhat optimum conditions for the synthesis of the bulk materials have been evolved. Either the nominal compositions in the form of oxides/peroxides (Tl_2O_3; CaO/CaO_2; BaO/BaO_2, CuO) or the matrix method (pre-reacted $Ba_2Ca_{n-1}Cu_nO_y$ + Tl_2O_3 mixtures) can be employed. Some of the recommended recipes are given below [14].

5.1 Preparation under flowing oxygen gas (1 atm. pressure)

The constituent oxides/peroxides or pre-reacted $BaCuO_2$ or ($Ba_2Ca_{n-1}Cu_nO_y$) and Tl_2O_3 are mixed thoroughly and pelletized and the pellets are wrapped in gold/Pt-foil and heated in a horizontal tubular furnace under flowing oxygen gas. Cooling is done at 10°C/min maintaining the oxygen flow. Temperatures and times are: (i) $Tl_2Ba_2CuO_6$ (2201): 890°C; 5 min.; cool at 10°C/min. (T_c ranges from 0-90K depending on the subsequent Ar- or vac- treatment). (ii) $Tl_2Ba_2Ca_1Cu_2O_8$ (2212): 905°C; 7 min.; cool at 10°C/ min. (T_c = 98K). (iii) $Tl_2Ba_2Ca_2Cu_3O_{10}$ (2223): 910°C; 7 min.; cool at 10°C/min. (T_c = 114K).

The above procedure has also been employed with some success for the synthesis of phase-pure m = 1 series (1212, 1223 etc.) in the Ba- and Sr-containing (with Pb partly substituted at the Tl-site) HTSC compounds.

5.2 Preparation in hermetically sealed containers

Either hermetically sealed gold tubes or evacuated and sealed quartz tubes with pellets wrapped in gold or Pt-foils have been employed in this method. The pellets are made up of either the reactant oxides/peroxides or pre-reacted matrix with the added Tl_2O_3. The temperature and heating times are given below:

(i) 2201: 875°C, 3 hr; slow cool (T_c = 83K; optimized after subsequent heating at 900°C for short times)
(ii) 2212: 900°C; 6 hr; slow cool (2°C/min.) (T_c = 110K)
(iii) 2223: 890°C; 1 hr; slow (T_c = 122K) (Single crystals obtained from nominal 2234: 920°C, 3 hr and cooled to 300°C at 5°C/min.; $T_{c,onset}$ = 125K)
(iv) 1212: 780-820°C; 5-6 hr (T_c = 65K after subsequent Ar-anneal at 780°C)
(v) 1223: 750-830°C, 5 hr (T_c ~110K)

For the Pb-substituted Sr-containing (instead of Ba) phases in the system, $(Tl_{0.5}Pb_{0.5})Sr_2Ca_{n-1}Cu_nO_y$, the conditions for both 1212 (T_c = 80-90K) and T_c = 120K) are: 850-915°C, 3-12 hr with the starting materials as Tl_2O_3, PbO_2, SrO_2, CaO_2 and CuO.

6. Conclusions

Preparative techniques for the 123, Bi- and Tl- cuprate HTSC have been discussed. The technique for the synthesis of polycrystalline rare earth and Y-based 123 compounds is now well established and standardized to give the optimum 90K superconducting behavior. Due to the complexity of phase relationships, intergrowth and substitutional disorder, synthesis of phase-pure Bi- and Tl- cuprate HTSC in polycrystalline bulk form is slightly difficult. However, empirically desived recipes are now available which will yield the optimum superconductivity behavior eventhough hundred percent phase purity is not guaranteed.

Acknowledgement

Thanks are due to Dr. U.V. Varadaraju for helpful discussions. Thanks are due to PMB-DST superconductivity project, Govt. of India and EEC project for financial support.

References

1. C.N.R. Rao (Ed.), "Chem. of Oxide Supercond.", Blackwell Sci. Publ., Oxford, UK (1988); C.N.R. Rao (Ed.), "Chem. and Struct. Aspects of HTSC", World Sci., Singapore (1988) (Progr. in HTSC Series, Vol. 7); A.K. Gupta, S.K. Joshi and C.N.R. Rao (Eds.), Rev. Solid State Sci., 2(2&3) (1988) (Proc. of Intl. Workshop on HTSC, Srinagar, India) (World Sci., Singapore).

2. Jpn. J. Appl. Phys. 26 (Supl. 3), 1987 (Proc. of 18th Intl. Conf. on Low Temp. Phys., Kyoto, Japan); Physica B, 148 (1987) (Proc. of 18th Intl. Conf. on Heavy Fermions and Highly Correl. Systems, Sendai, Japan).

3. J. Muller and J.L. Olsen (Eds.), Physica C153-155, 1-1774 (1988) (Proc. of the M^2 HTSC Conf., Interlaken, Switzerland).

4. K. Kitazawa and T. Ishiguro (Eds.), Adv. in Supercond., Springer-Verlag, Tokyo (1988) (Proc. of Ist Intl. Symp. on Supercond., Nayoga, Japan).

5. R. Nicolsky (Ed.), "Transport Properties of Supercond." World Sci., Singapore (Prog. in HTSC Series, Vol. 25); T. Akachi, J.A. Cogordan and A.A. Valladares, "High Temp. Supercond.", World Sci. Singapore (Prog. in HTSC Series, Vol. 20); H.W. Weber (Ed.), "High T_c Supercond.", Plenum Press, Newyork (1988).

6. C.N.R. Rao and B. Raveau, Acc. of Chem. Res., 22, 106 (1989); B. Raveau et al, Rev. Solid State Sci., 2, 263 (1988); C.N.R. Rao and T.V. Ramakrishnan, J. Phys. Chem., 93, 4414 (1989).

7. A. V. Narlikar (Ed.), "Studies of High Temp. Supercond.", Vols. 1-6, Nova Sci. Publ., New York (1989,1990,1991).

8. K. Bedell, D. Coffey, D. Meltzer, D. Pines and J.R. Schrieffer (Eds.), "Selected Experiments on High T_c Cuprates" (High Temp. Supercond.: The Los Alamos Symp., 1989), Addison-Wesley Publ. Co. (1990).

9. B. Batlogg, in "Proc. LT-19 Part III; Physica B (1990).

10. G.V. Subba Rao, in "High Temp. Supercond.", Eds., S.V. Subramanyam and E.S.R. Gopal, Wiley Eastern, New Delhi, 1989.

11. R.A. Laudise, L.F. Schneemeyer and R.L. Barns, J. Cryst. Growth, 85, 569 (1987); H.J. Scheel and Ph. Niedermann, J. Cryst. Growth, 94, 281 (1989); H.J. Scheel, J. Less-Comm. Metals, 15, 199 (1989).

12. U. Balachandran et al in "Proc. of Intl. Conf. on Supercond.", Bangalore, India, Jan. (1990); U. Balachandran et al, Mater. Lett., 8, 454 (1989).

13. R.R. Schartman and E.E. Hellstrom, Physica C, 173, 245 (1991); G.S. Grader, P.K. Gallagher and D.A. Fleming, Chem. Mater., 1, 665 (1989).

14. S.A. Sunshine and T.A. Vanderah, in "Chem. of Supercond. Mater.", Noyes Public., USA, 1991.

15. A. Ono, Jpn. J. Appl. Phys., 28, L 1372 (1989).

16. H. Sasakura et al, Jpn. J. Appl. Phys., 28, L 1163 (1989).

17. S. Koyama, U. Endo and T. Kawai, Jpn. J. Appl. Phys., 27, L 1861 (1988).

Some Aspects of Single Crystal Growth of High Temperature Superconductors

Katta Narasimha Reddy
ხ. Department of Physics, University College of Science
Osmania University, Hyderabad-500 007, India

1.INTRODUCTION

The field of high temperature superconductivity has matured considerably during the past four years. Despite of this progress the mechanisms of super conductivity have continued to remain elusive and general characteristics of recently discovered families of high temperature super conductors have shown that we are still far from having an even a satisfactory model of this most exiting phenomena of Solid state materials. Inspite of the fast growth of research in the field of high temperature superconductors(HTSC) the ultimate limits for transition temperature (T_c) in copper oxide based materials are still uncertain. The continued growth of the field now requires careful and well directed studies of various materials and physical issues vital to the understanding and applications of superconductor materials. One of such important areas of study is the preparation of oxide superconductors in single crystal form. There is an urgent need to quantify the physical and mechanical properties of these materials in order to make sensible judgments on the practical usage of these materials and define their limitations.

The presence of spurious phases in small amounts in the powder samples of HTSC materials as a rule constitutes a serious hindrance for many contemplated applications of these materials. Further, the results of measurements on such polycrystalline samples are not likely to be reliable and reproducible. One can overcome these drawbacks by employing single crystal samples. Obviously, the single crystals are of single phase and hence preferred in any high T_c superconductor materials applications. Besides the other important advantage with the single crystal is the investigation of anisotropic properties which are otherwise averaged out in the polycrystalline sample whose grains are more or less randomly oriented. The macroscopic nature and the mechanism responsible for the superconducting behaviour are strongly linked to this anisotropy. This emphasizes the immediate need of the single crystals of new superconductors of size suitable to allow reliable physical investigation.

335

Owing to the reactivity of the compounds of $YBa_2Cu_3O_{7-\delta}$ (YBCO) with water, hydrothermal growth is not possible. YBCO melts incongruent by precluding Bridgman-stockbarger, Czochralski or other melt techniques. The combined effects of thermal and chemical instabilities of YBCO lead to severe consequences for growing crystals because they exclude conventional crystal growth techniques, like direct growth from melts or by vapour halide transport, growth from high temperature solution using typical solvents for oxides, and hydrothermal growth. Only two ways have been found so far to achieve, with marginal success, crystal growth of YBCO by solid state reaction and crystal growth from high temperature solution of non-stoichiometric melts.

2. CRYSTAL GROWTH OF YBCO BY SOLID STATE REACTION (SSR)

This process involves long term annealing of ceramic pellets having stoichiometric or nearly stoichiometric composition. It usually led, possibly by liquid-assisted grain growth, to crystallites from few tens to hundred of microns. Intrinsic limitations of the method like high density of nucleation sites, and occurrence of crystal development with in the solidified mass, do not allow production of large and perfect crystals. But as a consequence of nearly isothermal, low rate growth process, with locally varying supersaturation, isometric and well developed cubic crystals were occasionally obtained.

3. CRYSTAL GROWTH FROM HIGH TEMPERATURE SOLUTION (HTS)

Crystal growth from high temperature solution of crystalline material from slow cooled supersaturated solutions and should in principle allow control of nucleation and achievement of large and perfect crystals. The main problems encountered in crystal growth of YBCO from HTS are to find proper crucible materials, proper high temperature solvents and to achieve flux separation. These problems could partially be solved through systematic investigations and the details are given below.

a) Crucible materials

In the growth of high temperature superconducting crystals from HTS method, selection of crucible plays a very important role. The correct choice of crucible material is of prime importance to the growth of satisfactory crystals. In the case of HTSC (High Temperature Superconductors) single crystals, the choice is very limited and the crucible material may even dictate the method of growth. The obviously desired property of a crucible material is that it should not react with the melt. YBCO system has generally been found to be quite reactive and during the extended growth periods at elevated temperatures, considerable crucible attack can occur. For some metal crucibles this is further complicated by the tendency to form metal oxides in high oxygen partial pressures which must be maintained during growth cycle.

The container materials investigated and used in the past three years includes both metallic and oxide systems. Platinum and Gold have been successfully used as container material to produce high performance Yttrium barium cuprate crystals (1-3) with well defined superconducting transitions close to 90K. Beside these two metallic crucibles, iridium crucible is often used in growing HTSC materials. Platinum and Iridium were found to react with YBCO melt. Particularly platinum is severely attacked by flux and can extensively perforated during the growth period. Platinum absorption into the flux and subsequently into growing crystalline system was observed to lead to the production of platinum rich phases (4,5) including both $Y_2Ba_2CuPtO_8$ and $Y_2Ba_3Cu_2PtO_{10}$. While gold is relatively corrosion resistant, and was used in the studies of Kaiser et al (6) which produced crystals with $T_c = 92.3$ ($\Delta T_c = 0.2K$) after oxygenation.

A wide range of oxide materials have been used for the growth of YBCO from $CuO-BaCuO_2$ flux system; including alumina, Zirconia (ZrO_2,YSZ,MgSZ etc), MgO, $MgAl_2O_4$, SiO_2(quartz), Mullite and more recently SnO_2 and Y_2O.The use of alumina with the cuprate fluxes known to lead to alumina contamination and poisoning of the YBCO crystals for extended growth periods (7,8) due to flux crucible interaction and the formation of barium alumina cuprates $BaAl_{2-x}Cu_xO_4$. Sadowski et al (9) were successful in growing large perfect single crystals of YBCO in alumina crucibles. They claimed that their grown crystals have good superconducting properties. They observed that the superconducting properties were unaffected by alumina impurity and T_c is found between 88 to 91K, with $\Delta T_c <$ 1K. Sadowski and Scheel (10)prepared single crystals of YBCO using both Al_2O_3 and ZrO_2 crucibles and found that the crucibles are corroded, and this led to crystallization of undesired phases and to incorporation of typically 2-3.5 at% of aluminium and 0.25 at% of zirconium in YBCO. The contamination effects from the crucibles indicated that the problem of crucible corrosion must be solved to achieve maximum T_c (because this corrosion significantly affects the oxygen kinetics of YBCO).

Keester et al (11) were successful in growing centimeter sized crystal plates of YBCO with T_c(zero) = 80K, using a magnesium oxide (MgO) porous crucible. They reported that Mg was not detected in the crystals. Debinsky et al (12) described a new method of growing YBCO crystals using an Yttria (Y_2O) consumable crucibles. The yttria crucible supplies yttrium to the melted mixture of BaO–CuO and YBCO crystallizes as the primary phase. The large superconducting crystals with $T_c \sim$ 40K were obtained by this method. The T_c of these crystals was found to be enhanced to 88K on 60h annealing at 450^0C. Zirconia and various stabilized modifications of Zirconia have been used extensively in connection with the YBCO (123) materials, not only as crucible materials for crystal growth but also as diffusion barrier layers in thin film production and substrate materials for high current thick films. This success of this material probably arises from a number of causes. The very high melting temperature (2700^0C) leads to a priori to a greater stability, however the most important reason almost certainly is associated with the formation of a complex interface layer based on Barium Zirconate ($BaZrO_3$). This material not only appears to be relatively stable against subsequent attack by $CuO–BaCuO_2$ flux but also appears to prevent Zirconium diffusion into (123) phase in contact with it.

Tin oxide (SnO_2) has also been used successfully in the growth of single domain YBCO crystals with critical temperatures close to 90K (13) and apparently no evidence for tin contamination. Scheel et al (13)however, showed in their study of flux attack on crucible material that a Barium stannous cuprate $BaSn_{1-x}Cu_xO_3$ was extensively produced as a reaction product at crucible surface. In attempting to encourage progress in this difficult area, Sheel et al (14) also suggested the possible use of BeO, TnO_2 and other highly refractory compounds. Clearly refractory materials other than the oxides must also be investigated, however Abell et al (15) reported that BN crucibles are unsatisfactory for crystal growth.

b) Melt composition

The reproducible growth of single crystals of YBCO strongly depends on the optimum stoichiometry of the constituents.

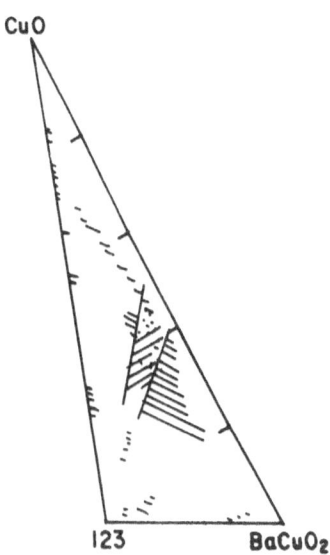

Figure 1: The starting compositions for the growth of (123) YBCO crystals shown shaded on the Cuo-123-BaCuO$_2$ phase field. The hatched area in this diagram represents the common area between the primary (123) crystallization field and the (123)+L field.

Thermal instability of YBCO does not allow growth from melts, vapour or high temperature solutions using conventional molten salts as solvents. YBCO can be grown only in small dimensions by solid state reactions and in some what larger dimensions from high temperature solutions using excess Bao-CuO as solvent (16-19). Folowing Scheel and Licci (16), the location of the starting compositions for some of the many crystal growth studies are shown in **Figure** 1.Since his study, Sadowski et al (9) have systematically explored compositions along the (123) el line, where they have taken BaO:CuO ratio to be 28:72 molar. This work included variations in several other parameters, including processing temperatures, crucible materials and flux decanting temperatures. The experimental data and the results of selected crystal growth techniques containing the melt composition is given in Table 1.

Balestrino et al(20) noticed (at the surface of thepellets of YBCO) the evidence of the formation of a liquid phase, which was likely to promote the growth of crystals. They have assumed this liquid to consist essentially of the eutectic of the two lowest melting components (namely CuO and BaO), and determined the composition (CuO-72%, BaO-28%) and the melting point as 870 + 5^{0}C, of this eutectic. They then performed a few trials of "Flux growth " using the above mentioned eutectic as solvent and a few

TABLE - 1

EXPERIMENTAL DATA AND RESULTS OF SELECTED CRYSTAL GROWTH EXPERIMENTS

Sample	Wt.%	Crucible	Soak temperatures/time °C/h	Cooling rate °C/h⁻¹	Decanting temperature °C	Flux separation	Crystal Area (mm²)	Crystal Thickness mm	Remarks
Y-Ba-Cu-O	5.8	AN	1020/30	2	920	V	5x3	0.2	
"	26.2	AC	1020/24	1	910	V	5x4.5	0.07	
"	5.9	AC	1020/24	1	910	V	4x2	0.15	
"	a	AC	980/5	2	915	V	3x2	0.1	a
"	10.2	ZC	1020/20	1.5	935	V	4x4	0.2	
"	7.8	AC	1000/about 24	1	920	V	3x3	0.1	Soak insufficient
"	24.7	AC	1030/16	0.6	915	V	2x2	0.4	h0l faces
"	30	AC	1030/16	0.6	915	V	2x2	0.4	h0l faces
"	20	AC	1030/16	0.6	915	V	2x2	0.4 inter-grown	h0l faces
"	6.7	AC	1020/28	1	920	V	15x11	0.4	T3x3x0.2
"	8.9	AC	1020/26	0.5	920	V	5x4.5	1.3	6x6x0.8; 3x3x1
"	8	AN	1020/20	1	930	V	Var.	Var	0.1 wt% dopant b
"	8.7	ZN	1020/about 15	0.4	920	M	3.15	0.08	
"	8.7	ZN	1020/about 15	0.4	920	M	2x1	0.05	Max. 10x5 c
"	6	AC	1020/about 15	0.4	920	V	8x3	1	T3x2x1

1. Wt. % of YBCO in the solvent.
2. Crucible material & shape : A, Al_2O_3; Z, ZrO_2; C, Cylindrical ; N, Normal shape.
3. Soaking temperature and time.
4. Cooling rate.
5. Temperature of decanting residual flux.
6. Degree of flux separation : V, Complete; M, medium.
7. Crystal dimensions, P, typical dimensions of atleast five crystals.

a) Decanted flux from earlier experiments at 930°C.
b) 0.1% of dopants; La_2O_3, In_2O_3, SnO_2, V_2O_5, Nb_2O_5, MoO_3, K_2Co_3, Li_2Co_3, B_2O_3, TiO_2; Crystal dimensions variable.
c) Surface : dendritic structure.

molar% of the third component of YBCO namely Y_2O_3 as solute. The representation of starting molar composition used by Balestrino et al (20) is shown in **Figure 2.**

Keester et al(11)reported the growth of single crystals of YBCO with the charge composition of $1/2Y_2O_3.4BaO.10CuO$. Their crystals showed a zero resistance at 80K, and showed complex twinning and crack patterns.Hidaka et al (21) grown crystals using flux method from molten oxide compound with BYCO (60 mole%) and CuO (40 mole%). Black plate shaped YBCO single crystals ranging from 0.5x0.5x0.03 mm^3 to 1x2x0.1 mm^3 were found together with insulating BaY_2CuO_5 greenish needles when the melt was cooled down. Das et al (22) reported the growth of single crystals by melting mixtures of CuO and a master alloy of "$Ba_3Cu_5O_x$" a mixture of $BaCuO_2$ and CuO. "$Ba_3Cu_5O_x$" is hypothetical compound lying at

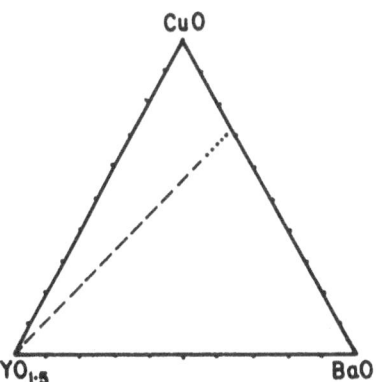

Figure 2. Representation of triangular diagram of the starting molar composition used for the flux of YBCO crystals. The dashed line represents the binary BaO-CuO Eutectic

the intersection of CuO-BaO pseudo binary and the extension of equilibrium line between 1:2:3 and Y_2BaCuO_5. They obtained fairly good and large single crystals of YBCO, when the charge selected lie along the line connecting 1:2:3 and Ba_3Cu_5O at a 30-70% ratio by weight. They reported a fairly broad composition range in the ternary phase field in which primary crystal formation is 1:2:3. Their results suggested that larger crystals can be grown from

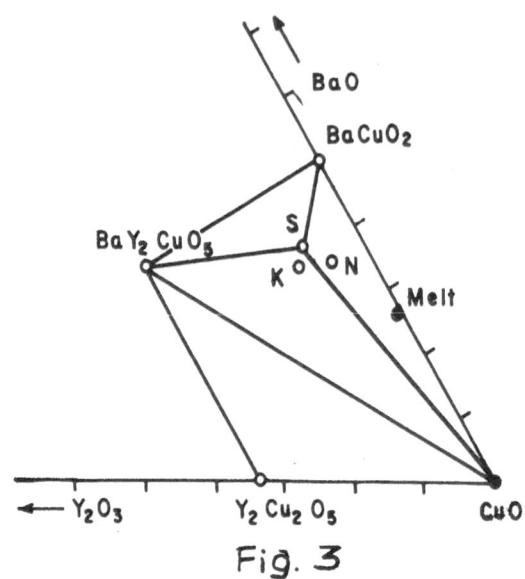

Fig. 3

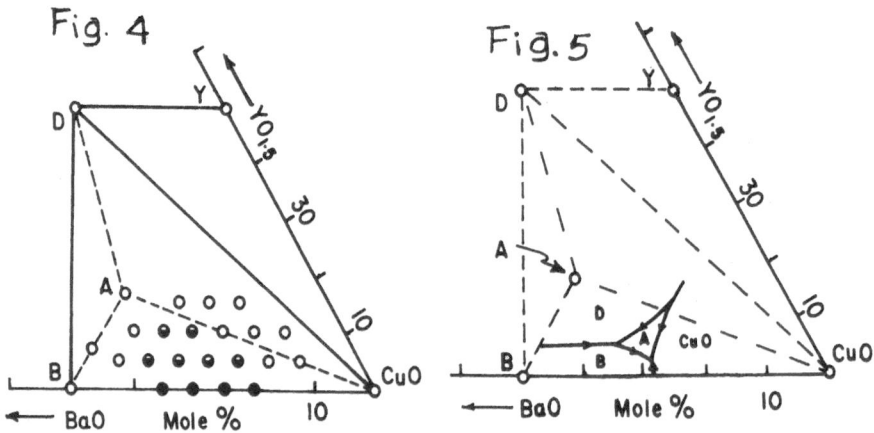

Figure 3: Composition of the starting materials in Ba-Y-Cu-O system's S: $Ba_2YCu_3O_{6.5+x}$

Figure 4: Part of phase-relations in the system $BaO-YO_{1.5}-CuO$. A: $Ba_2YCu_3O_x$ B: $BaCuO_2$ D: BaY_2CuO_5. Y:$YCuO_2$.

Figure 5: Tentative boundary lines separating the primary crystallization fields in the system

czochralski and Bridgeman techniques using similar compositions. Several attempts at pulling large crystals of YBCO were made. But runs were unfortunately short (about one hour) because of unavoidable reactions between the liquid and crucible (Al_2O_3). This reaction altered the liquid composition, viscosity and melting temperature. Success of the growth of single crystals of YBCO in the present case was attributed to finding suitable crucible material.

Aggregates of small crystals of YBCO about 20-80x20-80x5-25μm in size were synthesized by Ono and Tanaka (23) in air by heating mixtures just above and below the solidus temperature. They used three kinds of the mixtures. Their compositions are plotted on the $BaO-Y_2O_3-CuO$ diagrams (Figure 3) as circles of S,K and N. S is the stoichiometric composition, $Ba_2YCu_3O_{6.5+x}$ and the filled circle is the composition of melt in equilibrium with the superconductor at $940^{o}C$ in air. Single crystals of $Ba_2YCu_3O_7$ were grown from non-stoichiometric melts having composition of $BaO:YO_{1.5}:CuO$ = 12:1:26 by Takekawa and Iyi (24). They described the growth of single crystals of tetragonal $Ba_2YCu_3O_x$ (designated as A-phase) from non-stoichiometric melts in the YBCO system. The phase relation in the system $BaO-YO_{1.5}-CuO$ were given in Figure 4 on the basis of their data tentative boundary curves separating the primary crystallization fields were drawn as solid lines and shown in Figure 5, where arrows indicate the direction of falling temperature using the obtained diagrams, they tried several crystal growth experiments from the non-stoichiometric melts.

c) Fluxes

The growth conditions varyvery rapidly with detailed compositions of flux. The chosen composition can be formed either by the use of unreacted constituent oxides or by the use of processed (123) powders along with an unprocessed (BaO+CuO) or processed ($CuO+BaCuO_2$) flux.

Watanabe (25) attempted the growth of single crystals of superconductor materials by slow cooling technique. Numerous fluxes, chemical reagents of (Li,Na,K) fluoride, chloride,carbonate and borate, $PbO-PbF_2$, Li_2O-MoO_3, $BaCo_3$,fluoride, peroxide etc were used as flux. when alkali halides were used as a flux, needle CuO crystals of 1mm diameter and 20mm length were plentifully obtained. In the case of alkali or Ba carbonates, many large prismatic $YBa_4Cu_2Pt_2O_x$ single crystals which belong to the triagonal crystal system grew at the wall of the crucible. a flux containing heavy metal compounds (Pb,Bi or Mo) yielded three kinds of crystal needles of CuO, irregularly formed $Ba_2Y_2O_5$ or octohedral $BaMoO_4$.

Lin et al (25) and used an equimolar eutectic mixture of NaCl and KCl as flux and grew YBCO crystals. They obtained largest crystals for YBCO- flux ratios of approximately by 5:1 molar. In growing large single crystals consideration must always be given, however to the risk of flux- crucible interactions at elevated temperatures, and if growth is to extend over very long periods, the use of pre-reacted flux powders is to be preferred. When choosing a flux for superconducting materiallikeYBCOthe following considerations should be kept in mind

i) It should have low melting point

ii) It should be a good solvent, dissolving between 5 and 30 wt% of solution at the maximum temperature intended . The solubility should decrease with temperature.

iii) It should not form a compound with the solute, nor a solid solution beyond a degree which depends on the intended use of the crystal, 1% of flux present substitutionally can be tolerated for many experiments.iv) It should be compatable with the crucible material over the intended temperature range.v) For growth of slow cooling, it should be of low volatility. A volatile flux necessiates a low temperature range, a faster range of cooling and often results in nucleation at the melt surface. vi) It should be of low viscosity. vii) The flux should be easily prepared and easily separatedfromthecrystal.viii) Toxity should be low.

d)   Thermal   cycle

Accurate temperature control is essential for stable crystal growth, the temperature stability at the crucible may be improved by surrounding it with ceramic material to increase the thermal capacity. A stability of $+ 0.1^{0}C$ at the crucible is desirable, but better regulation is often pointless because of the temperature oscillations due to unstable convention. If the temperature and solute gradients are such that appreciable convection does not occur, closer temperature regulation may have benificial effects. In YBCO single crystal growth from flux, the detailed thermal processing sequence varies from laboratory to laboratory and clearly as long as a limited number of conditions are satisfied the precise sequence is not important; these conditions are

i) Heating rate

In the range of 800^{0}C, the heating rate should be slow enough to allowfull calcination of residual carbonates without the risk of trapped Co_2 bubbles.

ii) Soak temperature

Above the liquidus for long enough time to allow complete mixing and liquification (i.e. $1000-1040^{0}$C).

iii) Soaking time: 20-30 hours

iv) Cooling rate

Liquidus- 930^{0}C (oxygen)————————→crystal growth

$780-680^{0}$C ————————→T-O phase change

$600-300^{0}$C ————————→for oxygenation
or generally
$0.5-10^{0}$C per hour

v) After growth annealing: 450^{0}C/10-100 hours.

For optimized crystal growth of YBCO the thermal cycleshown in Figure 6 can be emplyoed.In this, a soak at a temperature in the range close to 1020^{0}C for 15-20 hours is adequate for homogenisation and liquification, while restricting the crucible reactions.

Hidaka et al (21) grown YBCO crystals from CuO flux. The temperature gradiant at the crucible was maintained at about 2^{0}C/cm by monitering the temperatures at the top and bottom of the crucible with two pairs of pt/Rh thermocouples. The temperature fluctuation in the middle of the furnace was within 0.5^{0}C. They heated the stoichiometric ratios of YBCO in the flux of CuO powder at 1400^{0}C for 2h and then slowly cooled down at 3.3^{0}C/h as shown in Figure 7.

The growth experiments of Balestrino et al (20) were performed as follows. Powders of $BaCo_3$,Y_2O_3 and CuO, in the proper amounts to form 75 to 94 molar percent of the eutectic and 25 to 6 molar percent of YBCO upto a total weight of about 50g, were throughly mixed and placed, with slight pressure, in a $50cm^3$ platinum crucible. The crucible was placed in a vertical tubular furnace in 99.99% oxygen atmosphere and rapidly heated to 975^{0}C, soaked at this temperature for about 15h, and then cooled at the rate of $3^{0}Ch^{-1}$ to 850^{0}C, namely slightly below the melting temperature of binary eutectic. Afterwards the crucible was furnace cooled to 300^{0}C and finally extracted. Such crystals were reported to be platelets upto 4mm edge and 0.1mm thick. To increase the size of the grown crystals they suggested several

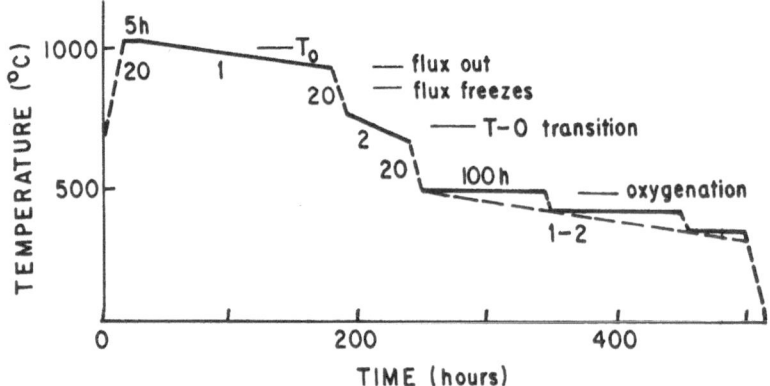

Figure 6: An ideal thermal processing cycle for crystal growth from $CuO-BaCuO_2$ flux, assuming that flux occurs at $T = 930°C$.

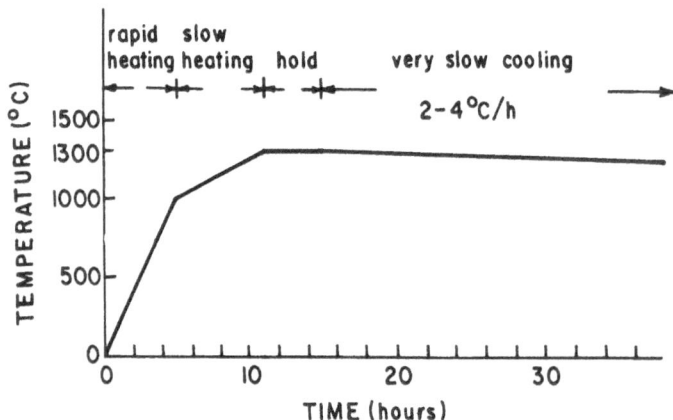

Figure 7: Temperature variation at the centre of the furnace.

technical improvements. In particular, the solubility curve of YBCO in the eutectic should be detected, and the growth should be performed by lowering the temperature from above the saturation point.

Kleester et al (11)have grown YBCO single crystals in a vertically mounted, single zone, tube furnace resistively heated with programable time and temperature control. Crystal growth was achieved in MgO crucibles completely filled with loose powder nexted inside 1.25 inch ceramic porcelain crucible to a temperature of 1000^{0}c and then allowed to soak at temperature from 2 to 36h. Judging from the behaviour of the melt, the starting material was reported to be partly carbonate free.

Soaking time is very important for the crystal growth.Without soaking no large crystals are obtained. A melt soaked for 4h at 1000^{0}C produce only one or two small crystals on the melt surface longer crystals upto 1cm (11). Most of the crystals obtained by the method of flux reported to be of tetragonal phase. Initially depending on the oxygen pressure (air or oxygen) and on cooling rate, the initially grown tetragonal crystals either transforms into superconducting or orthorhombic phase or have to be annealed after growth in oxygen at intermediate temperatures.

e) Crystal separation

Below 930^{0}C there is relatively little crystal growth in YBCO system and subsequent cooling simply allows solidification of the flux at approximately 890^{0}C. During freezing, the flux contracts severely leaving voids from which the YBCO crystals may be subsequently removed. These occur both within the body of YBCO flux system and between the flux and the crucible walls. Individual crystals always difficult to remove without damage and their enclosed location leads to incomplete and often very low oxygenation.

It is often convenient to remove the liquid flux from the crucible in the temperature interval 900–930^{0}C so as to leave the crystals self supporting and free from flux contamination. This can be achieved readily by one of two methods in these systems involving CuO based fluxes, either by physically decanting within the furnace or by use of flux pump (26) in which a piece of porous ceramic (fire brick) is used to soak up the liquid flux from the crucible at fixed temperature. The risk of damage to the crystals from thermal shock is greatly reduced using the latter technique as also is the problem of losing crystals with the decanting liquid. Provided care is taken in introducing the flux pumping ceramic there will be only a minor chance of breaking crystals by physical contact.

Once the flux is removed, by either means of residual liquid drains from the grown crystal and eventually solidifies as the system is cooled. Oxygenation is now much more readily achieved, and both the T-O phase, change and the oxygen absorption stages must now be included in the thermal cycle.

f) Flux creep

Under certain condition, the $CuO-BaCuO_2$ flux shows a strong tendency to creep over the surface of the crucible material, an effect which has been used to provide alternate means of reparting the crystal. In the early work of Kiser et al (6) using gold crucibles. Crystal growth took place in a cavity formed between the crucible bottom and a gold support plate, as the liquid moved over both the inner and outer walls of the crucible. This concept has been extended to oxide crucibles by a number of workers who have used pieces of broken ceramic to define growth cavities within the crucible. Finally an unusual use of flux creep at elevated temperatures has been to initiate crystal growth by melting a mass mixed YBCO and flux on a flat ceramic or gold plate. The flux transport across and off, the plate results in the crystal growth at various nucleation centers over the surfaces.

Careful manipulations of growth conditions through the knowledge of phase diagram, has recently led to the growth of very large, well formed single crystals. Unfortunately these are invariably in tetragonal anti-ferromagnetic state and there are no known examples in which these big crystals have been brought to a uniform oxygen concentration close to $\delta < 0.1$ by means of oxygen annealing. Indeed in view of known diffusion rates it seems likely that the surface degradation may occur to unacceptable levels before oxygenation is complete.

Alternative methods of crystal growth employing small amounts of alkalichlorides added to the YBCO melt have resulted in the successful growth of polycrystalline boules of materials from which three dimensional crystals of approximately $1cm^3$ can be cut or cleaved. These crystals are not' of extremely high perfection of those grown with the CuO flux and contain a few volume percent of (211) phase in the form of void inclusions. While these samples are not fully oxygenated on removal from the boul, they can be annealed to $\delta = 0.1$ in a relatively short time.

In the specific field of crystal growth establishment of relaible phase diagrams, development of suitable non-contaminating crucibles, and reaching optimum conditions for stable growth rate seem among the most urgent problems to be solved. However, due to the enormous complexity of preparing high purity ceramics and high quality crystals and epitaxal layers. Significant results will probably be achieved only after innovation in preparative methods combined with sophisticated characterization efforts.

Acknowledgements

I take this oppertunity to thank Prof.T.Navneetha Rao, Vice-chancellor of Osmania University, Hyderabad, India who gave me unfailing encouragement and support over a number of years.

I also thank Profs.K.Rama Reddy and A.A.Kamal of Department of Physics Osmania University, Hyderabad, India and Prof.S.Radha-Krishna of The Department of Physics, IIT madras and presently at The Institute of Advanced studies, University of Malaya, Kualalumpur, Malaysia for their encouragement, comments and suggestions based on their vast experience in research, which resulted in the substantial improvement of this work.I also acknowledge the help extended by ICTP Trieste, and DST New Delhi for their help.

REFERENCES

1. Holtzberg F; Kaiser D.L; Scott B.A; Mc Guire T.M; Jakson T.N; Kleinsasses A and Tozer S. Chem.HTSC ACS 351,79(1987)
2. Iwata C; Tazimo Y: and Hikita M; J.Cryst.Growth 91,274(1988)
3. Darlington C.N.W; O'Connor D.A; and Hollian C.A; J.Cryst.Growth 91, 308 (1988)
4. Katsui A; and Ohtsuka H; J.Cryst.Growth 91,264 (1988)
5. Menken M.J; and Menovky A.A; Cryst.Growth 91,264 (1988).
6. Kaiser D.L; Holtzberg F; Chisholm M.F and Worthington T.K; J.Cryst.Growth 85,593 (1987)
7. Shields T.C; Welhoffer F; Abell J.S Taylor K.N.R; and Holland D; MZS-HTSC stanford, Physica (c),1989 (in press)
8. Bailey A; Town S.L; Alvaraez G; Taylor K.N.R and Russell G.J; Physica C (1989) in press
9. Sadowski W; Walker E; and Triscone; Personal communication.
10. Sadwski W; and Sheel H.J; J.less Common Met 150,219 (1989).
11. Keester K.L; Hobley R.M and Marshall D.S J.Cryst.Growth 91,295 (1988).

12. Dembinsky k; Gervair M; and Coutures; Mat.Sci.Eng.B. Solid state Materials Adv.Tech B-5,345 (1990).
13. Thomson C; Cardona M: Gugenheimer B; and Liu R; Phys.Rev. B-37, 9860 (1988).
14. Sheel H.J; Sadowski W; and Schellenberg L: J.Super con.Sci & Tech 2,17 (1989).
15. Abell H.J; Private communication.
16. Sheel H.J and Licci F; J.Cryst.Growth 85,607(1987).
17. Sheel H.J Physica C,153, 44(1988).
18. Sheel H.J and Licci F; Mat.Res.Soc. Bull, 73,56(1988).
19. Damento M.A; Gschniedner Jr. K; and McCalhen R.W. Appl.Phys.Lett 51,690 (1987).
20. Balestrino G; Gambardella U; Liu Y.C; Marinelli M; Paroli P; Paoletti A and Paterno G; Vuto XVIII,222(1988).
21. Hidaka Y; Enonoto Y; Suzuki M; Oda M; and Murakami T; J.Cryst.Growth 85,581(1987).
22. Das B.N; Toth L.E; Smith A.K; Bonder B; O'Sofsky M; Pande C.S; Koom N.C and Wolf S: J.Cryst.Growth 85,588 (1987).
23. Ono A; and Tanaka T; Jpn.J.Appl.Phys 26,L825 (1987).
24. Takewa S; and Iye N; Jpn.J.Apll.Phys 26,L851(1987).
25. Watanabe K; J.Cryst.Growth 100, 293 (1990).
26. Lin S.H.et al J.Am.Ceram.Soc. 64,881 (1988).

Thermally Stimulated Luminescence Studies in High Temperature Superconductors

Katta Narasimha Reddy

Department of Physics, University College of Science
Osmania University, Hyderabad-500007, India

1.INTRODUCTION

Thermally Stimulated Luminescence (TSL) is generally credited as a potentially useful trap level spectroscopic method of investigating materials. The electronic and ionic transport phenomena in solids observed during non−isothermal temperature scans are often studied employing the TSL technique. Though the phenomenon of TSL has been well known for a considerable time, in recent years, interest in TSL has continued to increase, because the phenomena of TSL provide an ultra sensitive micro probe into the solid state.

In recent years many researchers have used TSL and Photoluminescence methods for investigating High Temperature Super conductors (HTSC). These methods have been employed in order to use the luminescence signals as characterization probes, and to obtain fundamental knowledge necessary for the development of new concept devices.The PL studies provide information about a microscopic field surrounding the ions and where as the TSL measurements will be performed with a particular emphasis to get insight into the energetic structure of defects and their site occupancy symmetry. The main problem with novel high temperature superconducting materials is their capacity allowing to observe only a weak fluorescence in the conventional range of optical spectra. Yet some researchers have succeeded recently to detect photoexcited visible fluorescence of some rare−earth ions such as Pr^{3+} and Eu^{3+} incorporated simultaneously into the $La_{1.85}Sr_{0.15}CuO_4$ ceramic compound. They have used rare earth ions as probes for studying the local environment of lanthanum sites.

Unlike pure metallic superconductors the new high temperature superconducting materials which are composed of ceramic oxides have a band gap [1] and defect sites (oxygen vacancies). Therefore they are highly susceptible to radiation damage. Because of this reason many researchers presently working in the field of high temperature superconductors (HTSC) are attracted towards the TSL measurements of HTSC materials.

2.PROCESS OFTHERMALLY STIMULATED LUMINESCENCE

Thermally Stimulated Luminescence (TSL) is the emission of light from an insulator or semiconductor (with a certain band gap), when heated this is not to be confused with the light spontaneously emitted from a substance when it is heated to incandescence. TSL is the emission of light following previous absorption of energy from radiation. Three essential ingradients are necessary for production of TSL, they are

 i)The material must be an insulator or semiconductor.Metals do not exhibit exhibit luminescent properties.
 11) The material must have some time the absorbed energy during exposure to radiation.
 iii)The luminescence emission is triggered by heating the material.

The fundamental principle which govern the production of TSL are essentially the same as those which govern all luminescence processes; and in this way TSL is merely one of the large family of luminescence phenomena. The "Family tree"of luminescence phenomena is shown in **Figure** 1.The prefix to the term luminescence distinguishes between the modes of excitation, whilst the delay between the excitation and emission τ_c distinguishes between fluorescence and phosphorescence. The essential features of the phenomena of TSL are as follows. When the radiation is incident on an insulating material, some of the deposited energy is stored inside the material at the defect sites, colour centres etc. Upon heating the crystals, this energy is released and fraction of it may be emitted as visible light, prior to the onset of the black body radiation. The pattern of this emitted visible light (luminescence output) versus temperature is called the TSL glow curve. The TSL glow curve exhibits many peaks depending upon the nature of material and these are called TSL glow curves. The peak positions in the glow curve are fairly constant for a given material and the amount of the light measured is found to be proportional to the initial excitation.

The phenomena of TSL is very sensitive to a variety of factors such as impurities in the material, pre and post irradiation treatments given to the material, type of radiation used, the rate of heating while recording TSL glow curves etc.

Inspite of all these complications, the TSL phenomenon finds successful practical applications in many disciplines to day. The very fact that the TSL is influenced by small changes in the state

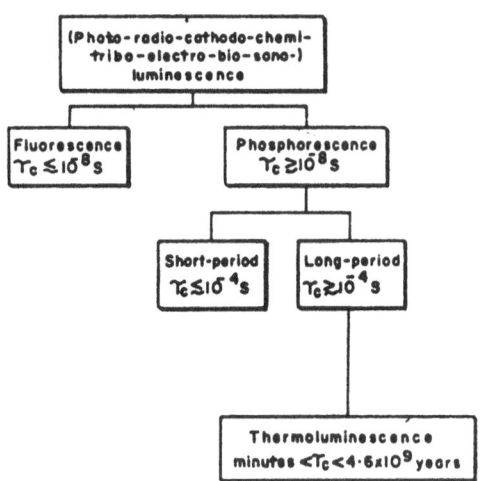

Figure 1. Family Tree of luminescence phenomena

of imperfections suggests that with a better understanding it may effectively be used as a powerful tool in the investigation of HTSC materials

3.TSL STUDIES IN Re–Ba–Cu–O SYSTEM

Cooke et al [2] have reported the first TSL and emission spectra measurements of X-ray induced defects in Re–Ba–Cu–O system (where Re= Eu,Ho,Gd,Tl etc) in the temperature interval of 80-300K. They observed a single TSL glow peak at 195K with concomitant single emission band at 500nm in the case of $GdBa_2Cu_3O_x$ and two TSL glow peaks (160 and 172K) in the case of $Ho_{1.5}Ba_{1.5}Cu_2O_x$ system. The interesting feature that they noticed was that the loss of luminescence sensitivity with time. Since oxygen defect perovskites have a priority to readily lose or gain oxygen, they suggested this loss of oxygen, contributes to the decrease in luminescence intensity. Thus TSL measurements may represent a very sensitive way to investigate the problem of oxygen stability in these materials. They also felt that the surface rather than the bulk properties determined the TSL glow curves.

Cooke et al [3] have reported the existence of X-ray induced TSL in single phase $GdBa_2Cu_3O_7$ and two phase $Re_{1.5}Ba_{1.5}Cu_2O_x$ (Re= Ho,Eu) compounds. They have observed several TSL glow peaks in the 80-675K temperature range and proposed tentative explanations for the defect mechanisms involved, e.g the recombination of electrons trapped as F-type colour centres with thermally released v-type holes or deexcitations of rare earth ions. Unfortunately the samples used by cooke et al [3] were unstable with respect to their luminescent properties

Fujiwara and Kobayashi [4] performed TSL measurements on $Gd_xBa_{1-x}CuO$ system and obtained about six glow peaks for Gd composition (x) of o.3,0.33 and0.4 (see **Figure 2**). They also reported the cumulative x-ray irradiation time dependence of TSL glow curves obtained for x=0.3 samples. They noticed that with the increase in x-irradiation time a peak around 130K appeared and grew. This Peak disappeared when the x-ray'ed sample was kept in dark at RT for 15 hours. Similar result was observed even in repeated irradiated samples. These results suggest that the 130K peak is meta-stable and is relaxed by RT annealing. Further in samples of x=0.4 the cumulative x-irradiation is found to influence only A and B peaks. The increase in x-irradiation resulted in the increase of B-peak intensity. Though the origin of these peaks are not clearly discussed by Fujiwara and Kobayashi [4],these results can give a clue to clarify the mechanism of creation and annihilation of the traps.

TSL of x-ray induced defects in Re-Ba-Cu-O samples (Re=Ho, and Eu) was measured by Cooke et al [3] in the interval of $25-400^{0}C$. Ho and Eu samples exhibit similar TSL with peaks at $65,135$ and $185^{0}C$. All the TSL peaks have been attributed to the recombination of Fe_3^{+} and F_3^{-} centre electrons with V_3^{-}-type holes.

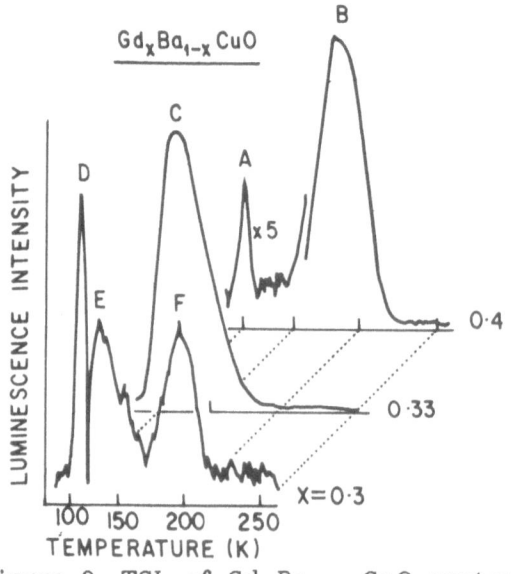

Figure 2. TSL of $Gd_xBa_{1-x}CuO$ system

354

Cooke et al [3] observed the TSL glow curves of rare earth based superconductors taken under identical conditions showed a dramatic decrease in intensity in just three days, and after two weeks exhibited no intensity at all. This was thought to be due to the degradation of rare earth doped sample before taking the TSL measurements, besides this the rare earth based samples were opaque and the measurements corresponded to the surface rather than bulk luminescence, thus most of the TSL studies made on rare earth based HTSC materials probably reflect a greater sensitivity to sample properties rather than the usual bulk measurements.

4.TSL STUDIES IN Bi AND Tl BASED SUPERCONDUCTORS

Fuziwara and Kobayashi [4] succeeded in the observation of TSL from new oxide superconductor systems with out rare earth elements i.e Bi–Sr–Ca–Cu–O and Tl–Ba–Ca–Cu–O (Figure 3).

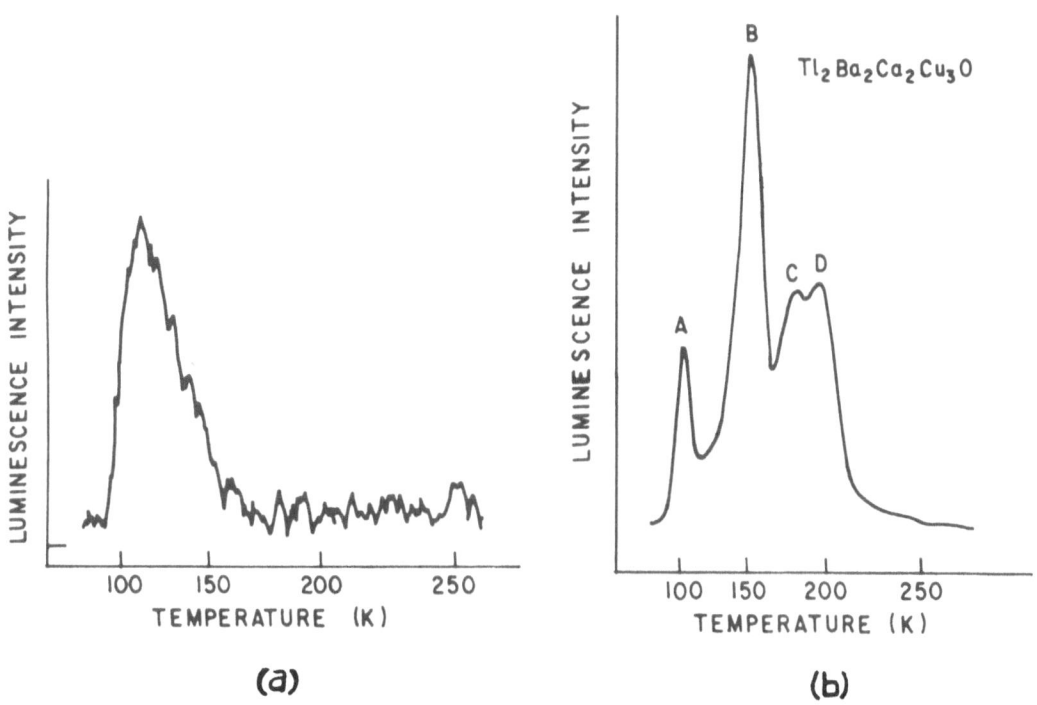

(a)   (b)

Figure 3: a)TSL spectra of Bi–Sr–Ca–Cu–O and b)Tl–Ba–Ca–Cu–O

It was found that in these two cases the glow curves significantly differ from each other, though they share crystallographic properties in common.

Fuziwara et al [5] studied TSL in TL based superconductors with the increase in x-irradiation time. They observed that a new peak around 120K grows on account of an increase in cumulative x-ray irradiation time (Figure 4). Inspite of these interesting observations Fuziwara et al [5] could not give proper explanation for their observations.

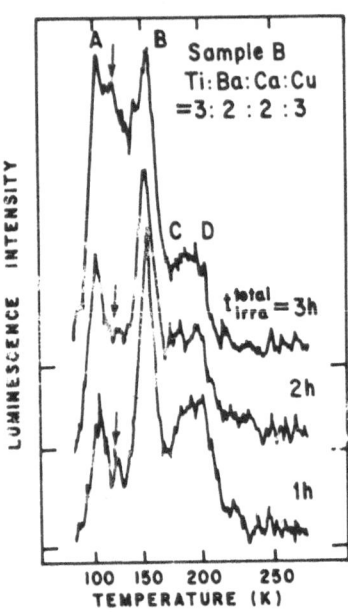

Figure 4. Cumulative X-irradiation time dependence of the glow curves observed in Tl based superconductors

5.DEPENDENCE OF TSL ON OXYGEN STOICHIOMETRY

It was shown by Tarascon et al [6] that the oxygen-defect perovskites have a propensity to readily lose er gain oxygen. This loss or gain in oxygen alters the number of oxygen vacancies and upon irradiation would change the concentration of F-type centres affecting the luminescence intensity.

Thermally stimulated luminescence of nominally doped and undoped $YBa_2Cu_3O_7$(YBCO) ceramics x-irradiated at low temperature (20K) has been observed by Roth et al [7]. High purity YBCO exhibited two glow peaks, at 90 and 195K. They have also noticed that the spectral composition of the 90K and 195K glow peaks were identical. They attributed 195K peak to F-centres. Comparing their TSL results with those of Cooke et al [3], Roth et al [7] have concluded that 90 and 195K peaks are due to intrinsic defects rather than rare earth or any other trace impurity present in the sample.

The author and his co workers [8] have studied TSL of YBCO samples as a function of quenching temperature (with increase in oxygen deficiency). They reported three glow peaks around 90,145 and 195K in unquenched samples.With the increase in quenching temperature they observed the increase of magnitude of 145K peak (**Figure 5a**). Annealing the irradiated samples for 5 days at room temperature resulted in the decrease of the magnitude of the 145K peak and the total elimination of 90K peak (**Figure 5b**). In view of these observations they tentatively assigned the 90K peak to back ground trace impurities.

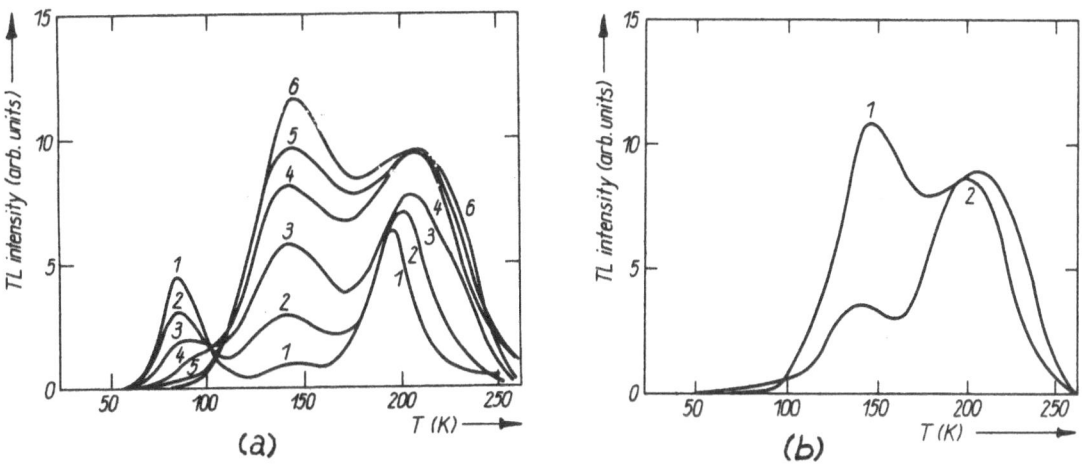

(a) (b)

Figure 5. a)TSL glow curves of $YBa_2Cu_3O_{7-\delta}$ ceramic samples(1)unquenched; (2)quenched from $200^{\circ}C$; (3)$400^{\circ}C$; (4)$500^{\circ}C$; (5)$600^{\circ}C$; (6) $700^{\circ}C$
b) 1) quenched from $700^{\circ}C$ (2) annealed at RT for 5 days and TSL recorded

Bartyakhtar et al [9] in their recent paper studied the oxygen losses from YBCO at different quenching temperature and concluded that $\delta(T_g)$ (where δ is the oxygen content in $YBa_2Cu_3O_7$) depends linearly on quenching temperature, and they have approximated this by the formula

$$\delta = 1.273 \times 10^{-3} T_q - 0.409$$

where T_q is the quenching temperature. This relation shows how oxygen deficiency takes place in YBCO at different quenching temperatures. In the present case it is observed that as the quenching temperature increases (i.e as the oxygen deficiency increases) the intensity of 145K and 195K peaks increases. When the sample is quenched from 700^0C and annealed at RT for 5 days(as shown in Figure 5b) the intensity of 145K peak decreases, that means 145K peak is sensitive to the oxygen content in the sample. Therefore the author and his co-workers assigned this peak to oxygen vacancy centers. Finally They attributed 145K and 195k to F^+-centres (Two electrons trapped at an oxygen vacancy and F-centres (one electron trapped at the oxygen vacancy). It is note worthy from this work that the quenching of HTSC material like YBCO should lead to an increase in oxygen vacancies which upon irradiation would yield a larger concentration of F-centres.

6.DETECTION OF INSULATING PHASES USING TSL STUDIES

Cooke et al [10] demonstrated that TSL is an important method for detecting insulating impurity phases that commonly occur in high T_c-superconductors. They have shown that all insulating phases, Y_2O_3, Ba_2CuO_4, $BaCuO_2$, $BaCO_3$ and Y_2BaCuO_5 all exhibit characteristic TSL glow curves with well defined maxima(Figure 6). In addition they observed strong luminescence from a sample which has been depleted of large fraction of oxygen content (see curve labeled $YBa_2Cu_3O_{6.2}$) which is characteristic of very weak TSL. This result demonstrated that not only can TSL detect the impurity phases present, but it can also detect the non-superconducting tetragonal phases, which results from poor oxygenation of $YBa_2Cu_3O_x$. No TSL was reported from superconducting orthorhombic phase (x>6.5) as for all metals there is no band gap of sufficient magnitude to trap the radiation induced changes. This feature of TSL studies of high T_c materials is important because no pure superconductor will exhibit TSL. Any observed signal is a characteristic of the insulating surface impurity. Thus TSL serves as a quick and inexpensive method for determining the quality of high temperature superconductors.

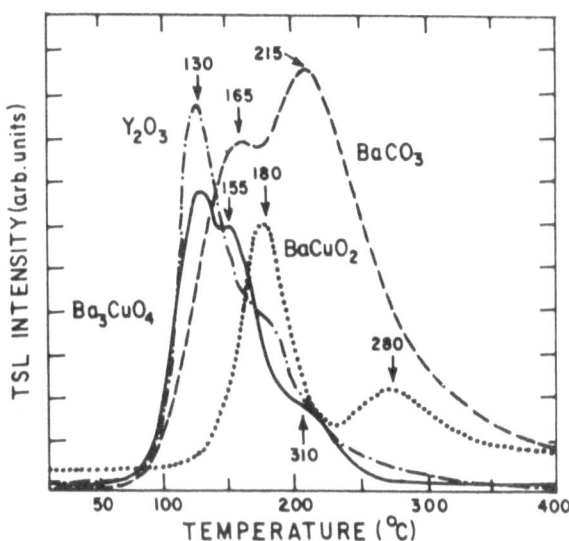

Figure 6. TSL glow curves of $Y_2O_3, Ba_2Cu_3O_x, BaCuO_2, BaCo_2$.

Cooke et al [12] in their another communications pointed out that a correlation can be established between TSL and radi frequency surface resistance (R_s) and suggested that TSL can be used as quick and reliable method for estimating R_s of bulk superconductors. their data also showed that for selected substrates one can use TSL method for estimating R_s of thin films as well. Their results indicated that the intensity of luminescence related to the magnitude of the rf surface resistance.

γ-ray induced surface defects in YBCO have been investigated by Jahan et al [12] using TSL. Irradiated samples stored in either vacuum or oxygen environments showed that no insulating chemical species were formed on the surface of the superconductors. In contrast the exposure of YBCO samples to the humid environments produces various chemical components on the surface and consequently TSL yield was enhanced. Their TSL emission spectra results indicated that more than one chemical insulating species were formed on the YBCO surface as a result of the reaction with water vapour. $BaCO_3$ was identified as one of these species. Y_2BaCuO_5 and oxygen depleted $YBa_2Cu_3O_x$ (x= 6.2) are two other species that were reported to have been formed.

Narasimha Reddy and Prasad [13] studied the effect of

microstructure on thermally stimulated luminescence (TSL) spectra of YBCO. They studied TSL in as prepared, quenched and water treated conditions. They reported that YBCO in as prepared condition gives a weak TSL signal. A tempering treatment (consisting of heating, the sample at 500^{0}C for specific time and then cooling to RT in about 10 hours) enhanced the TSL output and two peaks (around 190 and 225^{0}C) were observed even for small radiation exposure time (Figure 7).

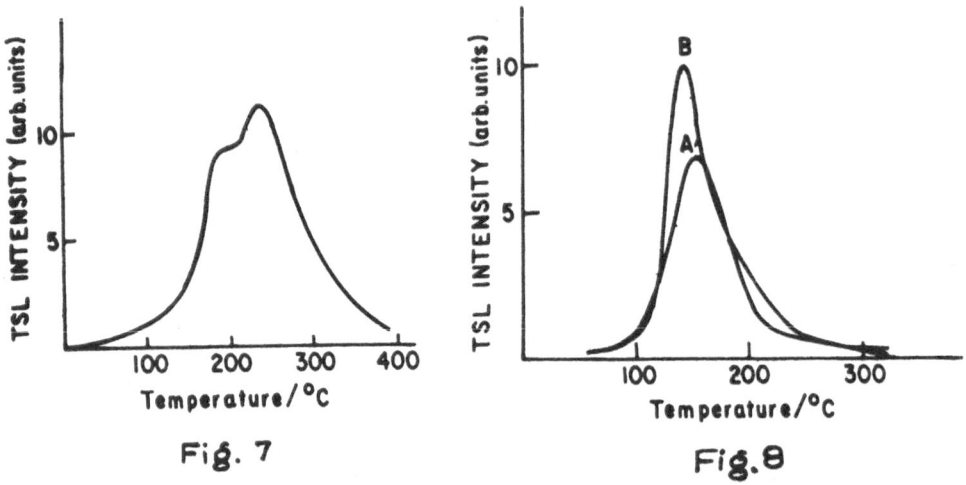

Fig. 7

Fig. 8

Figure 7. TSL spectrum of tempered YBCO sample (γ-rayed for 10 min

Figure 8. Curve A:TSL of YBCO quenched from 1050^{0}C and γ-rayed, curve B: TSL of water treated YBCO sample after γ-irradiation

The TSL glow curve of YBCO samples, quenched from 1075^{0}C (curve A) and samples exposed to water environments for longer periods (>10h or so)(curve B) are shown in Figure 8 . In as quenched sample only one glow peak around 150^{0}C was observed and where as in water treated sample a glow peak was observed at 145^{0}C.In the sample quenched and annealed at RT for 10 days, two peaks at 130 and 180^{0}C are observed (Figure 9 curve A). Water treated samples kept at RT for 5 days and subsequently exposed to radiation showed two TSL peaks at 160 and 215^{0}C (Figure 9 curve B).

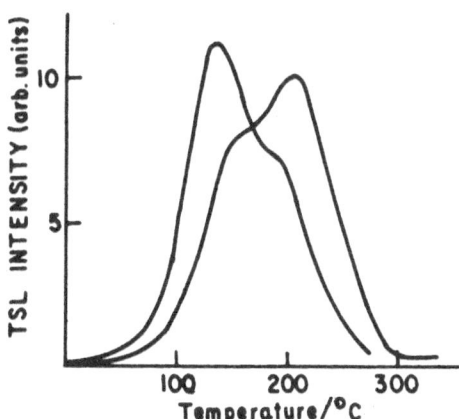

Figure 9. Curve A: TSL spectrum of quenched YBCO kept at
RT for 10 days and exposed to γ-irradiation
Curve B: TSL spectra of water treated YBCO
kept at RT for 5 days and exposed to
γ-radiation

These results could be explained basing on the observations of
Murakami et al [14] that when YBCO quenched from and above $1050^{0}C$
decomposes according to the following reaction

$$YBa_2Cu_3O_6 \longrightarrow Y_2BaCuO_5 + 3BaCuO_2 + 2CuO$$

$BaCuO_2$ is not stable at high temperatures and further decomposes
to BaO and CuO. Y_2BaCuO_5 is also not stable and over a period of
time further decomposes to Y_2O_3+BaO and CuO mixtures. The TSL
signal taken with a gap is different(see Curve A of Figure 8 and
curve A of Figure 9) indicating that the compound formed
immediately after quenching is different from the one formed after
10 days of annealing at RT. Comparing the results of Curve A and
Curve B of Figure 9, one can conclude that the insulating phases
responsible for these two TSL curves are same and hence the
quenching and water treatments produces the similar insulating
phases of YBCO.

7. Analysis of TSL Glow Curves and TSL kinetics

The analysis of TSL glow curves provides information on
recombination kinetics, which is essential for complete
understanding of thermally stimulated luminescence process. Many
methods have been advocated in the literature by several authors.
Most of these methods depend rather heavily on peak temperature
and some other special features of the glow peaks (such as half

width, full width, lower and higher half width temperatures) and have limited applicability. As the observed TL is the net result of multiple processes taking place on heating the material, methods based on the shape of entire glow peak are more useful than others. The analysis of Kelly and Braunlich [15] based on general order kinetic equations have shown that the glow curve shapes depend upon parameters such as the ratio of retrapping and recombination rates, the number of thermally activated traps, the number of deep traps and trap occupancy, as well as on experimental factors such as initial temperature and and heating rate. Since these parameters are generally available from glow peaks alone, the determination of activation energies and frequency factors from glow peaks can not be strictly valid. However, the determination of such parameters could be useful for a comparison of experimentally determined values of these activation parameters with those estimated by other researchers using different methods. In most of the cases the glow curves are complex ones and considerable overlap exists between the various glow peaks, therefore the peaks must be isolated by using some method or other. The glow curves are usually analysed for better results by the method of numerical curve fitting [16]. In this method the following version of Randall-Wilkins function [17] is used to fit the observed light intensities at various temperatures:

$$I(T) = P_0 h_0 \exp\left[-(\emptyset/T)-(T/T_g)^2 \exp(\emptyset/T_g-\emptyset/T)\right.$$
$$\left. \times\{1-2(T/\emptyset)+6(T/\emptyset)^2-24(T/\emptyset)^3+120(T/\emptyset)^4\}\right]$$

where $\emptyset = E/k$, E is the trap depth, k =Boltzmann constant, h_0=the number of electrons at temperature T_0, P_0=attempt frequency factor and T_g=glow peak temperature.

The Randall-Wilkins [17] equation for intensity of TL emission reduces the above form by using the approximation

$$\int_{T_0}^{T} \exp(\emptyset/T) dT \approx T e^{-\emptyset/T} \sum_{n=1} (T/\emptyset)^n (-1)^{n-1} n!$$

the frequency factor P_0 is calculated with the known values of $\emptyset$ with the help of the condition for maximum glow peak intensity

$$P_0/q = (\emptyset/T^2)\exp(\emptyset/T_g)$$

the calculated values of E and P_0 from the present glow peaks and

362

reported values in the case of other superconductors are cited in Table 1.

Table 1

sample	peak position (K)	activation energy (eV)	frequency factor (s^{-1})	kinetic order
$Tl_2Ba_2CaCu_3O_7$	100	0.12	11.2	1.12
[5]	150	0.24	67.5	1.11
	190	–	–	–
	200	0.27	59.1	1.11
$Ho_{1.5}Ba_{1.5}Cu_2O_x$	160	0.26	21.4	1.00
[7]	172	0.29	20.5	1.80
$GdBa_2Cu_3O_x$	195	0.42	61.6	1.3
[7]				
$YBa_2Cu_3O_7$	90	0.16	–	–
[4]	195	0.42	–	–
$YBa_2Cu_3O_{7-\delta}$	90	0.18	12	1.2
(present	145	0.23	65	1.1
study)	195	0.42	66	1.1

8.ACKNOWLEDGEMENTS

I take this opportunity to thank **Prof.T.Navaneetha Rao,** Vice-chancellor of Osmania University, Hyderabad, India who gave me unfailing encouragement and support over a number of years.

I also thank **Profs. K.Rama Reddy** and **A.A.Kamal** of Department of Physics Osmania University, Hyderabad and **Prof.S.Radhakrishna** of Department of Physics,IIT Madras and Presently at The Institute of Advanced Studies,University of Malaya, Kuala Lumpur, Malaysia for their encouragement, comments and suggestions based on their vast experience in research, which resulted in the substantial improvement of this work.

I acknowledge the financial support extended by The International centre for Theoretical Physics (ICTP) Trieste,in the form of a research fellowship to work at Italy and to **Department of Science and Technology,**New Delhi for a research support.

REFERENCES

1. Matheiss L.F; Phys.Rev.Lett $\underline{58}$,1028(1987)

2. Cooke.D.W; Rempp.H;Fisk.Z; Smith.J.L; and Jahan.M.S Thermally Stimulated Luminescnce from rare–earth dopedBarium Copper Oxides; Phys.Rev.B $\underline{36}$(4),2287 (1987)

3. Cooke.D.W; Rempp.H; Fisk.Z; Smith.J.L; and Jahan.M.S Luminescent properties of irradiated rare–earth doped Cu– Oxides; J.Mater.Sci $\underline{6}$,(2),1115 (1987).

4. Fujiwara.Y and Kobayashi.T; Characterization of high T_c–super conductors by luminescence methods, IEEE Transactions on magnatics $\underline{25}$(2),2563 (1989)

5. Fujiwara.Y; Tonakhi.M and Kobayashi.T; TSL from high T_c–Super conducting Tl–Ba–Ca–Cu–O system; Jap.J.Appl.Phys $\underline{27}$(9),L1706 (1988).

6. Tarascon ; Green L.H; Mckrnnon.W.R; Hull G.W and Geballe T.H; Science 235,1373 (1987).

7. Roth M; Halperin.A and Katz.S; Thermally Stimulated Luminescence of $YBa_2Cu_3O_7$ high Temperature Superconductors; Solid State Communications $\underline{67}$,(2),105 (1988).

8. Narasimha Reddy.K;Prasad.T.S.P.L.N;Laxmi Narsaiah.E; Subba Rao.U.V;and Paroli.P Thermoluminescence in $YBa_2Cu_3O_{7-\delta}$ samples; Phys.Stat.Sol (a) $\underline{119}$,655(1990).

9. Bartyakhtar. J.T et al; Super conductivity and crystal structure pecularities of the oxygen deficient $YBa_2Cu_3O_{7-\delta}$ compounds; IEEE Trans.Magnetics $\underline{25}$, 2262(1989).

10. Cooke.D.W. et.al ; Detection of surface (~1μm) impurity phases in high T_c–super conductors; Appl.Phys.Lett $\underline{54}$(10),960(1989).

11. Cooke D.W.et.al; Correlation of TSL with Radio frequency surface resistance of high–temperature superconductors; Appl.Phys.Lett $\underline{55}$(10), 1038(1989).

12. Jahan M.S; Cooke D.W; Shenberg H; Smith J.L; and Lianos D.P; J.Mat.Res $\underline{4}$,759 (1989).

13. Narasimha Reddy K; Prasad T.S.P.L.N; and Subba Rao U.V; The Effect of Microstructure on Thermally Stimulated Luminescence of Y–Ba–Cu–O system; Cryst.Res.Technol $\underline{26}$(4),465(1991).

14. Murakami,M; Morita,M; Doi,K; Niyamoto.K; and Hamada.H; Microstructural study of the Y–Ba–Cu–O system at high temperature; Jpn.J.Appl.Phys

3,L399(1989).

15. Kelly P; Laubitz M.J; and Braunlich P; Exact solutions of kinetic equations governing TSL and conductivity; Phys.Rev.B4, 1960 (1971).

16. Mohan N.S; and Chen R; Numerical curve fitting for calculating glow curve parameters: J.Phys.D; Appl.phys 3,243 (1970).

17. Randall J.T and Wilkins M.A.F, Phosphorescence and electron traps I: The studies of trap distributions Proc.R.Soc.Lond A184, 365 (1945).

Oxide Ion Conductors for Solid Oxide Fuel Cells

Osamu Yamamoto, Takayuki Kawahara, Kazushige Kohno,
Yasuo Takeda and Nobuyuki Imanshi
Department of Chemistry, Faculty of Engineering, TSU, 514 Japan

This paper reviews the electrical properties of the stabilized zirconia with high mechanical strength and toughness. These oxide ion conductors are attractive for the electrolyte of the self–supported planar type solid oxide fuel cells(SOFC). The electrical conductivity of the tetragonal stabilized zirconia(TZP), which has high bending strength of 1200MPa, was measured to be 6.5×10^{-2}S/cm at 1,000°C. The electrical and mechanical properties of the composites of cubic stabilized zirconia and alumina were examined. The 20 weight% Al_2O_3 and 80 weight% stabilized zirconia with 8 mole% Y_2O_3 showed improved mechanical strength and only slight decrease in electrical conductivity.

1. INTRODUCTION

High temperature solid oxide fuel cells have become of great interest for electric power generation system, because of the high energy conversion efficiency, the simplicity of system design and the availability of high quality by-product heat. However, the realization of a useful cell is depended on whether the many material problems arising from the high operating temperature of 1,000°C will be solved. The most of significant material limitation at present time is imposed by the electrolyte. The oxide ion conducting solid was observed in ZrO_2–9mol%(m/o) Y_2O_3 by Nernst[1] as early as 1899. In 1937, Bauer and Preis[2] constructed the first SOFC using this electrolyte. During the past three decades, many oxide systems have been examined as oxide ion conductors. Figure 1 shows the important solid oxide ion conductors reported previously. The electrolyte for SOFC has to meet the following requirement; high ionic conductivity, low electronic conductivity, chemical and physical stability under reduced and oxidized atmospheres at a higher temperature, and the easy preparation of densed films. Most of the oxide ion conductors crystallized in the cubic fluorite structure. For zirconia the fluorite phase is stabilized only when it is doped with di- or tri-valent metal oxides like CaO, Y_2O_3, and Sc_2O_3.

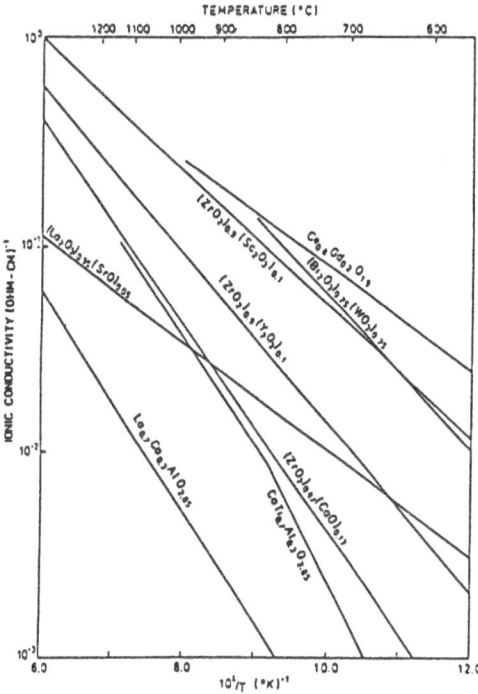

Fig. 1 Ionic conductivity(Arrhenius) plots for selected oxides.

367

The systems based CeO_2 have a high ionic conductivity, but Ce^{4+} ions are easily reduced to the +3 valence state at low oxygen pressure. Measurement of ionic transport number, t_i, in the CeO_2–11m/o La_2O_3 system indicated that $t_i=0.92$ at $Po_2=10^{-1}$atm, and was only 0.54 at $Po_2=10^{-8}$ atm at 1,000°C (3). Despite much research activity over the 30 years, stabilized zirconia, the first material discovered which shows oxide ion conductivity, remains one of the best conductors for SOFC.

Many types of configuration of SOFC, such as tubular, planar, and monolithic types, has been proposed and designed(4,5) as shown in Fig.2. For the tubular type SOFC, where the electrolyte film was deposited on the porous support tube with the anode or cathode, the mechanical strength of the electrolyte is not a significant requirement . While, the self–supported planar configuration,which could be expected to have exceptional potential for high electric power density, requires a more good mechanical strength and toughness for the electrolyte in order to construct a large size fuel cell. To obtain a high performance SOFC, the contribution of the electrolyte resistance should be less than 0.2–0.3 ohm/cm^2 at the operating temperature. Therefore, the thickness of the stabilized zirconia is required to be around 0.2–0.3mm. In this paper, the electrical properties of the stabilized zirconia with a high mechanical strength and toughness have been reviewed.

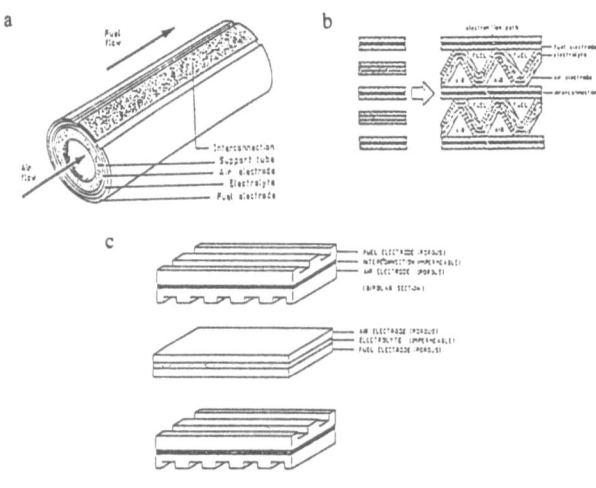

Fig. 2 SOFC configuration

a: Westinghous single tube configuration
b: Argonne multi–channel monolithic configuration
c: planar configuration

2. ELECTRICAL CONDUCTIVITY
OF TETRAGONAL ZIRCONIA

Zirconia has three polymorphic modification structures; monoclinic (up to about 1,200°C), tetragonal (about 1,200 to 2,370°C) and cubic (above 2,370°C). The transformation to the low temperature modification, however, can be suppressed by addition of some metal oxides such as CaO, Y_2O_3, and rare-earth oxides. The zirconia-rich portion of the ZrO_2–Y_2O_3 phase diagram by Pascula and Duran(6) is shown in Fig.3. The cubic zirconia phase(FSZ) is stabilized to room temperature by addition of >7.5m/o Y_2O_3. The tetragonal phase(TZP) has recently been stabilized by taking advantage of fine-particle technology and minor doping with Y_2O_3(7). This phase is expected to find a wide range of applications as a structural ceramics because of its high mechanical strength as shown in Table 1. If TZP has a high oxide ion conductivity, it is considerable interest as the electrolyte of the self-supported planar configuration SOFC. The first report on the conductivity of TZP containing 1.4m/o Y_2O_3 was presented by Gupta et al. in 1981(8),. The d.c.conductivity of TZP containing 3m/o Y_2O_3 was measured over the temperature range of 400 to 1,000°C by Badwal and Swain(9). The conductivity at 1,000°C was found to be 4×10^{-2} S/cm, which is one third that of FSZ. Move recently, Yamamoto et al.(10) reported the conductivity of the tetragonal zirconia of systems ZrO_2–M_2O_3 (M=Sc,Y,Yb) up to 1,000°C. Fig.4 shows the temperature dependence of the a.c. conductivity of TZP containing 3.0m/o Y_2O_3(3Y-TZP), 2.9m/o Sc_2O_3(2.9Sc-TZP) and 3.0m/o Yb_2O_3(3Yb-TZP). The highest conductivity of 1.0×10^{-1} S/cm at 1,000°C was found in 2.9Sc-TZP. In the cubic zirconia, the highest conductivity was also found in the system ZrO_2–Sc_2O_3(10). The conductivity data of TZP are summarized in Table 2 along with those of FSZ.

Table 1. Mechanical properties of tetragonal stabilized zirconia with 3Y-TZP and cubic stabilized zirconia with 8YSZ

	3Y-TZP	8YSZ
Bending strength (MPa)	1200	300
Fracture toughness ($MN.m^{-1.5}$)	8	3

Table 2. Conductivity ata of tetragonal and cubic zirconia

Electrolyte	Phase	Conductivity(S/cm) 1000°C	800°C	Activation energy(kJ/mol)
ZrO_2–3m/oY_2O_3	100%tetra	6.5×10^{-2}	1.8×10^{-2}	72
ZrO_2–8m/oY_2O_3	100%tetra	1.6×10^{-1}	4.5×10^{-2}	70
ZrO_2–2.9m/oSc_2O_3	100%tetra	1.0×10^{-1}	3.1×10^{-2}	62
ZrO_2–10.3m/oSc_2O_3	100%cubic	3.2×10^{-1}	1.3×10^{-1}	57
ZrO_2–3m/oYb_2O_3	100%tetra	6.3×10^{-2}	1.4×10^{-1}	80
ZrO_2–8m/oYb_2O_3	100%cubic	2.0×10^{-1}	5.6×10^{-1}	80

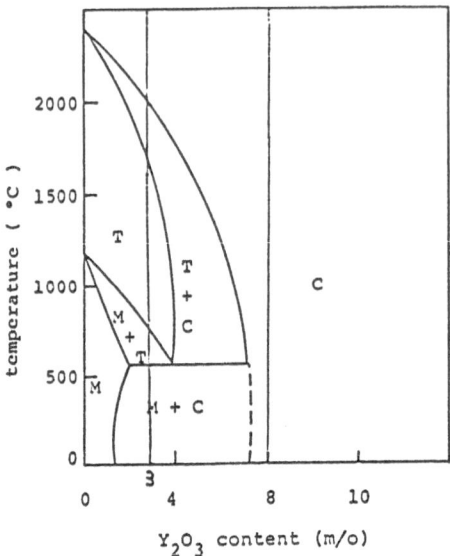

Fig. 3 Phase diagram of the ZrO_2–Y_2O_3 system

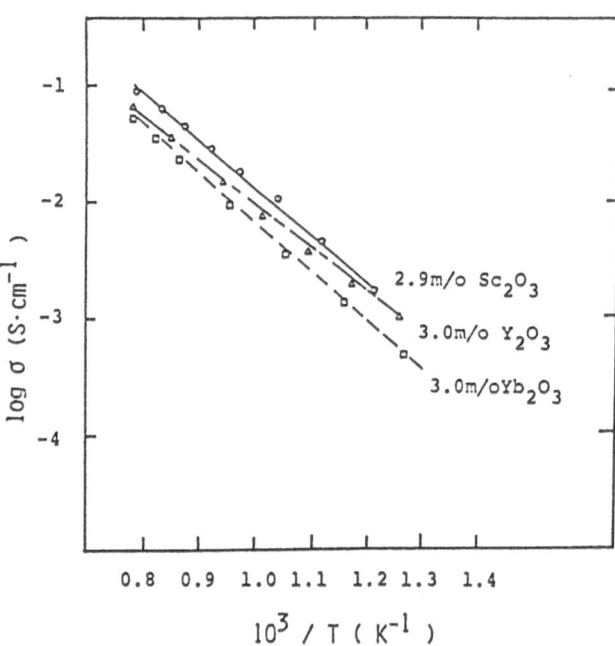

Fig. 4 Temperature dependence of ac conductivity for 3Y–TZP, 2.9Sc–TZP and 3Yb–TZP

Several reports have shown that the use of the ZrO_2-Y_2O_3 is accompanied by substantial deterioration of electrical conductivity by annealing at a higher temperature(11). It is important to study these changes when prolonged use is anticipated. Figure 5 shows the conductivity change of 3Y-TZP, 2.9Sc-TZP, and 3Yb-TZP on annealing at 1,000°C in air(12). The aging process for FSZ was first investigated by Carter and Roth(13) for the ZrO_2-CaO system, and it was concluded that defect ordering processes must be taking place. A square-root law for the conductivity change of 8m/o Y_2O_3 doped ZrO_2 at 875° was found by Kleitz et al.(14), indicating a diffusion-controlled process for the segregation of oversaturated impurities at the grain boundaries. The conductivity change for TZP shown in Fig.5 suggested that there may be two process contribution to the aging mechanism; the segregation of impurities and/or non-conductive second phase at the first stage and the ordering into tetragonal phase. Consequently, tetragonal zirconia is a promising electrolyte for SOFC, because of its excellent mechanical properties and long-term conductivity stability.

3. ELECTRICAL CONDUCTIVITY OF THE ZIRCONIA ALUMINA COMPOSITES

The composites of yttria stabilized tetragonal zirconia and Al_2O_3 have been extensively examined by Tsukuma(15) to improved the high temperature mechanical properties. It is well known that TZP exhibits extremely high bending strength at room temperature as shown in Table 1. However, the strength of

Table 3. The bending strength of Y-TZP and Y-TZP-Al$_2$O$_3$ composites

	Temperature	
	25°C	800°C
3Y-TZP	1200MPa	200MPa
3Y-TZP-20w/oAl$_2$O$_3$	2400MPa	700MPa

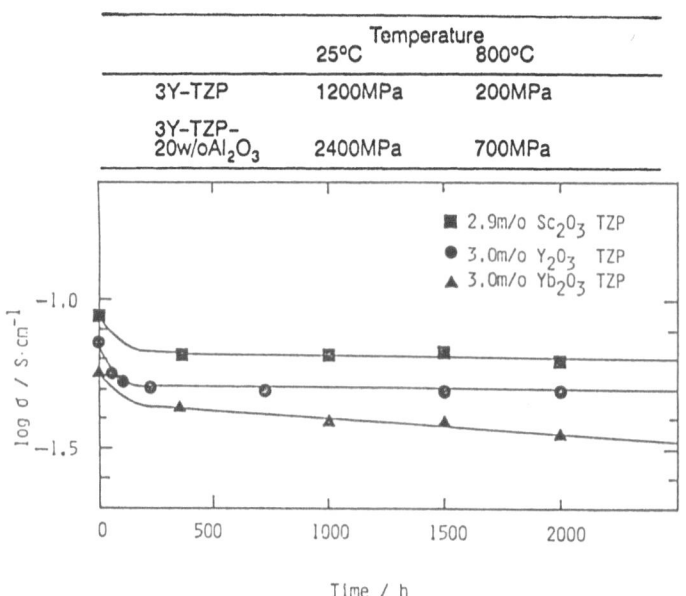

Fig. 5 Conductivity change of TZP on annealing at 1,000°C in air.

371

TZP decreased remarkably with increasing temperature. The addition of Al_2O_3 improved the high temperature strength of pure–Y–TZP, because the strength of alumina does not decrease so largely as that of zirconia with increasing temperature. In Table 3, the bending strengths of Y–TZP and the Y–TZP–Al_2O_3 composites are shown. The temperature dependence of electrical conductivity of the Y–TZP–Al_2O_3 composites is shown in Fig.6. The conductivity decreases with increasing Al_2O_3 content. The conductivity of the composites of 3Y–TZP with 20weight%(w/o) Al_2O_3(3Y20A) was 4.5×10^{-2} S/cm at 1,000°C. The conductivity value is slightly lower than acceptable one as the electrolyte SOFC. However, the composites are quite attractive, because they have a extremely high bending strength at higher temperatures. The tetragonal zirconia phase is metastable below 1,200°C as shown in Fig.3 and undergoes a phase transformation to monoclinic phase and a consequence degradation of the conductivity, when it was annealed at higher temperature, because of the grain growth. XRD analysis showed that no detectable amount of the monoclinic phase formed in 3Y20A annealed at 1,550°C for 3h, compared one mole% of the monoclinic phase formed in 3Y–TZP without Al_2O_3.

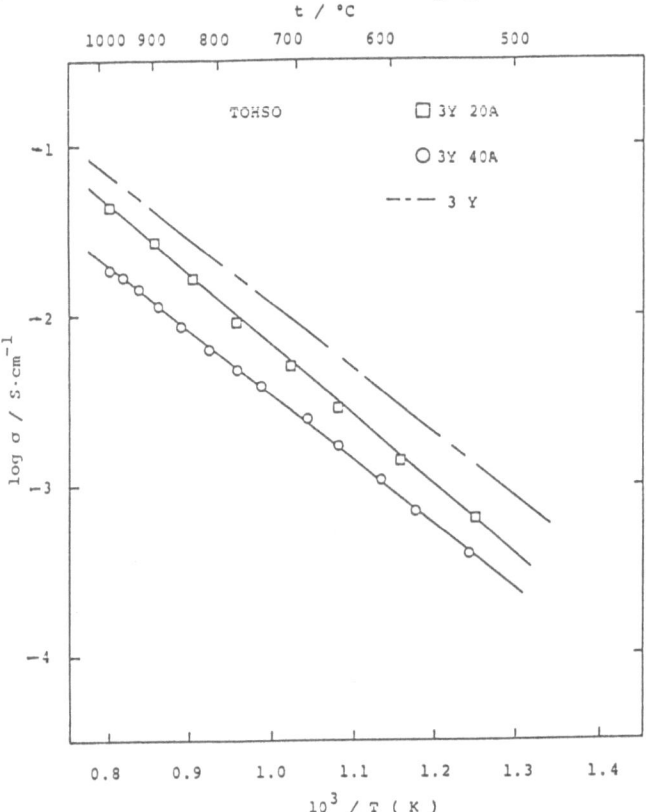

Fig. 6 Temperature dependence of ac conductivity for 3Y–TZP, 3Y–TZP with 20w/oAl_2O_3(3Y20A) and 3Y–TZP with 40w/o Al_2O_3(3Y40A)

Many papers have been reported on the effect of small Al_2O_3 addition on the microstructure and the conductivity of yttria stabilized cubic zirconia(YSZ). Small additions of Al_2O_3 are effective for densification of YSZ(16). Only few papers have been published on the mechanical and mechanical properties of the composites of YSZ with large amount of Al_2O_3. The composition dependencies of three−point bending strength of the 8YSZ−Al_2O_3 composites at various tem perature are shown in Fig.7. The bending strength of 250MPa at room tempera− ture in 8YSZ without Al_2O_3 is comparable to that reported(17). No result of mechanical properties of the 8YSZ−Al_2O_3 system have been reported, except of the preliminary results by Ishizaki et al.(18) . In the case of the 8YSZ−Al_2O_3 composites, the bending strength decreases at 500°C and then increases at 1,000°C. The bending strength at 1000°C is almost comparable to that at room temperature. The cubic zirconia exhibits a super plasticity at a higher tempera− ture(19). Therefore, the high strength of 8YSZ at 1000°C could be explained by the contribution of the plasticity. The bending strength of the 8YSZ−Al_2O_3 composites increases with increasing Al_2O_3 content up to 20wt% Al_2O_3. Further addition of Al_2O_3 is not so effective for strengthening. The addition of 20wt% Al_2O_3 gives the bending strength of 330MPa at room temperature. The similar composition dependence of the strength was observed in the Y−TZP−Al_2O_3 composites, where the maximum bending strength was given in the addition of 20wt% Al_2O_3. The enhancement of bending strength by addition of Al_2O_3 in 8YSZ could be explained simply by the replacement of 8YSZ with Al_2O_3,be− cause bending strength of Al_2O_3 is superior to that of 8YSZ. The effective of the particle size of Al_2O_3 on the bending strength was not remarkable. This result indicated that the homogeneous distribution of Al_2O_3 in 8YSZ is not so impor− tant for the mechanical properties.

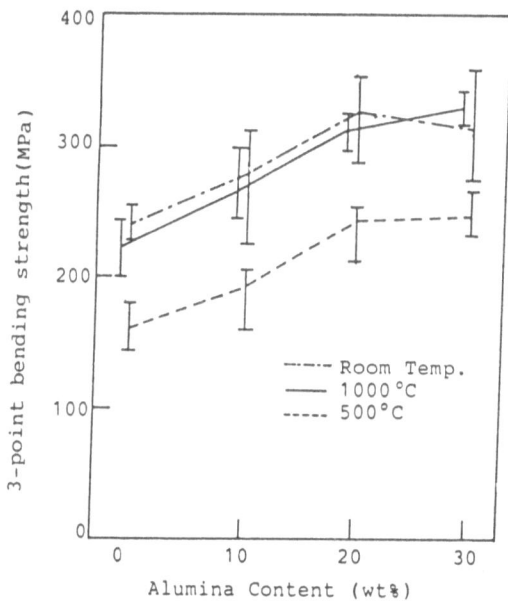

Fig. 7 Composition dependence of the three−point bending strength of the 8YSZ and Al_2O_3 composites (Error bar indicates the maximum and minimum vales of ten test samples)

The electrical conductivity of the composites is the most important requirement for the electrolyte in SOFC. The temperature dependencies of the conductivity are shown as a function of Al_2O_3 content along with the composition dependence of that at 1,000°C and 830°C in Fig.8. The conductivities at 1,000°C increases slightly with increasing Al_2O_3 content up to 1wt% and then decreased. The maximum conductivity of 0.13S/cm at 1,000°C was observed in the composite with 1wt% Al_2O_3. The electrical conductivity of 8YSZ with small amount of Al_2O_3, where Al_2O_3 is effective as a sintering agent, has been examined in the temperature range 400–600°C by many authors. The effect of Al_2O_3 addition on the conductivity of 8YSZ is rather complicated. Verkerk reported that additions of 0.78mol% Al_2O_3 decreased the bulk and grain–boundary conductivity in 8YSZ(16). On the other hand, Bernard found that 1.0wt% addition in 9YSZ increased the conductivity(20). These discrepancy may be due to the level of sample impurities ; silica has a marked negative influence on the conductivity(14). Butler suggested that Al_2O_3 in cubic stabilized zirconia acts as a scavenger for SiO_2, removing it from grain–boundary localities. Such a interactions of Al_2O_3 offers a possible explanation for the improvements in conductivity brought about by Al_2O_3 additions(21). In the composites, however, the content of silica was extremely low. The enhancement of electrical conductivity by addition of small amount of Al_2O_3 could not be explained with the scavenger effect of Al_2O_3. The decrease of the electrical conductivity with the Al_2O_3 content is observed above 3wt% Al_2O_3. The conductivity of 8YSZ with 20wt%

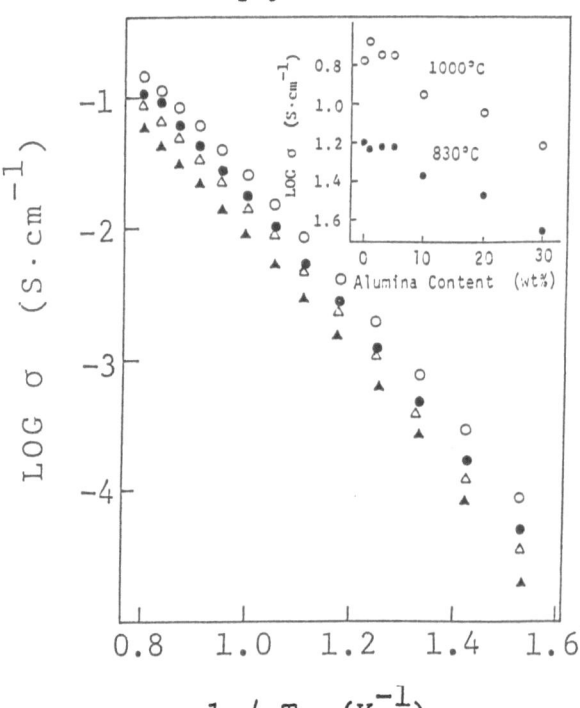

Fig. 8 Electrical conductivity of 8YSZ and Al_2O_3 composites
O, 8YSZ; ●, 10wt% Al_2O_3+ 8YSZ; △, 20wt% Al_2O_3+ 8YSZ;
▲, 30wt% Al_2O_3 +8YSZ

Al$_2$O$_3$ was around 0.10S/cm at 1,000°C, which is about 65% of that of pure 8YSZ. The conductivity decrease by addition of Al$_2$O$_3$ could be interpreted with several mechanisms; in the mixture phase of the isolated conducting and insulating particles, the conductivity should be proportional to the volume fraction of the conducting particles. and the blocking effects of ion diffusion at the grain-boundaries is proportional to the area of the boundaries and increases significantly as the grain size of the sintered material decreases(14). The decrease of the conductivity of the 8YSZ and Al$_2$O$_3$ composites could not be explained only by the decrease of the volume fraction of 8YSZ. The grain size of the composites was extremely smaller than that of pure 8YSZ. Therefore, the conductivity decrease may also result from the decrease of the grain size. The effect of the particle size of Al$_2$O$_3$ on the electrical conductivity is shown in Fig.7, where the content of Al$_2$O$_3$ was 20wt%. The composites with larger particle size of 0.68μm show a slightly higher conductivity at lower temperature. At higher temperature, no significant change in conductivity was observed. The composites with 0.68μm Al$_2$O$_3$ showed a homogeneous dispersion of Al$_2$O$_3$ particles in 8YSZ matrix, but those with 0.23μmAl$_2$O$_3$ a segregation of the Al$_2$O$_3$ particles at the grain-boundaries. At lower temperatures, the grain-boundaries play an important role in the conductivity. The segregation of the Al$_2$O$_3$ particles at the grain-boundaries could block the migration of oxide ions through the boundaries. The blocking effect vanishes at a higher temperature, which markedly depends on the level of sample purity(14).

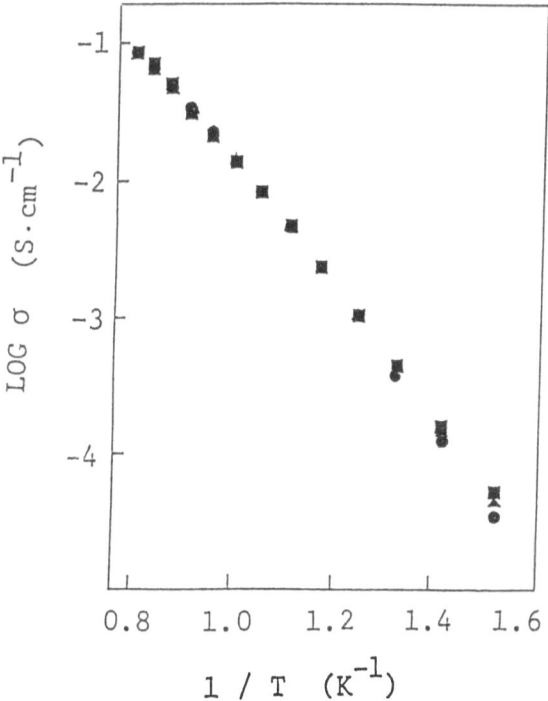

Fig. 9 Al$_2$O$_3$ particle size dependence on electrical conductivity of the 8YSZ with 20wt% Al$_2$O$_3$ ●,0.22μ;▲,0.39μm; ■,0.68μm 20wt% Al$_2$O$_3$ content

375

4. CONCLUSION

High temperature solid oxide fuel cells have a possibility to achieve extremely high current density as high as one ampere/cm^2, because of high operating temperature of 1,000°C. Especially, the planar type SOFC has exceptional potential for high electric power density. To obtain a high performance planar SOFC, however the electrolytes with good mechanical properties and high ionic conductivity should be developed. We could conclude that yttria stabilized tetragonal zirconia and the composites of yttria stabilized cubic zirconia were the best candidates for the electrolyte in self supported planar SOFC.

References

(1) W.Nernst, Z.Elektrochem., **6**,41(1899)
(2) E.Bauer and H.Preis, Z.Electrochem.,**43**,727(1937)
(3) T.Takahashi, K.Ito and H.Iwahara, Rev.Energy.Primaire,**2**,27(1966)
(4) O.Yamamoto, M.Dokiya, and H.Tagawa,Eds."proceedings of the International Symposium on Solid Oxide Fuel Cells", SOFC Soc. of Japan (1990)
(5) B.C.H.Steele, "Ceramic Electrochemical Reactors", Ceramionics (1987)
(6) C.Pascula and P.Duran, J.Am.Cerm.Soc.,**66**,23(1983)
(7) T.K.Gupta, J.H.Bechtold, R.C.Kuzunickl, L.H.Cadoft and B.R.Rossing,J.Mat.Sci.,**128**,929(1981)
(8) T.K.Gupta, R.B.Gvekila, and E.C.Subbarao, J.Electrochem.Soc.,**128**,929 (1981)
(9) S.P.S.Badwal and M.V.Swain, J.Mat.Sci.Lett.,**4**,487(1985)
(10) O.Yamamoto, Y.Takeda, R.Kanno, K.Kohno, and T.Kamiharai, J.Mat.Sci.Lett.,**8**198(1989)
(11) E.Shouler, G.Giround, and M.Kleitz, J.Chim.Phys.Phisiochim.Biol.,**70**, 1309(1973)
(12) O.Yamamoto, Y.Takeda, R.Kanno, and K.Kohno, J.Mat.Sci.,**25**,2805 (1990)
(13) R.E.Carter and W.L.Roth,"EMF Mesurement in High Temperature Systems." ed. by C.B.Alock,P125(1968)
(14) M.Kleitz, H.Bernard, E.Fernandes, and E.Shouler, Advance in Ceramics, vol.3, ed. by A.H.Heuer and L.W.Hobbs, P.310(1981)
(15) K.Tsukuma, Thesis, Osaka University(1986)
(16) M.J.Verkerk, A.J.A.Winnubst and A.J.Buggraat, J.Mat.Sci.,**17**,3113(1982)
(17) K.Kobayashi, H.Kawajima and T.Masaki, Solid State Ionics,**3/4**,489(1981)
(18) E.Ishizaki, T.Yoshida, and S.Sakurada, "Proceedings of Electrochem. Soc. Fall Meeting",P3(1989)
(19) K.Mastusue, Y.Fujisawa and T.Takuhara, Yogyo-Kyokai-shi,**91**,59(1983)
(20) H.Bernard, Rep.CEA-R-500, CEN-Saclay, France (1981)
(21) E.P.Butler and J.Drennan, J.Am.Ceram.Soc.,**65**,474(1982)